J.G.E.LEWIS

The biology of centipedes

CAMBRIDGE UNIVERSITY PRESS
Cambridge
London New York New Rochelle
Melbourne Sydney

CAMBRIDGE UNIVERSITY PRESS
Cambridge, New York, Melbourne, Madrid, Cape Town, Singapore, São Paulo

Cambridge University Press
The Edinburgh Building, Cambridge CB2 2RU, UK

Published in the United States of America by Cambridge University Press, New York

www.cambridge.org
Information on this title: www.cambridge.org/9780521234139

First published 1981
This digitally printed first paperback version 2006

A catalogue record for this publication is available from the British Library

ISBN-13 978-0-521-23413-9 hardback
ISBN-10 0-521-23413-1 hardback

ISBN-13 978-0-521-03411-1 paperback
ISBN-10 0-521-03411-6 paperback

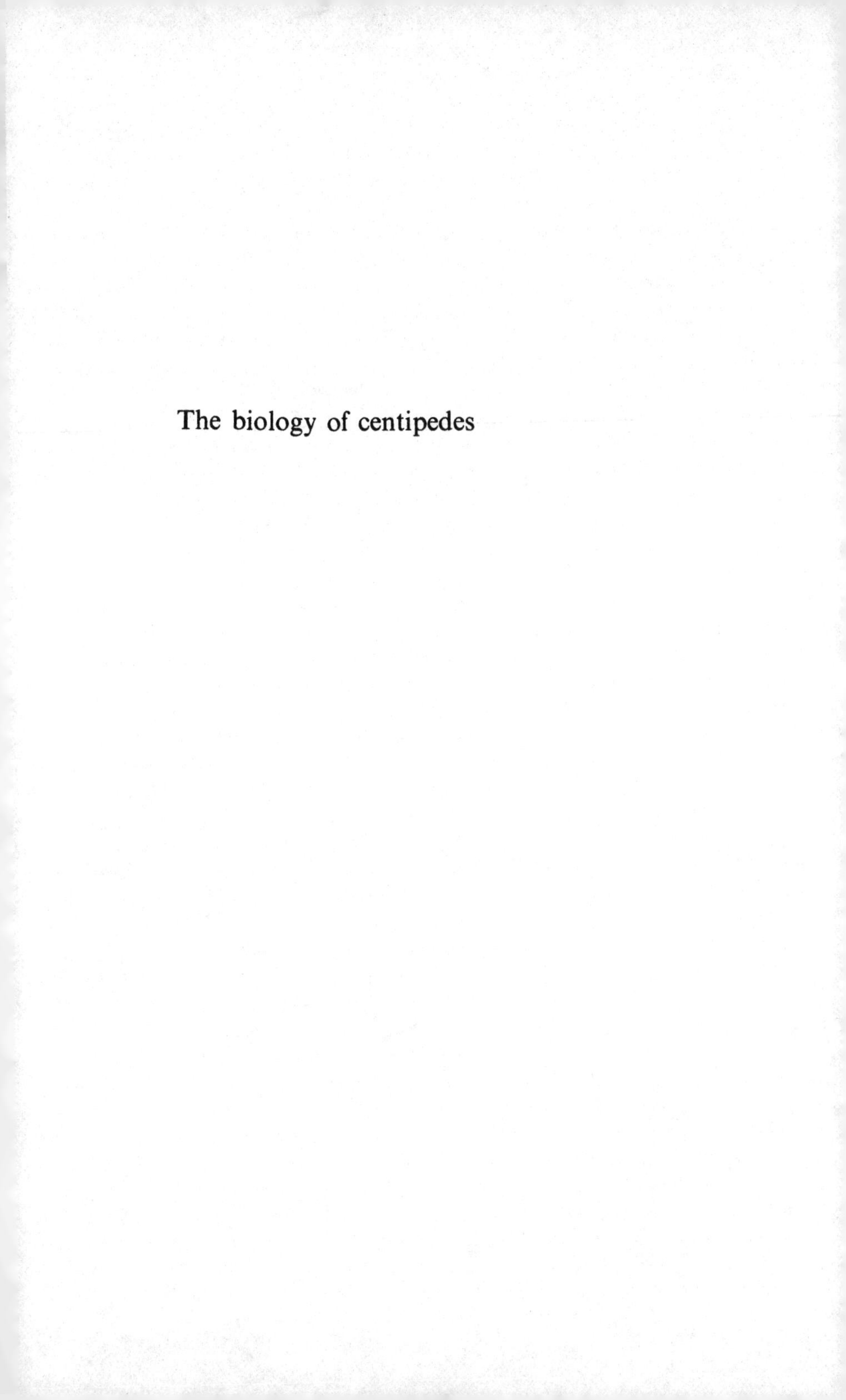

The biology of centipedes

CONTENTS

PREFACE

This book is an attempt to bring together information on centipedes hitherto to be found only in old texts or scattered through the literature and in unpublished theses.

My interest in the group is due to Sir Eric Smith, FRS who, as Professor J. E. Smith, suggested that I carry out postgraduate research on littoral centipedes. Since that time I have had a great deal of advice and encouragement from Mr. J. Gordon Blower, Dr E. H. Eason, Dr R. E. Crabill, Jr and Professor J. L. Cloudsley-Thompson. Over the years, I have received much help from members of the Arachnid Section of the British Museum (Natural History) most notably, Mr K. H. Hyatt, Mr D. Macfarlane, Mr F. R. Wanless and the late Mr D. J. Clark and also from the library staff of the Museum.

The Headmaster and Governors of Dover College generously allowed me a term's leave of absence to take up a School Teacher Fellow-Commonership at Sidney Sussex College, Cambridge, in order to complete the book. I am greatly indebted to the Master and Fellows of Sidney Sussex for their help and hospitality and to the librarians of the Cambridge Zoology Department Library for their assistance. I am grateful to Dr D. L. Gunn, CBE and the Royal Society Research in Schools Committee for their support during the preparation of this book.

I am likewise indebted to Miss K. E. Yule for help with French and German and to Mrs K. Stefan for translating the Russian papers. My thanks are also due to Mrs B. Bradford for typing the manuscript and to the staff of Cambridge University Press for their advice during its processing for publication.

Above all, my thanks are due to my wife, for checking the manuscript and for constant advice, support and encouragement.

to S.J.C.

1

Introduction

Centipedes (members of the arthropodan class Chilopoda) are common and relatively familiar animals which are found in soil and litter or under stones or bark. They are soft-bodied and dorso-ventrally flattened having from 15 to 181 pairs of legs, one pair to each trunk segment. Species from temperate regions are usually of moderate size, varying from 1 to 10 cm in length and of drab brownish or yellowish coloration. Many tropical species of the order Scolopendromorpha are large, one reaches a length of 26 cm, and are brightly coloured: red, black and orange, green or violet.

Like other arthropods, the centipedes are bilaterally symmetrical, metamerically segmented animals with a double ventral nerve cord, typically with a ganglion in each segment and concentrations of nervous tissue above and below the gut at the anterior end of the body. A circulatory system is present carrying blood forwards in a dorsal vessel and backwards in a ventral vessel. The body is covered by a non-living layer, the cuticle, which is secreted by the epidermis. The cuticle is in the form of relatively rigid sclerites separated by flexible arthrodial membranes; it is shed periodically to allow growth, a phenomenon known as moulting or ecdysis. The anterior part of the body is differentiated to form a head which bears a pair of antennae, a pair of jaws (mandibles) and two pairs of jointed legs modified to form mouthparts (the first and second maxillae). The legs of the first trunk segment are modified to form the characteristic poison claws which are used to seize prey.

The centipedes have often been grouped as the Myriapoda with the millipedes (Class Diplopoda) and two classes of soil-dwelling arthropods: the Symphyla with 12 pairs of legs and a pair of stout anal cerci and the minute Pauropoda with nine or ten pairs of legs and biramous antennae. Myriapods may be defined as arthropods with one pair of antennae, two or three pairs of mouthparts and numerous pairs of legs.

Opinions as to the status of the four groups of myriapods have varied considerably. They have sometimes been regarded as four classes showing little true relationship to each other. Manton (1970), however, has discussed the reasons for the reinstatement of the Myriapoda as a natural group. A study of the jaw and leg

Fig. 1. *Geophilus carpophagus*, body length 40 mm (after a photograph by Manton in Eason, 1964).

morphology have led her to the conclusion that the anatomical differences in head and trunk anatomy within the myriapods are associated with divergent habits of life: the myriapods could have evolved from a similar basic stock although no one class could have given rise to another.

The class Chilopoda comprises the orders Geophilomorpha, elongated, worm-like soil-dwelling forms with 31 to about 181 pairs of legs (Fig. 1), the Scolopendromorpha containing the large tropical and subtropical species with 21 or 23 pairs of legs (Fig. 2), the Lithobiomorpha, short-bodied animals with 15 pairs of legs (Fig. 3)

Fig. 2. *Scolopendra morsitans*, body length 88 mm (after Manton, 1965).

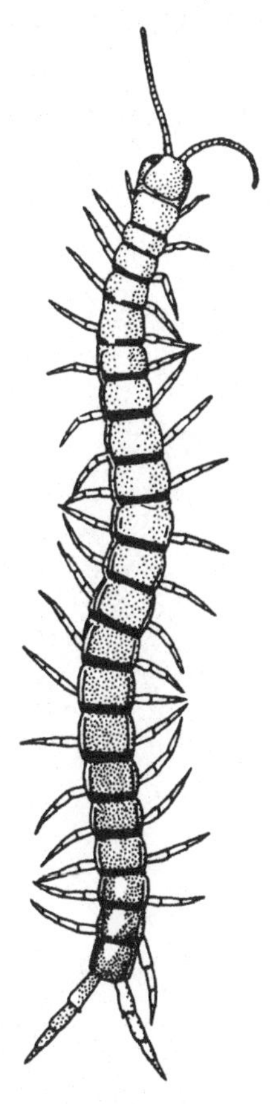

and the Scutigeromorpha which are mainly tropical and subtropical forms with 15 pairs of very elongated legs (Fig. 4). A fifth type with problematical affinities is the Australasian species *Craterostigmus tasmanianus* Pocock (Fig. 5) which shows similarities to both Lithobiomorpha and Scolopendromorpha.

Centipedes and millipedes are sometimes confused but although there is some superficial resemblance between lithobiomorphs and 'flat-backed' millipedes the two classes are readily distinguishable. Centipedes are opisthsogoneate (have a posterior genital opening), possess three pairs of mouth parts and a pair of poison claws, and one pair of walking legs per segment. Millipedes are progoneate (have an anterior genital opening), possess two pairs of mouth parts and two pairs of legs per segment with the exception of the first four segments. They lack poison claws. Centipedes are largely

Fig. 3. *Lithobius variegatus*, body length 30 mm (after Manton, 1965).

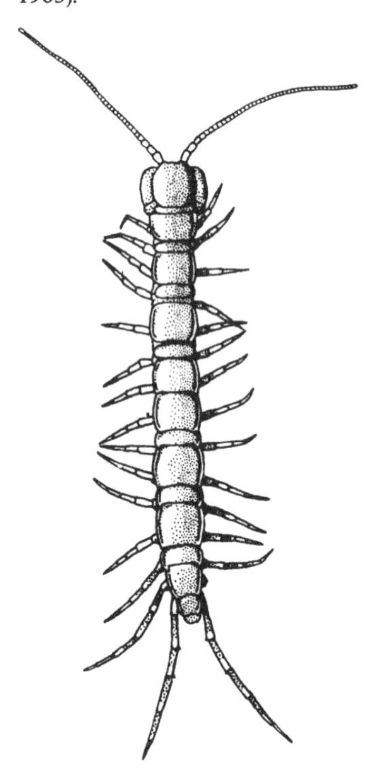

carnivorous, millipedes, largely herbivorous or saprophagous.

The most conspicuous terrestrial arthropods, the arachnids and insects, owe much of their success to the evolution of a waxy waterproofing layer on the cuticle and a uricotelic excretory system which enables them to get rid of their waste nitrogen as non-toxic and insoluble uric acid and thus save water. By contrast, the centipedes appear to have relatively permeable cuticles and excrete

Fig. 4. *Scutigera coleoptrata*, body length 30 mm (after Manton, 1952*b*).

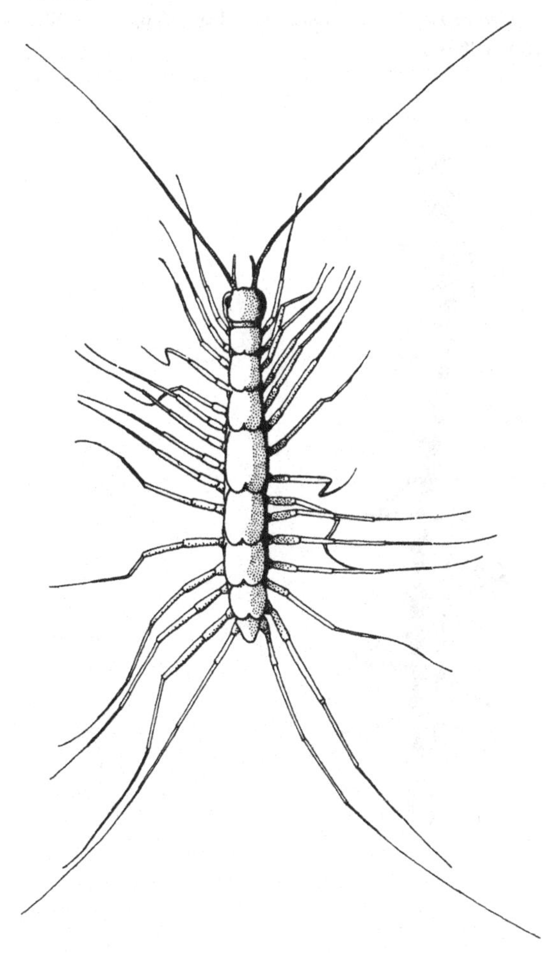

much of their waste nitrogen in the form of ammonia which is toxic and has to be diluted with large quantities of water to be excreted.

There are more than 850 000 species of insects but only about 3000 species of centipedes have been described, yet despite their apparent insignificance they are found from north of the Arctic Circle to the desert fringe. Some species are very widely distributed and may be successful in a wide variety of habitats. A number of geophilomorphs have adapted to life on the sea-shore, a habitat that has been difficult even for the insects to penetrate.

Fig. 5. *Craterostigmus tasmanianus*, body length 46 mm (after Manton, 1965).

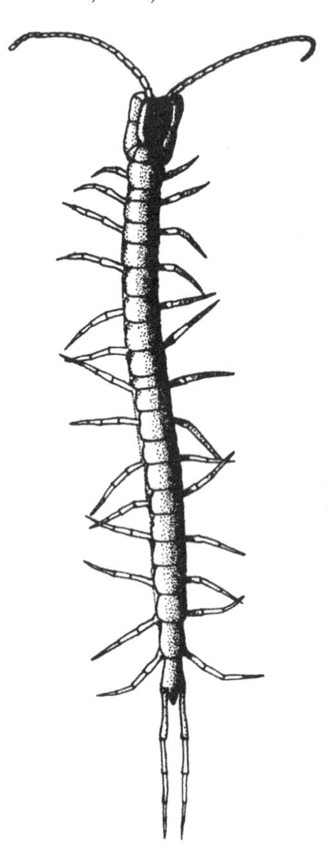

2

External morphology and functional anatomy

External morphology

The dorsal surface of the centipede head is covered by a more or less rigid head capsule which bears a pair of moniliform antennae. It may bear simple ocelli or paired compound eyes as in the Scutigeromorpha. The mouth is situated ventrally and is bounded in front by the labrum or upper lip, which is usually divided into a mid-piece with a side-piece on each side. The rigid chitinous plate on the ventral side of the head anterior to the labrum is termed the clypeus (Fig. 6). The mouth is bounded laterally by a chitinous 'pleurite' on each side. The appendages of the first three segments posterior to the mouth have been modified to form mouthparts and the dorsal skeletal elements (tergites) of these segments have been lost. The first pair of mouthparts are the mandibles, each of which consists of a trunk bearing a condyle which articulates with the skeleton of the head capsule and is

Fig. 6. Ventral aspect of the head of *Geophilus osquidatum* with the mouthparts removed, to show the clypeus and labrum. The setae are shown on the clypeus only (after Eason, 1964).

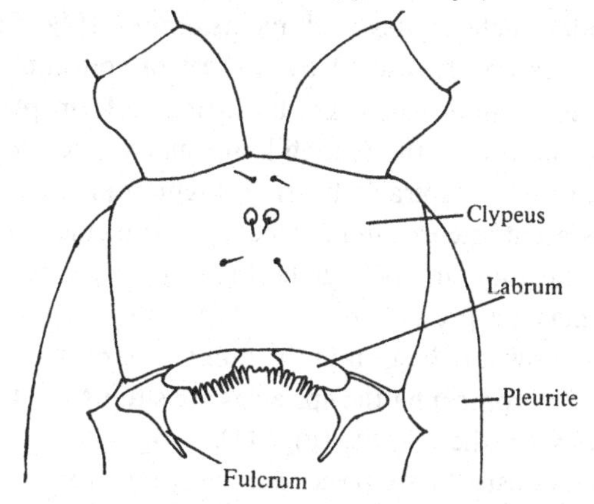

produced proximally into a curved shank affording attachment for the muscles operating the appendage (Fig. 7). The distal end of the mandible is broadened and armed with teeth and/or comblike structures.

The second pair of mouthparts, the first maxillae, are situated ventrally to the mandibles and obscure them when the head is viewed from the ventral side (Figs. 8, 17). This pair consists of a basal plate formed from the fused coxae (the basal segment) of each leg and the sternite (the ventral skeletal plate) of the segment and hence called the coxosternite. The coxosternite bears a pair of conical jointed appendages (the telopodites) between which is a second pair of conical processes, the coxal projections. The third pair of mouthparts, the second maxillae, partly cover the first maxillae and are recognisably leg-like in structure, each coxosternite bearing a telopodite consisting of three segments and an apical claw.

The head morphology of centipedes other than scutigeromorphs is correlated with head flattening sufficient for the animals to catch and eat prey within narrow crevices. The difficulty of obtaining transversely cutting mandibles and a suitable abductor mechanism have been resolved by the development of entognathy and the use of the endoskeleton (Manton, 1964).

Behind the head there is a variable number of trunk segments: there is no distinct thoracic region in centipedes as there is in insects. Typically, each trunk segment consists of a dorsal shield or tergite and a ventral shield or sternite. Tergites and sternites are sometimes divided transversely into narrow pretergites and presternites and wider metatergites and metasternites (Fig. 9). The tergites and sternites are connected by a sheet of soft cuticle, the pleural membrane, areas of which are hardened to form pleurites. These are best developed in the Geophilomorpha where there is a large prescutellum and a spiracle-bearing stigmatopleurite. More ventrally, there is a katopleure, and a procoxa and metacoxa which form an incomplete ring supporting the basal leg segment (coxa). Other pleurites may be present in geophilomorphs (Fig. 10). The remaining orders show a reduction in the number of pleurites (Figs. 12, 26, 39). In Scutigeromorpha the spiracles are situated not on the pleura but dorsally on the tergites (Fig. 11).

The legs, which consist of six basic segments, the coxa, trochan-

Fig. 7. Mandible of *Lithobius piceus* (after Brölemann, 1930).

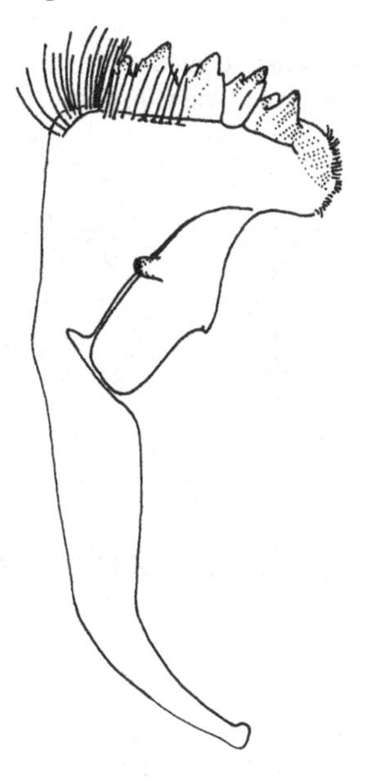

Fig. 8. Ventral view of the head, forcipules and first leg-bearing segment of *Lithobius forficatus* (after Rilling, 1968).

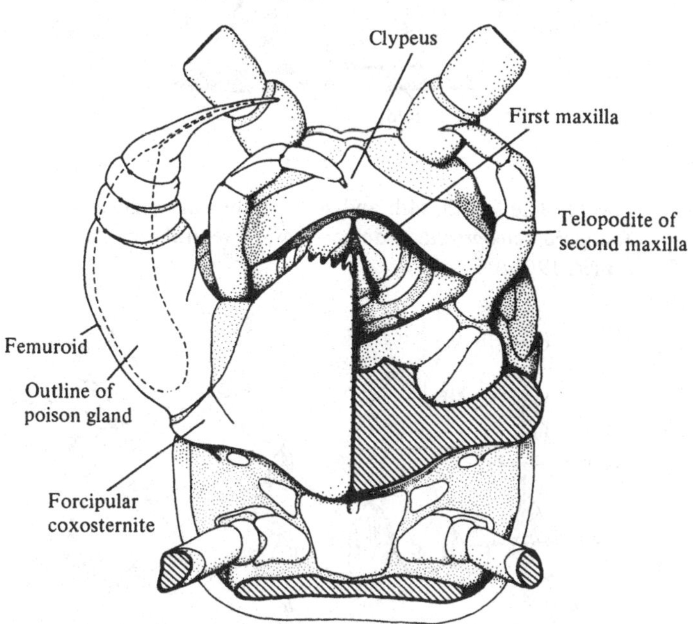

Fig. 9. The fifth and sixth leg-bearing segments of *Necrophloeophagus longicornis a*, dorsal view; *b*, ventral view (after Füller, 1963*a*).

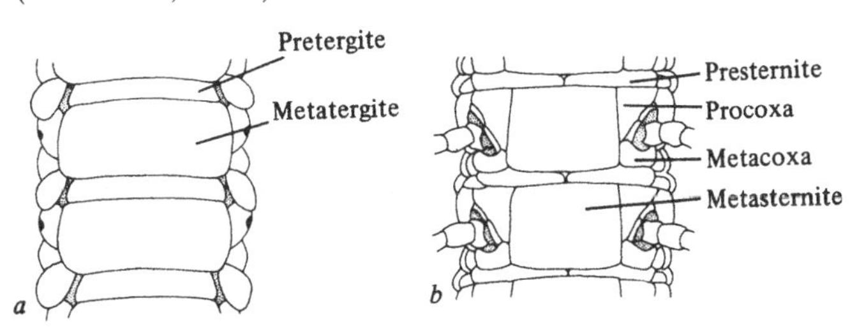

Fig. 10. Lateral view of two trunk segments of *Haplophilus subterraneus* (after Manton, 1965).

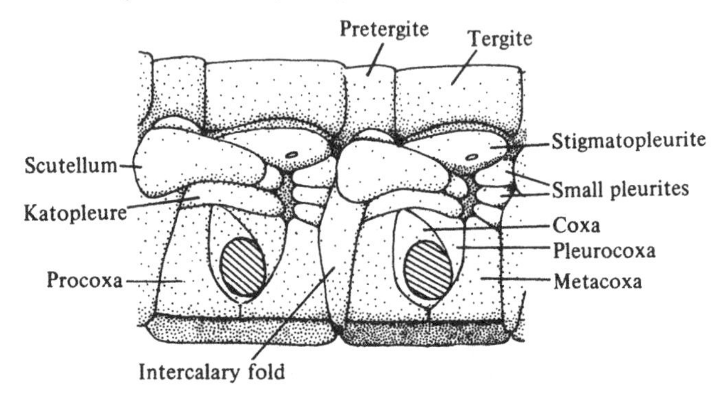

Fig. 11. The fourth, fifth and sixth leg-bearing segments of *Scutigera coleoptrata. a*, dorsal view; *b*, ventral view (after Füller, 1963*a*).

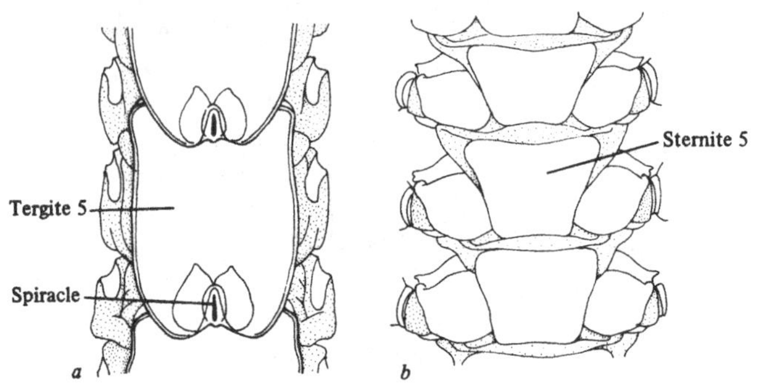

ter, prefemur, femur, tibia and tarsus are terminated by an apical claw and are inserted laterally into the pleural region (Fig. 12). The tarsus may be subdivided as in the Scutigeromorpha in which it consists of a very large number of segments (Fig. 13).

A marked exception to the normal constructional pattern of the trunk segments is seen in the first and to a lesser extent the last segments. The legs of the first trunk segment form large poison claws, variously termed prehensors, toxicognaths, maxillipedes and forcipules. The last term will be used here. The forcipules reach forward under the head and partially cover the mouthparts (Fig. 8) forming part of the feeding apparatus. The forcipular tergite may be free or fused with the tergite of the second trunk segment; the sternite is invariably fused with the coxae to form the forcipular coxosternite or prosternum. The anterior border of the coxosternite

Fig. 12. Anatomy of the trunk segments of *Lithobius forficatus*. *a*, transverse section; *b*, lateral view of segments 10, 11 and part of 12 (after Manton, 1965).

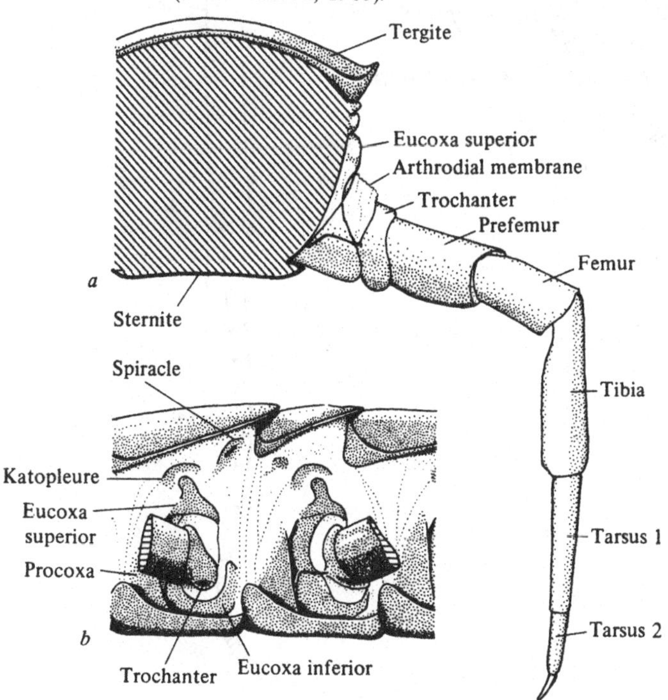

frequently bears spines or teeth. The telopodite consists of four segments, a very large basal segment, followed by two very short segments and an apical segment which bears a powerful claw. There is no agreement about the homology of the segments of the forcipule, Eason (1964) regarded the basal segment as a femuroid and the following three tibia, tarsus and apical claw. Crabill terms them the trochanteroprefemur, femur, tibia and tarsungula. The forcipule contains a poison gland which discharges by means of a

Fig. 13. Transverse section of the trunk of *Scutigera coleoptrata* showing one walking leg (after Manton, 1965).

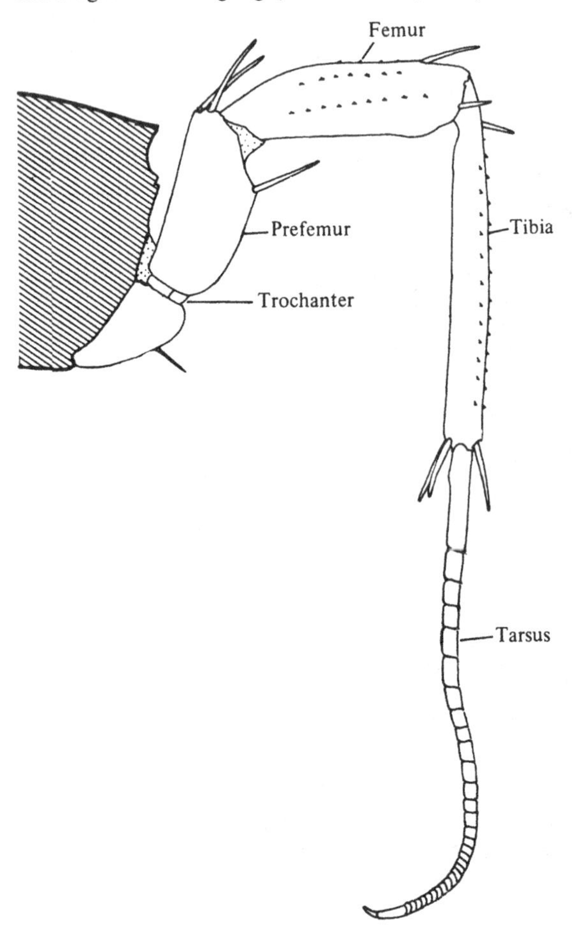

duct that opens just behind the tip of the claw. The second post-cephalic segment is normally regarded as the first trunk segment and the corresponding tergite and sternite are labelled tergite 1 and sternite 1.

The last leg-bearing segment differs from the preceding ones in that the spiracles are absent and the pleurites have been largely incorporated into the coxae of the legs. In Geophilomorpha and Scolopendromorpha the coxae are inflated and frequently, bear pores which are the openings of the coxal glands (Fig. 14). The coxae of the last four pairs of legs bear pores in many lithobiomorphs (Fig. 31*b*). The last pair of legs in centipedes may be modified for sensory or defensive functions and is not used in locomotion.

There has been confusion over the terminology relating to the

Fig. 14. Ventral view of the terminal segments of a male *Strigamia maritima* (after Lewis, 1960).

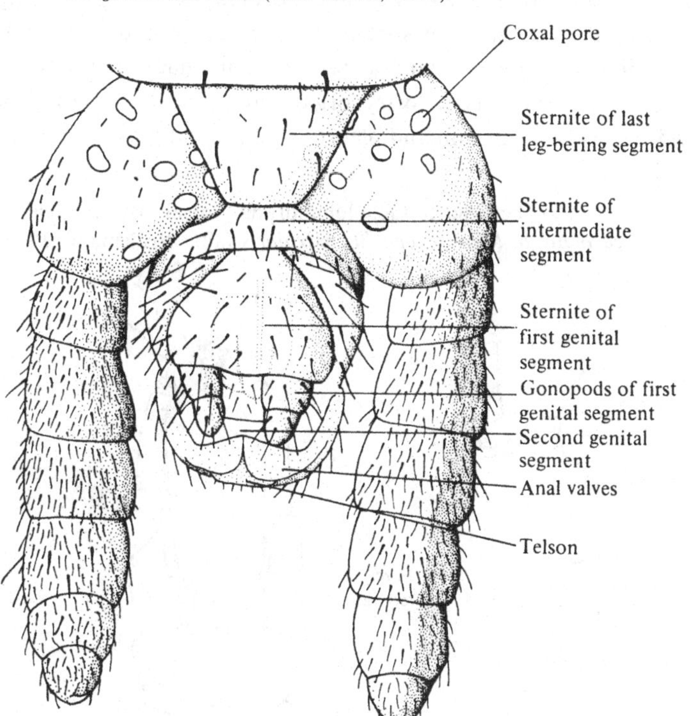

segments behind the last leg-bearing segment. That of Brölemann (1930) is adopted here. He recognised an intermediate segment, first and second genital segments and a terminal telson (Fig. 14). The first genital segment bears paired gonopods, the second carries the penis in males and the vulva in females. The telson carries a pair of anal valves flanking the anus.

The following descriptions of the orders are based to a large extent on Brölemann (1930) and Eason (1964).

Geophilomorpha

The geophilomorphs are long-bodied animals, the number of leg-bearing segments is always odd varying from 31 to 181. In most species there is some variation in leg number and females tend to have more leg-bearing segments than males. A notable exception is the family Mecistocephalidae in which the number of segments is fixed in each species.

The head capsule is flattened and normally lenticular in shape (Fig. 15*a*) though in some species it is elongated and rectangular (Fig. 15*b*). The head bears a pair of antennae each with 14 segments: eyes and Tomösváry organs are absent. Ventrally are found the clypeus, labrum and pleurites. The mandibles of geophilo-

Fig. 15. The head in Geophilomorpha, dorsal view. *a*, *Henia bicarinata*; *b*, *Mecistocephalus maxillaris* (after Brölemann, 1930).

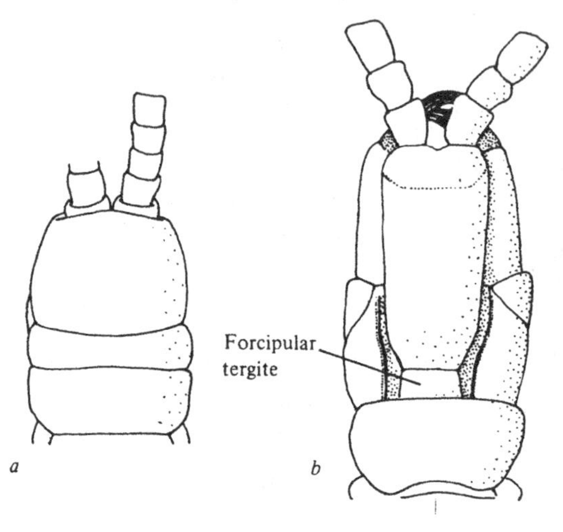

morphs are unusual in that they are very small and show considerable variation from family to family. In the family Himantariidae each mandible bears one dentate lamella and several pectinate lamellae (Fig. 16*a*); in the Schendylidae one dentate lamella and one pectinate lamella (Fig. 16*b*). The Oryidae and the Mecistocephalidae lack the dentate lamella but possess a number of pectinate lamellae (Fig. 16*c*). The remaining families (Geophilidae and others, see Chapter 24) have a single pectinate lamella of fine teeth (Fig. 16*d*).

The coxosternite of the first maxilla is occasionally divided medially. It bears the paired telopodites and the coxal projections and may carry two pairs of lateral maxillary palps (Fig. 17). The second maxillary coxosternite is undivided. The 'metameric pores' on the posterior border of the coxosternite are the openings of the second maxillary glands.

In geophilomorphs the forcipular coxosternite is undivided and a pleurite is present on each side. The second and third segments of the telopodite are very short and formed of incomplete rings. The forcipules vary considerably from genus to genus (Fig. 18) and may

Fig. 16. Mandibles of geophilomorphs. *a*, *Meinertophilus arcisherculis* (Brölemann); *b*, *Haploschendyla europaea* (Attems); *c*, *Mecistocephalus cephalotes* Meinert; *d*, *Geophilus pyrenaicus* Chalande (*a* & *d* after Brölemann, 1930; *b* & *c* after Attems, 1929).

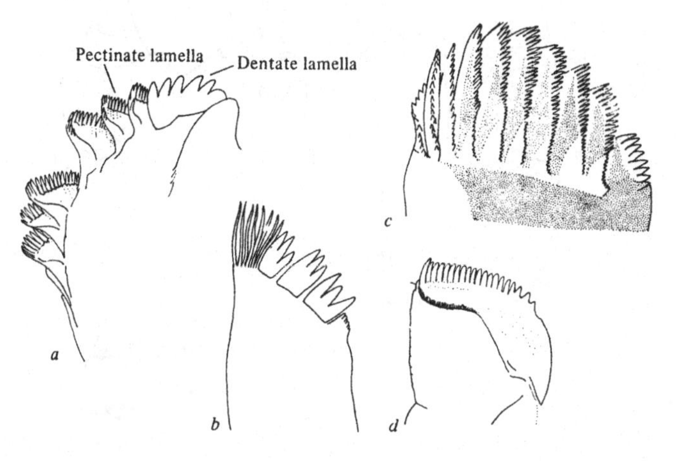

Fig. 17. The first and second maxillae of *Pleurogeophilus mediterraneus* (Meinert), ventral view (after Brölemann, 1930).

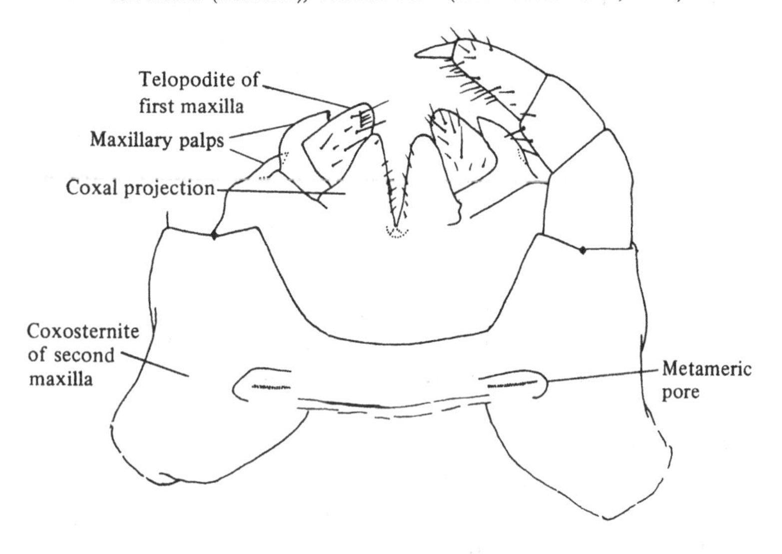

Fig. 18. Forcipules of geophilomorphs. *a*, *Arctogeophilus inopinatus* (Ribaut); *b*, *Henia bicarinata*; *c*, *Eriphantes telluris* (*a* & *b* after Brölemann, 1930; *c* after Crabill, 1970).

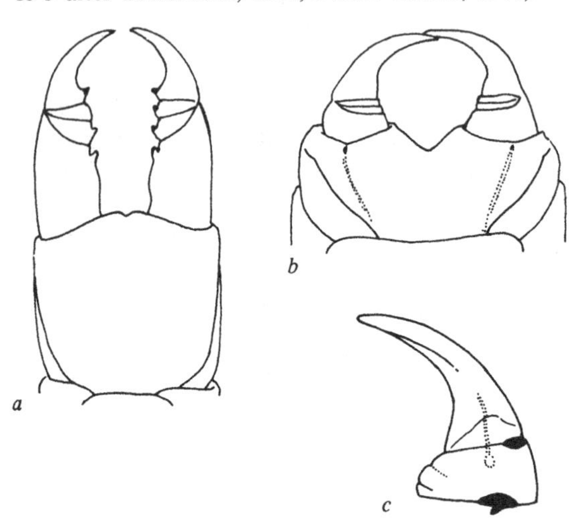

be very small. In *Eriphantes telluris* Crabill the telopodite consists of only two moveable parts, the claw and the fused femuroid, tibia and tarsus. In geophilomorphs two large posterior extensions of the coxosternite sink into the body, well into the second segment in *Orya* (Fig. 72) and further back in some other geophilomorphs (Manton, 1965).

Presternites and pretergites are well developed in the order. The metatergites usually bear paired paramedian longitudinal sulci and the sternites often bear groups of pores which are the openings of unicellular glands and may be diffuse or arranged in well-defined depressed areas (Fig. 19). These 'sternital pore fields' are of considerable taxonomic importance. They are frequently present only on the anterior segments.

The sternites of the anterior third of the body of a number of geophilomorphs show median or lateral pit like structures. In *Geophilus carpophagus* Leach (45 to 55 pairs of legs) strongly sclerotised 'carpophagus pits' are found from segment 4 and fade out after segment 11 being best developed on sternites 7 and 8. Situated medially on the anterior edge of the sternite, each carpophagus pit articulates with a peg on the middle of the hind border of the preceding sternite (Fig. 19*a*). *Haplophilus subterraneus* (Shaw) (69–89 leg-bearing segments) has strongly sclerotised pits, the 'virguliform fossae', immediately behind the anterior angle of the sternite from about segment 25 to segment 40 (Fig. 19*b*). *Nesoporogaster souletina* (Brölemann) (99–107 pairs of legs) has large median pits on the posterior borders of tergites 44 to 52 (Fig. 19*c*) whilst *Nothobius californicus* Cook (153 pairs of legs) has wide, deep parasternital clefts between the lateral tergite margins and the adjacent subcoxal pleurites from segment 45 to 75 (Crabill, 1960*a*) (Fig. 19*d*).

In *Mecistocephalus* spp. there is a Y-shaped suture on the sternites of anterior segments associated with the prolongation of the posterior border of the tergites into endosternites (Fig. 19*e*).

Spiracles are borne on all leg-bearing segments except the last.

Eason (1964) pointed out that there is sometimes a change in the structure of geophilomorphs about two-fifths of the way along the trunk. This may be striking and abrupt as in *Nesoporogaster*, spread over several segments as in *Geophilus*, or be absent alto-

Fig. 19. Sternites of geophilomorphs. *a*, *Geophilus carpophagus*, segment 7; *b*, *Haplophilus subterraneus*, segment 34; *c*, *Nesoporogaster souletina*, segment 46; *d*, *Nothobius californicus*, segment 45; *e*, *Mecistocephalus maxillaris* (Gervais), segments 11 & 12 (*a* after Manton, 1965; *c* after Eason, 1964; *d* after Crabill, 1960; *b* & *e* after Brölemann, 1930).

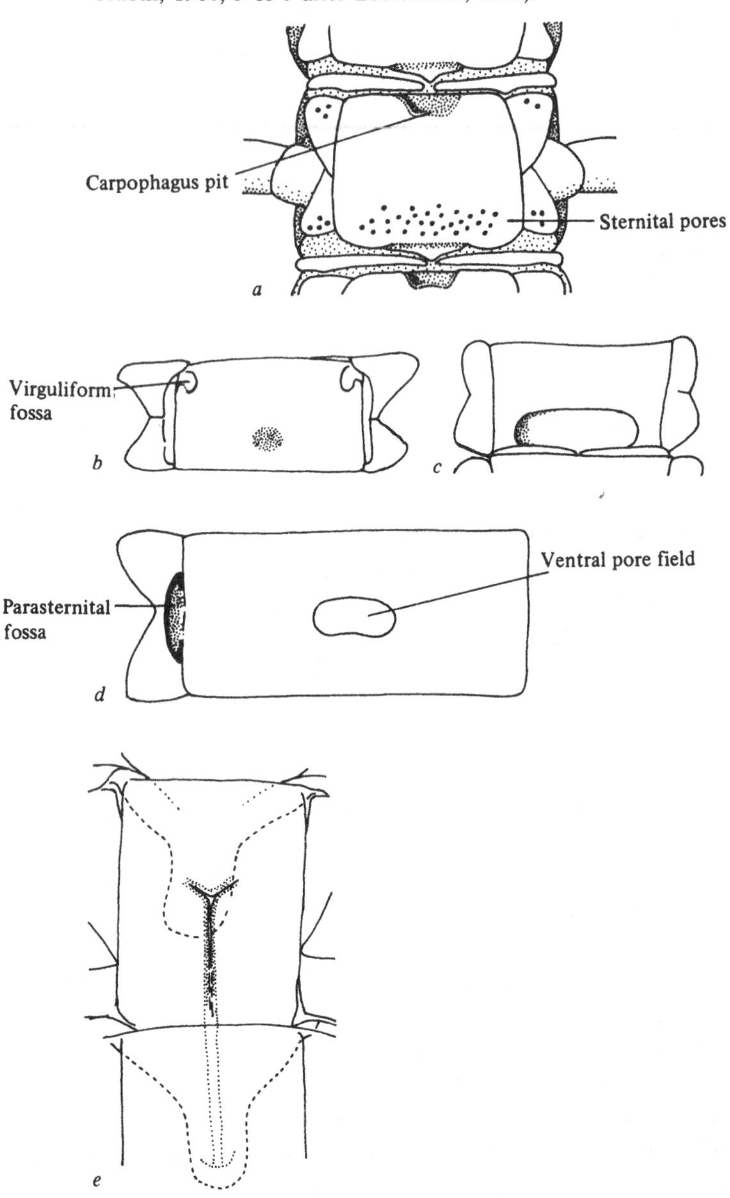

gether as in *Strigamia.* At this 'transition' the body becomes broader and each segment becomes longer, the legs more slender, the arrangement of the pleurites alters and they become less well sclerotised. The sternites lose much of their reticular pattern and the arrangement of the sternal pores often changes. It is around this point that the sternal fossae of Himantariidae are found and the carpophagus pits of *Geophilus* terminate.

The first pair of legs is considerably reduced in the Mecistocephalidae but otherwise the legs do not show marked change in size along the trunk. The tarsus and metatarsus of the legs are fused.

The last leg-bearing segment shows a number of modifications: with one exception pleurites are absent, probably having been incorporated into the coxa. The single pleurite is usually fused with the pretergite. The last pair of legs lies almost parallel to the body axis. The inflated coxae usually bear few to the many coxal pores (Fig. 14). Sometimes these open into pits. In *Nothobius californicus* there are two groups of pores in two concealed cavities: dorsally the majority of the pores are concealed in a hidden trench extending in a sinuous course into the penultimate leg-bearing segment (Crabill, 1960*a*). The telopodites are sometimes swollen in both sexes, more often just in males, and may be densely setose ventrally. The tarsus and metatarsus are separate and the claw may be absent or rudimentary.

The genital region is covered dorsally by the large tergite of the intermediate segment. In females the sternite of this segment is very narrow and is followed by the first genital sternite. This is fused with its pleurites and bears a narrow, transversely elongated unsegmented appendage (Fig. 20*a*). Sometimes, as in *Stigmatogaster gracilis* (Meinert), this unsegmented appendage is replaced by a pair of two-segmented gonopods (Fig. 20*b*). The first genital segment lacks a tergite. The oviduct opens behind this segment on the second genital segment. Behind this is the telson consisting of a small tergite beneath which there is a pair of anal valves.

In males the intermediate segment has distinct pleurites (Figs. 14, 20*c*) as does the first genital segment. The latter bears a pair of two-segmented gonopods in between which there is a protrusible structure frequently termed the penis. The telson has the same structure

Fig. 20. Genital region of geophilomorphs, ventral view. *a*, *Chaetechelyne vesuviana*, female; *b*, *Stigmatogaster gracilis provincialis* Chalande & Ribaut, female; *c*, *Chaetechelyne vesuviana*, male (after Brölemann, 1930).

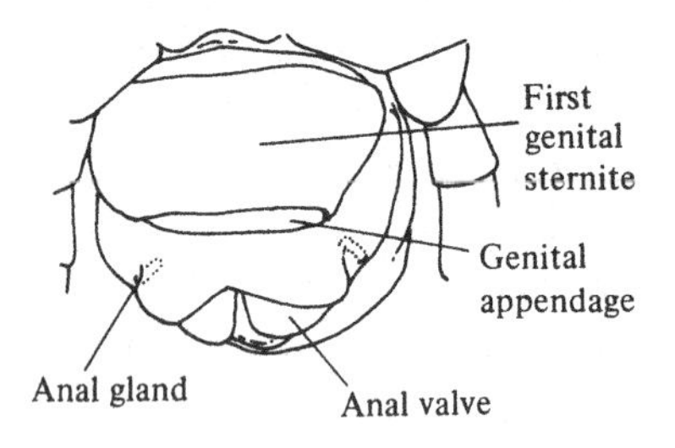

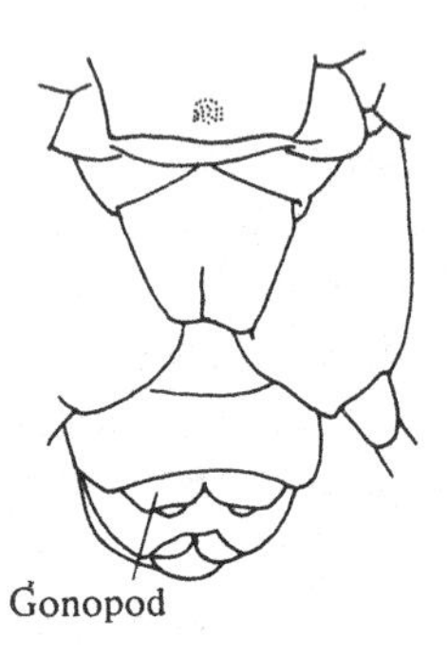

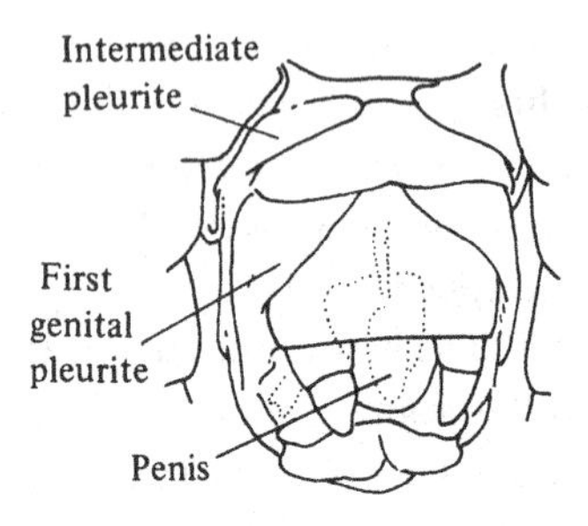

as in the female. In *Chaetechelyne* there is a transverse sclerite just in front of the anal valves which appears to be a post-genital segment.

In many geophilomorphs a pair of 'anal glands' opens laterally just anterior to the anal valves. These probably belong to the second genital segment.

Scolopendromorpha

There are usually 21 leg-bearing segments in the Scolopendromorpha but in the scolopendrid genus *Scolopendropsis* and the cryptopid subfamily Scolopocryptopinae, there are 23. The more or less rounded head capsule bears antennae composed of 17 or more segments and either four ocelli on each side or none. Tömösváry organs are lacking. The mid-piece of the labrum is small, projecting posteriorly as a single tooth, the side pieces are well developed and fringed with setae. The mandibles are well developed, the trunk of each bearing a cruciform suture. The apical ridge bears a dentate lamella with five tricuspid teeth on one mandible, four on the other. The dorsal angle bears a tuft of fine setae, the ventral angle bears several pectinate lamellae (Fig. 21). The coxosternite of the first maxillae is divided medially, lacks lateral palps but bears coxal projections (Fig. 22). The

Fig. 21. Mandible of *Cormocephalus violaceus* Newport (after Attems, 1930*a*).

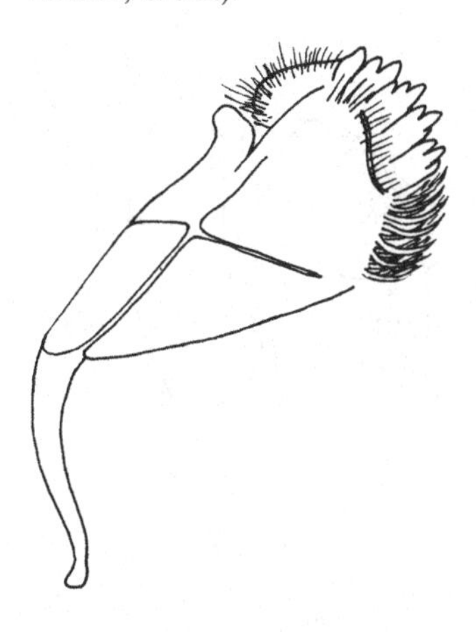

distal segment of the telopodite is terminated by a pad bearing very fine hairs. The structure of the second maxillae is shown in Fig. 22.

The forcipular tergite is fused with the first trunk segment. In many Scolopendridae the anterior border of the coxosternite bears a pair of tooth plates and a single pleurite is present on each side. The tibia and tarsus are shorter than in geophilomorphs and form incomplete rings. The femuroid may bear a lateral tooth (Fig. 23). As in the Geophilomorpha a pair of posterior extensions from the coxosternite sinks into the first trunk segment (Manton, 1965).

The trunk tergites show a varying degree of heteronomy, segments 2, 4, 6, 9, 11, 13, 15, 17 and 19 being slightly shorter than the remainder (Fig. 2). This is noticeable in the anterior segments, barely perceptible in the posterior ones. Spiracles are present on segments 3, 5, 8, 10, 12, 14, 16, 18, 20 and sometimes on segment 7. In species with 23 pairs of legs spiracles are present on segment 22 also. In *Plutonium*, spiracles are present on segments 2 to 20.

Separate pretergites are wanting but transverse sutures at the front of each tergite delimit an area that probably corresponds to the pretergite. Longitudinal paramedian sutures are usually present. In some *Cryptops* species there is a cruciform suture on the first tergite (Fig. 24). In scolopendrids the more posterior tergites

Fig. 22. Ventral view of the first and second maxillae of *Scolopendra cingulata* (after Brölemann, 1930).

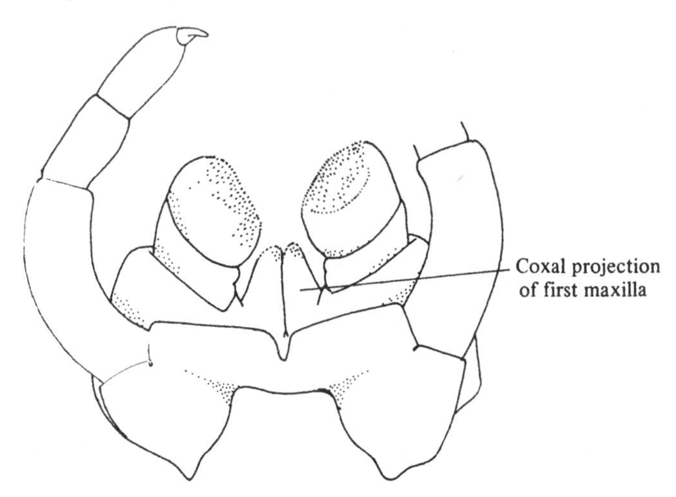

Fig. 23. Ventral view of the head and forcipules of *Cormocephalus calcaratus* Porat (after Manton, 1965).

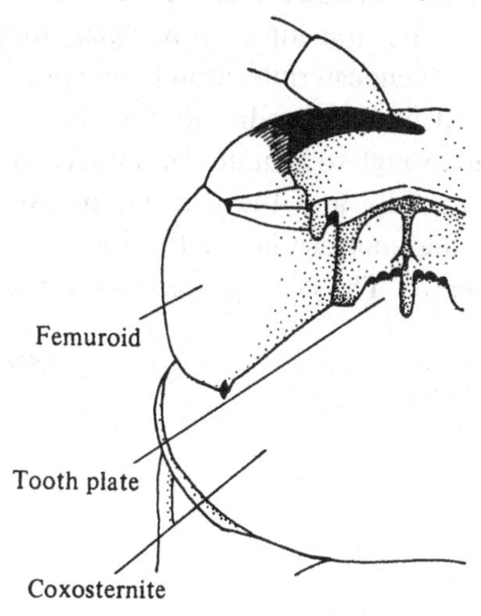

Fig. 24. Head capsule and first three segments of *Cryptops anomalans* (after Eason, 1964).

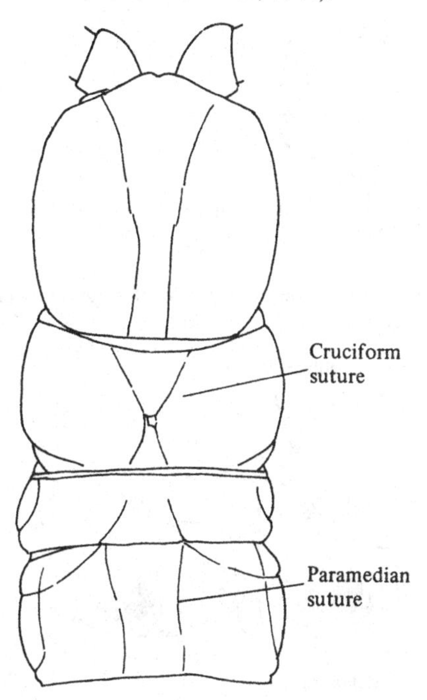

have a ridge along each lateral margin and are then said to be marginate. Presternites are present and similar to those of geophilomorphs. The metasternites are sometimes prolonged above the subsequent metasternite as an endosternite, as in *Cryptops* (Fig. 25).

The pleurites are reduced in scolopendromorphs; the stigmatopleurites are only just large enough to contain the spiracles (Fig. 26). The coxa is incomplete dorsally and the legs are of five or six segments and show a gradual increase in length from the anterior region of the trunk backwards. The last leg-bearing segment has a

Fig. 25. Sternites of segments 1 and 2 of *Cryptops savignyi* (after Brölemann, 1930).

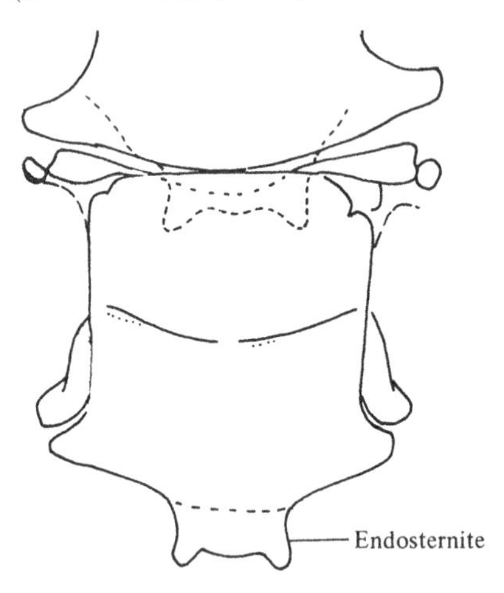

Fig. 26. Lateral view of segments 5 and 6 of *Cryptops hortensis* (after Füller, 1963*a*).

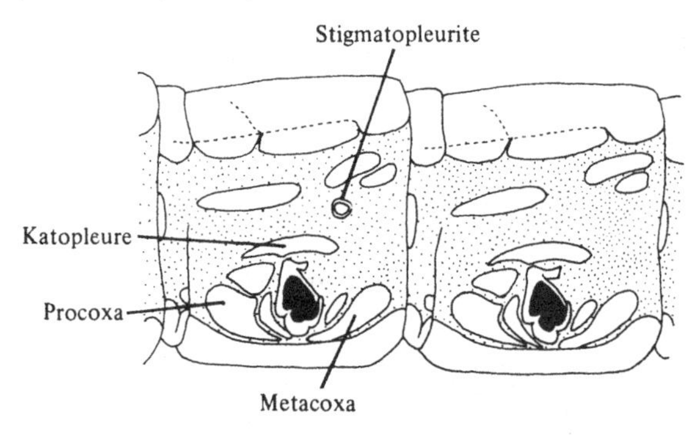

tergite which is narrower than the preceding ones as is the sternite. The pleurites are fused with the inflated coxae of the terminal legs which frequently bear numerous pores and in many scolopendrids there is a posteriorly projecting coxopleural process armed with spines (Fig. 27). The telopodite of the last pair of legs consists of five segments, the trochanter being absent. These legs are usually thicker than the normal walking legs. In scolopendrids they frequently bear spines on the prefemur (Fig. 27) but in *Cryptops* the prefemur is without spines. Instead there is a row of saw-like teeth on the ventral margin of the tibia and tarsus (Fig. 28*a*) which meet when the legs are flexed and may be used as prehensile organs. In *Newportia* there are 23 pairs of legs and the femur of the last pair usually carries from one to three teeth: the second tarsus is divided into a large number of secondary segments (Fig. 28*b*).

The terminal segments are strongly retracted beneath the twenty-first tergite and are scarcely sclerotised. Seen from below they hardly project beyond the sternite of the last leg-bearing segment. Bücherl (1942) described the terminal segments of several

Fig. 27. Ventral view of last leg-bearing segment and terminal segments of *Trachycormocephalus mirabilis* (Porat) (after Lewis, 1967).

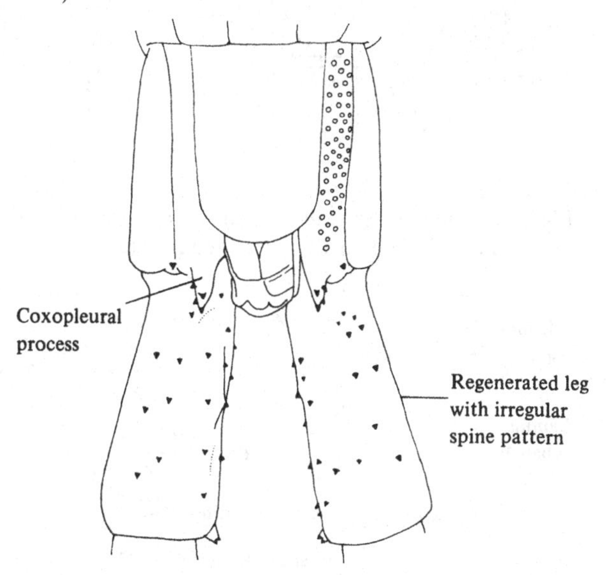

Brazilian species of *Scolopendra*; Jangi (1956, 1957) gave a clear account of their structure in *Scolopendra morsitans* Linn. In the female *S. morsitans* the genital segment (first genital segment) is represented by a sternite, the post-genital segment (second genital segment) is introverted to form the genital atrium and the anal

Fig. 28. Terminal legs of cryptopids. *a*, *Cryptops anomalans* (Newport) (after Eason, 1964); *b*, *Newportia longitarsis longittarsis* (Newport) (after Attems, 1930*a*).

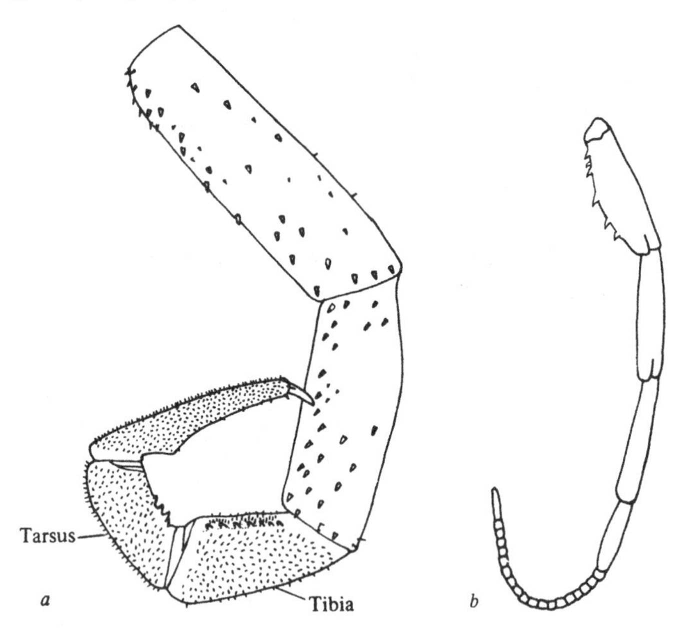

Fig. 29. Terminal segments of *Scolopendra morsitans*. *a*, female (after Jangi, 1957); *b*, male (after Jangi, 1956).

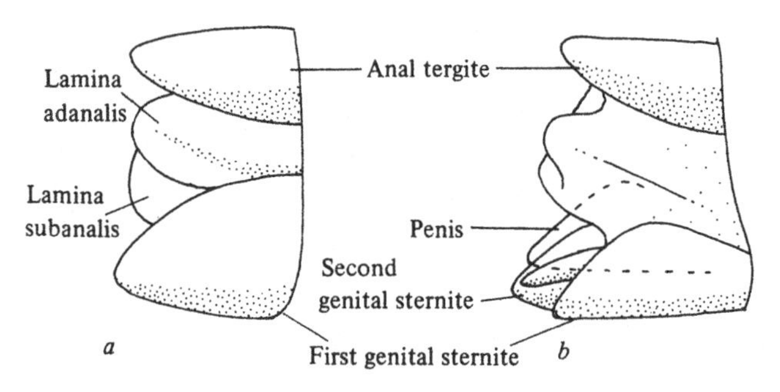

segment (telson) by the anal tergite, lamina subanalis and two laminae adanales (Fig. 29*a*).

In the male the sternite of the first genital segment is well sclerotised and carries a pair of narrowly conical genital appendages. These are not present in all species, for example *Scolopendra valida* (Lucas) (Demange & Richard, 1969). The second genital sternite is represented by a transversely elongated sternite and the 'penis' consisting of two triangular valves (Fig. 29*b*). The anal segment or telson is as in the female. Anal glands are absent in both sexes.

Lithobiomorpha

The lithobiomorphs have 15 leg-bearing segments. The head capsule is approximately heart-shaped with a distinct marginal ridge posteriorly and half-way along each side (Fig. 30*a*). A transverse suture crosses the anterior third of the head giving rise at its lateral extremities to anterior and posterior running longitudinal sutures. The paired antennae consist of from 18 to more than 100 segments. There is a group of from one to thirty ocelli on each side of the head capsule: ocelli are absent in some species. Immediately antero-ventrally to the ocelli there is a single Tömösváry organ on each side. This is a small spherical sensory structure with a central pore (Fig. 30*b*).

Fig. 30. The head of *Lithobius*. *a*, *L. variegatus*, dorsal view; *b*, *L. forficatus*, lateral view (after Eason, 1964).

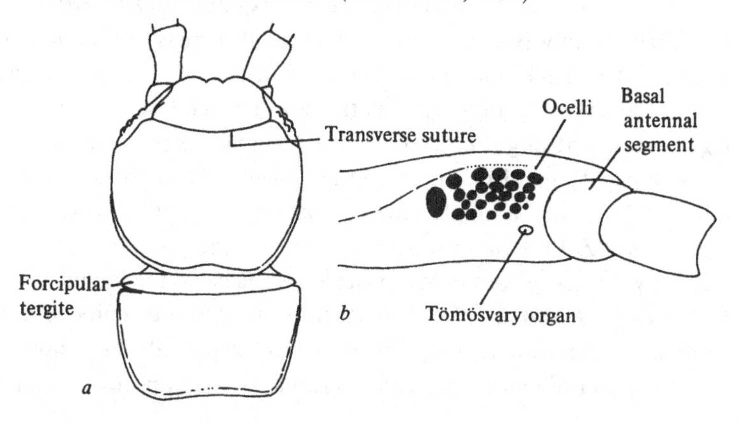

The clypeus bears an anterior median conical projection. The labrum is similar to that of the Scolopendromorpha. The structure of the lithobiomorph mandible is shown in Fig. 7. The maxillae are similar to those of the scolopendromorphs but the coxosternite shows definite signs of a median division.

The forcipular tergite is small (Fig. 30*a*). The coxosternite has a median longitudinal cleft and usually carries a number of teeth along its anterior border (Fig. 8). In some species of *Lithobius* there are two teeth on each and in *Bothropolys* as many as 11. The pleurites do not encroach on the ventral border of the coxosternite as in the previously described orders. The tibia and tarsus form complete rings (Fig. 8).

The lithobiomorphs show a very marked heteronomy, tergites 2, 4, 6, 9, 11 and 13 being very much shorter than the others (Fig. 3). A sulcus runs along the lateral and posterior margins of tergites 1 and 3 but is limited to the lateral margins of the more posterior tergites. Tergite 7 and the more posterior short tergites may carry lateral triangular projections (Fig. 218*f*). There are no pretergites or presternites. Between tergite 14 and the last large tergite there is a small tergite, only visible in relaxed and extended specimens, which belongs to segment 15 (Fig. 31*a*). The large tergite following it is that of the intermediate segment. The pleurites are reduced in number; stigmatopleurites which spiracles are present on segments 3, 5, 8, 10, 12 and 14 in most species and in addition on segment 1 in the Henicopidae. The other pleurites are insignificant (Fig. 12*b*).

The legs consist of a coxa, a telopodite of five or six segments and an apical claw. In some genera, for example *Esastigmatobius*, the tarsus is subdivided. There is a gradual increase in leg length from front to back and this is accompanied by an increase in the size of the coxae. In most species the coxae of the last four pairs of legs have a ventral gutter in which are located one or more rows of coxal pores (Fig. 31*b*). In the genus *Pseudolithobius* the last five pairs of legs bear a single row of coxal pores, and in *Dakrobius krivolutskyi* Zalesskaja coxal pores occur on the last three pairs of legs only. In many species the posterior surfaces of the femur, tibia, tarsus and metatarsus of the fourteenth and fifteenth pairs of legs sometimes the twelfth and thirteenth and rarely all legs show a concentration of telopodal glands opening by small pores. In many

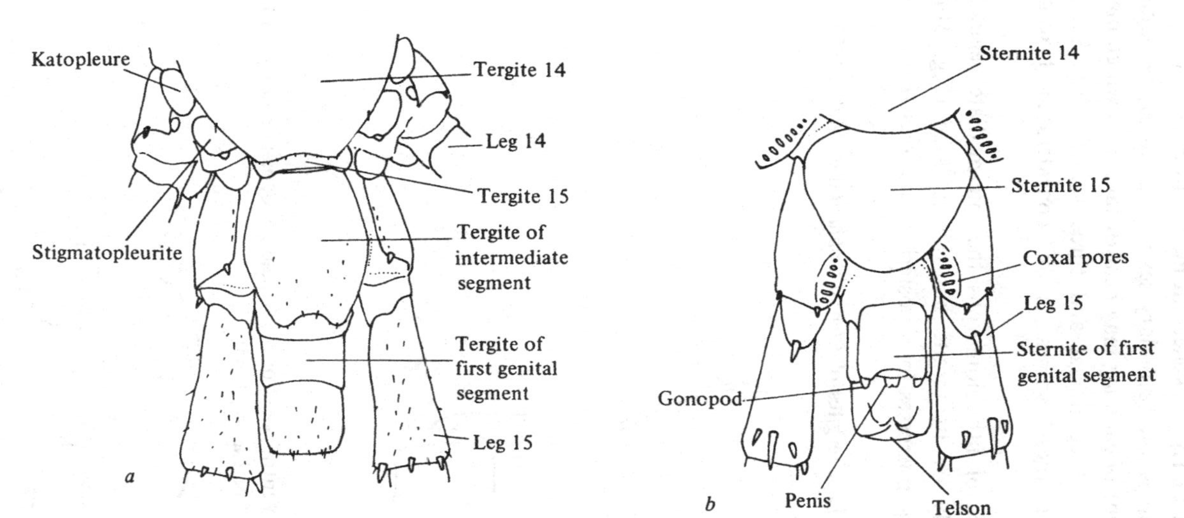

Fig. 31. Terminal segments of male *Lithobius pilicornis* Newport. *a*, dorsal view (after Brölemann, 1930); *b*, ventral view (after Brölemann, 1930 and Eason, 1964).

species and particularly in males the fourteenth and fifteenth pairs of legs are particularly thickened. In males either or both of these pairs of legs may show secondary sexual structures (see Chapter 17). Each segment of each leg may bear distal spines which may be as many as six, three dorsal and three ventral (Fig. 93). The arrangement of these spines is of considerable taxonomic importance.

In both sexes of lithobiomorphs the intermediate segment is represented only by its tergite but this is large (Fig. 31*a*): in *Pleurolithobius* it possesses long posterior processes (Fig. 32). The

Fig. 32. Last tergites of *Pleurolithobius* (after Matic, 1962).

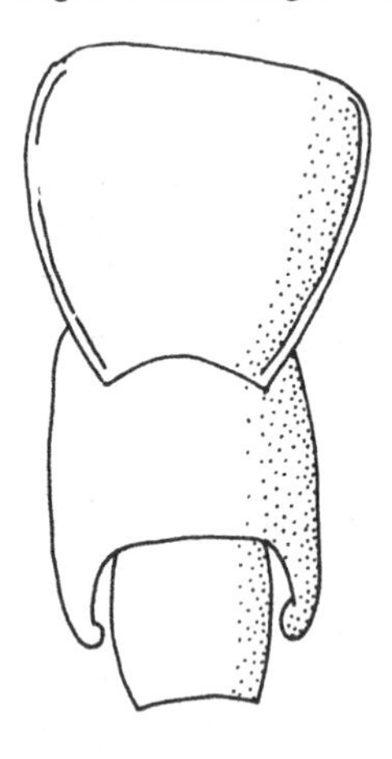

Fig. 33. Ventral view of the female gonopods of *Lithobius forficatus* (after Eason, 1964).

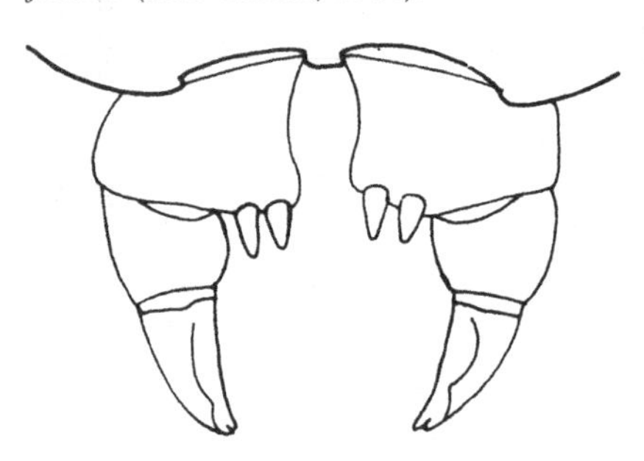

first genital segment is represented by a tergite and a trapezoidal sternite with an embayed hind border on each side of which are the genital appendages. In males these are reduced to small humps in *Lithobius* but are relatively long and two-segmented in *Bothropolys*. In females the gonopod consists of a broad basal segment, which usually carries two or three large flattened spines on its inner (posterior) edge, and a second segment which carries a large claw (Fig. 33). The second genital segment is represented by the penis in males and by a small transverse chitinised band bearing two small lobes in the female. The telson consists of a rectangular tergite and a pair of valves. In larval *Lithobius* a pair of anal glands is present: these persist in adult *Lamyctes* (Brölemann, 1930).

Scutigeromorpha

The scutigeromorphs have 15 pairs of legs. The body is fusiform in shape and not dorso-ventrally flattened. The head capsule is approximately hemispherical and the clypeus and pleurites are not folded underneath it as in other orders. The origin of the antenna is lateral rather than frontal: the two basal segments (the scape) are large and the first bears the opening to a sensory organ (the *Schaftorgan*) (Fig. 80). The rest of the antenna consists of the long flagellum made up of numerous annulations. Along the flagellum there are one or two 'nodes', which are considerably longer than the annulations (Fig. 34). The compound eyes are situated behind the antennae (Fig. 35) and the small Tömösváry organs are found between the antennae and the eyes.

The mandibles and first maxillae are similar to those of the Lithobiomorpha. On the dorsal surface of the coxosternite of the first maxillae there are two areas of deeply invaginated cuticle covered with hairs and spindle-like processes (Fig. 36). These were termed maxillary organs by Haase (1884*b*). He regarded them as sensory organs. They are, in fact, distensible cleaning organs through which the legs and antennae are passed repeatedly. The second maxilla has a very narrow coxosternite. The telopodites consist of four long slender segments, the first three bearing two to four long spines. The last segment lacks a claw (Fig. 37).

The forcipular tergite is short; the coxites, each of which carries three or four long spines on its anterior border, are separate and

Fig. 34. Basal segments of the antenna of *Scutigera coleoptrata* (after Verhoeff, 1902–25).

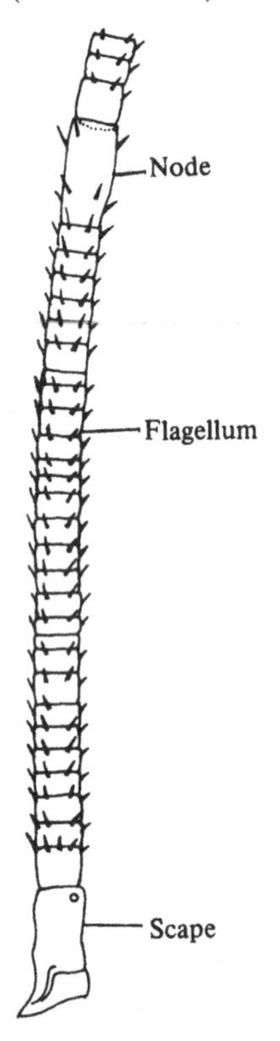

Fig. 35. Dorsal view of head capsule of *Scutigera coleoptrata* (after Attems, 1926).

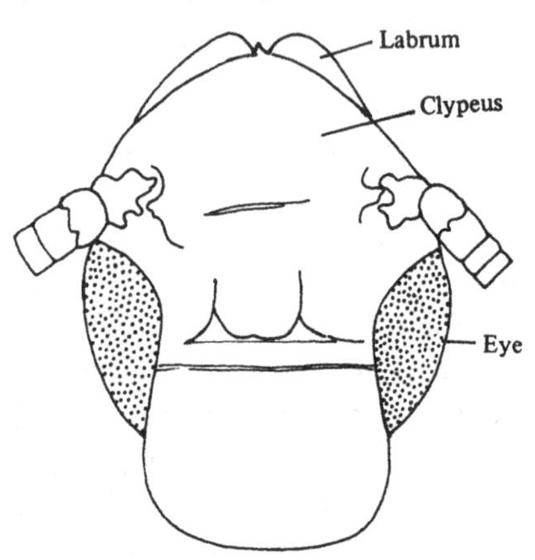

the sternite is vestigial. The segments of the forcipules are much longer and more slender than in other orders and the poison claw is relatively small. The femuroid bears a long spine (Fig. 38).

Each leg-bearing segment possesses a sternite but there are only seven tergites formed by the fusion of atrophied short tergites with

Fig. 36. Ventral view of first maxilla of *Scutigera coleoptrata* (after Brölemann, 1930).

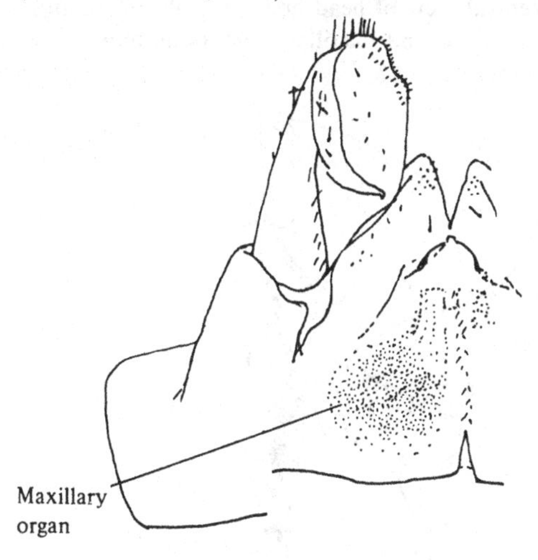

Fig. 37. Second maxilla of *Scutigera coleoptrata* (after Brölemann, 1930).

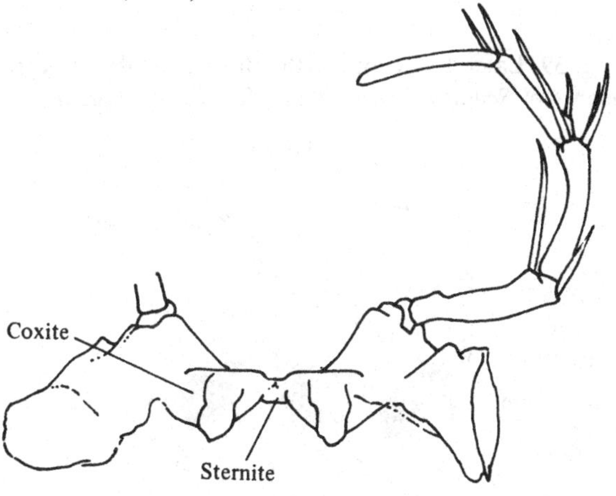

the subsequent long tergite. Tergite 1 corresponds to segment 1, tergite 2 to segments 2 and 3, tergite 3 to 4 and 5, tergite 4 to 6, 7 and 8, tergite 5 to 9 and 10, tergite 6 to 11 and 12, and tergite 7 to segments 13 and 14. The tergite of segment 15 is absent (Eason, 1964). At the anterior end of each large tergite there is an indistinct line marking off a poorly chitinised band which represents the fused small tergite (Fig. 39). The hind border of each long tergite is

Fig. 38. Ventral view of head and forcipules of *Scutigera coleoptrata*, the second maxillary telopodite omitted (after Manton, 1965).

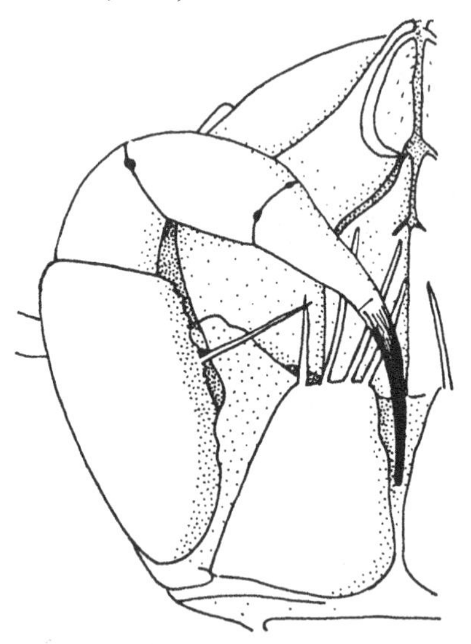

Fig. 39. Lateral view of fourth, fifth and sixth leg-bearing segments of *Scutigera coleoptrata* (after Füller, 1963*a*).

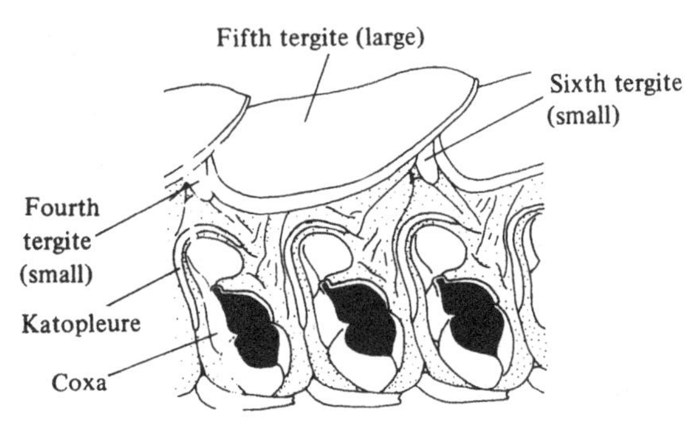

notched and in front of this is a slit-shaped spiracle opening into the so-called tracheal lungs (Fig. 11*a*). The pleurites are very weakly sclerotised. Murakami (1959*a*) considered that the eight long tergites belong to the odd leg-bearing segments and that each even segment has a small covered tergite. Therefore, according to him, tergite 4 does not correspond to segments 6, 7 and 8.

The legs are very long and increase in length from the front to the back of the body. The coxa is well developed and bears a ventral spine, the trochanter is very much reduced. The prefemur, femur and tibia bear two or three longitudinal ridges beset with spines or teeth and the ends of these three segments bear spines homologous with those of the Lithobiomorpha. The tarsus and metatarsus are subdivided into annulations (Fig. 13); there is a well-developed apical claw. On the ventral surface of the annulations there are numerous small setae and small pegs (tarsal papillae) giving a firm grip on the substratum, (Fig. 219*g*). The last leg-bearing segment lacks a tergite but a normal sternite is present. The last pair of legs is much longer than the preceding pairs. The coxae are parallel to the longitudinal axis of the body. The prefemur, femur and tibia have weak longitudinal ridges; their distal spines are reduced in number. The tarsus and metatarsus are not distinct. They are divided into more than 500 annulations bearing ventral hairs and pegs but lacking a terminal claw.

Brölemann (1930) gave a clear description of the structure of the

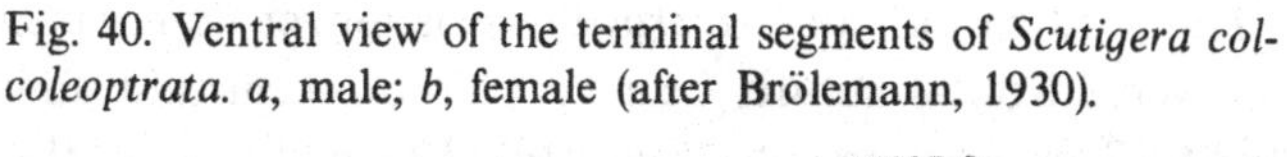

Fig. 40. Ventral view of the terminal segments of *Scutigera coleoptrata*. *a*, male; *b*, female (after Brölemann, 1930).

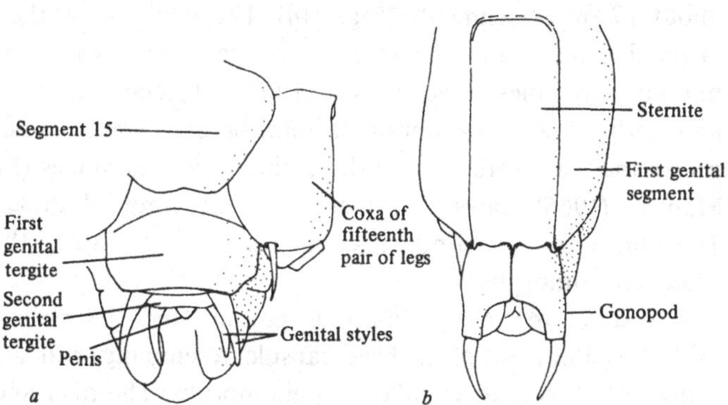

genital region of *Scutigera coleoptrata* (Linn.). The four terminal segments are covered by tergites 8 and 9. Tergite 8 is that of the intermediate segment which lacks a sternite. The tergite of the first genital segment is membranous but the sternite is well-developed. In males it is a trapezoidal structure which bears a pair of attenuated styles (gonopods). The second genital segment bears a similar pair of styles between which there is a penis (Fig. 40*a*). In the female the ventral part of the first genital segment is more developed than it is in the male and almost entirely covers the following segments (Fig. 40*b*). It is a coxosternite with a central sternite and lateral coxites. A condyle on the posterior border of each coxite articulates with the two-segmented gonopod, the second segment of which is style-like. The second genital segment is not sclerotised.

Tergite 9 is that of the anal segment. There are no anal glands.

Craterostigmomorpha

Craterostigmus tasmanianus is a moderate sized centipede; the type specimen is 37 mm long. It occurs in Tasmania and New Zealand and was first described by Pocock (1902). The systematic position of the species has been much discussed and it has been considered to be both a lithobiomorph and a scolopendromorph. Pocock placed it in a separate order the Craterostigmomorpha. There is still no large measure of agreement as to its affinities. Its relationships are discussed in Chapter 23.

The head capsule of *Craterostigmus* is longer than wide: it has a transverse frontal suture, bears a pair of ocelli and antennae of about 17 or 18 segments (Fig. 41*a*). The mid-part of the labrum bears five teeth and the side pieces are fringed with setae. The minute mandibles have a thick fringe of setae posteriorly and anteriorly a large membranous lobe bearing small hairs. In the centre there are three sets of three horny teeth or spines (Fig. 41*b*). Manton (1965) states that there are only three of these spines. The first and second maxillae are much like those of scolopendromorphs.

A distinct forcipular tergite is present and the poison claws extend well in front of the head capsule, resembling in arrangement those of the mecistocephalid geophilomorphs. The tibia and tarsus are complete rings as in the Lithobiomorpha and the inner face of the femuroid bears a group of five teeth. The anterior border of the coxosternite has a tooth plate of five to seven teeth on each side (Fig. 42).

There are fifteen leg-bearing segments as in lithobiomorphs but

Fig. 41. *Craterostigmus tasmanianus*. *a*, dorsal view of head region; *b*, mandible; *c*, fourteenth leg; *d*, ventral view of terminal segments; *e*, dorsal view of anterior segments (after Pocock, 1902).

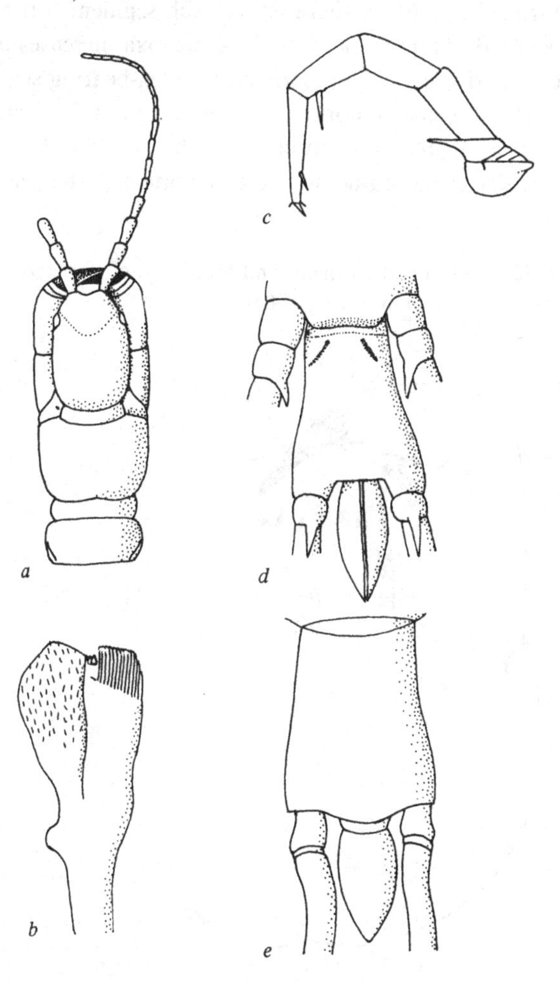

since these are rather long and narrow the animal resembles a *Scolopendra* in shape (Fig. 5). There appear to be 21 trunk tergites plus an anal tergite. This is because tergites 3, 5, 7, 8, 10 and 12, which correspond to the long tergites of scolopendromorphs and lithobiomorphs, are subdivided. Pocock maintained that there were no pretergites but Manton has shown that these are present on all but tergites 1 and 14 (Fig. 43). There are 15 sternites, each with presternites and two transverse hinge lines (Fig. 43*e*).

Spiracles are present on segments 3, 5, 8, 10, 12 and 14 which therefore bear a stigmatopleurite behind which there is a post-stigmatopleurite. In addition there is, on each segment a procoxa, katopleure and other minor pleurites. The precoxa increases in size towards the posterior end of the body and on leg-bearing segments 13, 14 and 15 it forms a continuous plate almost, or entirely covering the pleural area and fused with the sternite (Fig. 43*d*).

The legs consist of six segments: the trochanters of the last three

Fig. 42. Ventral view of head and forcipules of *Craterostigmus tasmanianus* (after Manton, 1965).

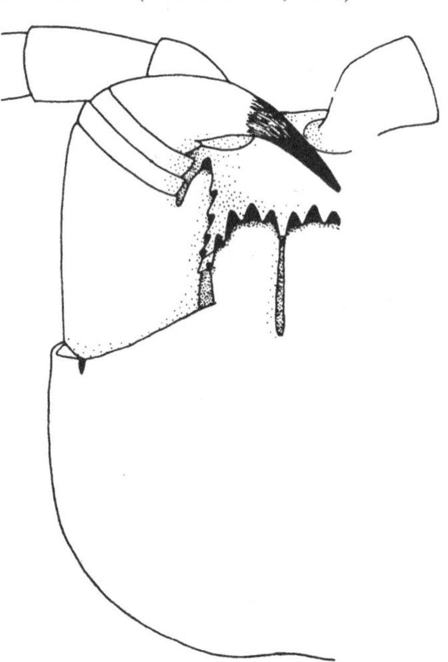

pairs of legs are each armed with a long spine (Fig. 41*c*). The last pair of legs are long and slender and lack spines apart from the one on the trochanter.

The terminal segments are represented by a sclerite formed of two valves which are fused dorsally but meet to form a longitudinal

Fig. 43. Anatomy of *Craterostigmus tasmanianus a–d* lateral view showing segmentation; *e*, ventral view of segment 3 (after Manton, 1965).

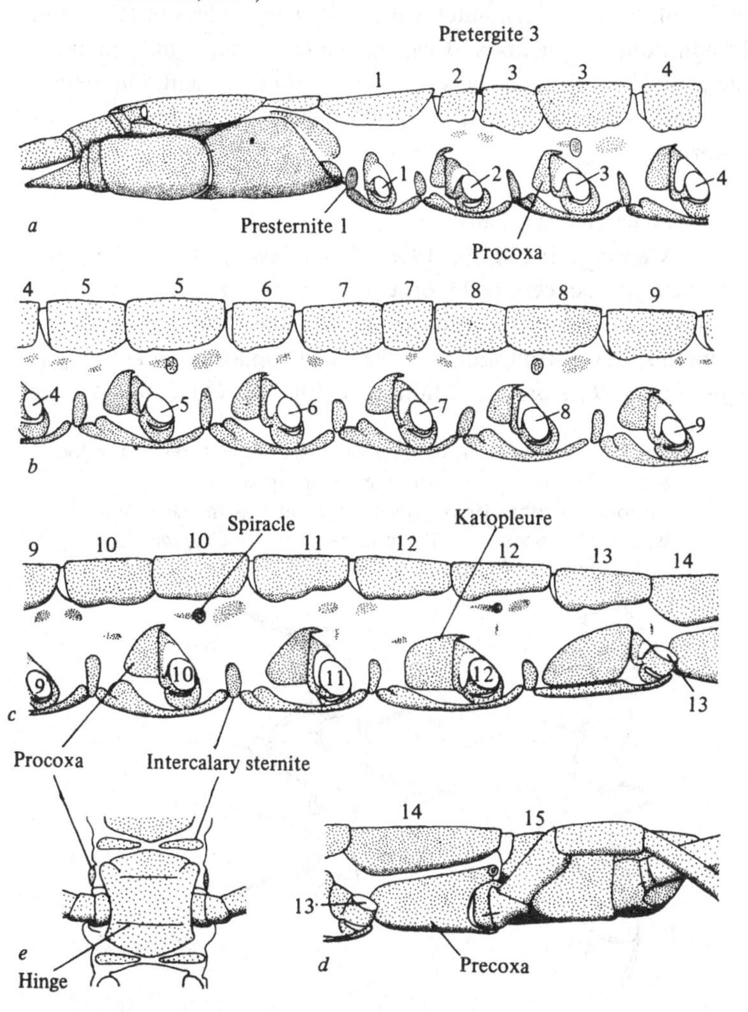

slit ventrally (Fig. 41 *d*, *e*). Pocock considered it to be the homologue of the anal sternite (telson). There is no trace of the genital sternite or gonopods.

Epimorpha and Anamorpha

Haase (1880) suggested that there was a basic subdivision within the Chilopoda into the Epimorpha (Geophilomorpha and Scolopendromorpha) which hatch from the egg with a full complement of legs and the Anamorpha (Lithobiomorpha and Scutigeromorpha) which hatch with seven pairs of legs or fewer and add additional segments and legs in the early stages of their post-embryonic development. This subdivision is convenient. Discussion continues as to whether *Craterostigmus* is anamorphic or epimorphic.

Functional anatomy

Manton (1952*a*, *b*, 1958, 1965) sought to explain the characters of the centipede orders in terms of their functional anatomy, particularly locomotory mechanisms. She first examined the method of locomotion of the onychophorans *Peripatopsis sedgwicki* and *P. moseleyi* (Manton, 1950). The Onychophora are

Fig. 44. Diagrams illustrating the essential features of the locomotory mechanism of Scolopendromorpha and Lithobiomorpha. Heavy lines represent legs in the propulsive backstroke, open lines those in recovery. *a*, *Cryptops*; *b*, *Lithobius* (after Manton, 1952*b*).

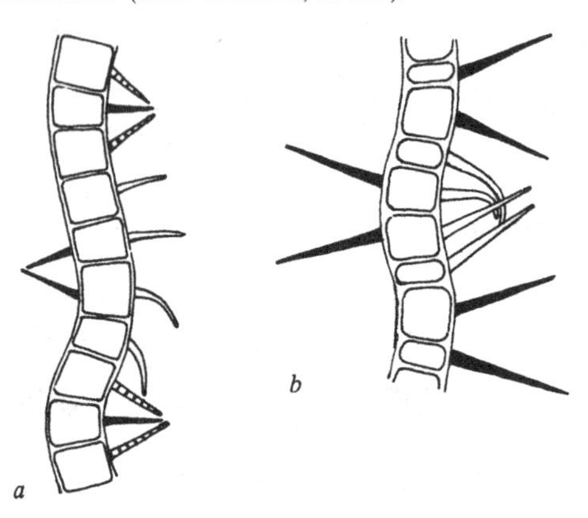

probably the most primitive extant arthropods: they are caterpillar-like animals with short conical legs extending out ventro-laterally from the body. Locomotion in *Peripatopsis* involves a variety of 'gaits'. In the bottom gear gait, which is associated with low speeds of movement, the forward stroke of the leg is relatively fast, the backward (propulsive) stroke slow. In middle gear which is used for moderate to fast walking the duration of forward and backstrokes is the same, while in top gear, which is used for easy walking when momentum has already been achieved, the forward stroke is slow, the backstroke fast. Manton (1952*b*) demonstrated that in the epimorphic centipedes the middle gear gait has been developed and since the phase difference between the legs is always greater than 0.5 the successive propulsive legs always converge (Fig. 44*a*). In the Anamorpha the top gear gait has been developed and as the phase difference between successive legs is less than 0.5, the propulsive legs diverge (Fig. 44*b*).

In centipedes, as in many other arthropods, the body is hung from the legs (Fig. 45*b*); such a stance is more stable than a 'standing-up' position such as is seen in *Peripatopsis* (Fig. 45*a*) since it keeps the centre of gravity low and at the same time allows an increase in leg length (Manton, 1952*a*). Moreover, it allows the legs to cross over each other during running, provided that is, that the legs become progressively longer towards the posterior end of the animal.

Geophilomorpha

The main morphological characteristics of the Geophilomorpha, namely the small head, tapered anterior end, large number of segments with separate intercalary tergites and smaller intercalary sternites (pretergites and presternites) and relatively short legs are associated with the methods of locomotion in

Fig. 45. Cross section of *a*, *Peripatopsis* and *b*, *Cryptops* (after Manton, 1952*a*).

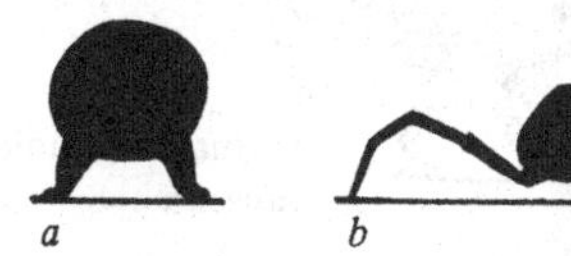

the order. Geophilomorphs employ two methods of locomotion, namely walking and burrowing. In walking, increased speed is achieved by decreasing the relative duration of the leg's backstroke. As there is little variation in the length of legs it is impracticable for them to cross over appreciably during the propulsive phase and in most gaits the angle of swing is constant. Burrowing is brought about by elongation and contraction of the body. Telescoping is facilitated by the presence of the intercalary tergites and sternites which slide under the tergites and sternites and is brought about by a strong longitudinal musculature (Fig. 46). A tendency towards a shortening of the segments and their increase in number is also correlated with the burrowing mechanism. In a large *Orya barbarica* (Gervais), a subterranean species found in the Mediterranean region, the segments are at least 6 mm wide and only 1.5 mm long in the longitudinally contracted condition and there are 107 to 125 pairs of legs. In species of *Geophilus* found under stones there are usually only 40 to 50 leg-bearing segments. The small head and tapered anterior end assist entry into crevices and holes.

The fact that the anterior third of the trunk is traversed by a narrow oesophagus, the mid-gut lying behind this region, means that no large internal food masses will hinder anterior mobility and leaves more room for locomotory muscles. The heart, oesophagus and nerve cord are centrally placed in this anterior region, thus

Fig. 46. Posterior view of a section through segment 7 of *Geophilus carpophagus* (after Manton, 1965).

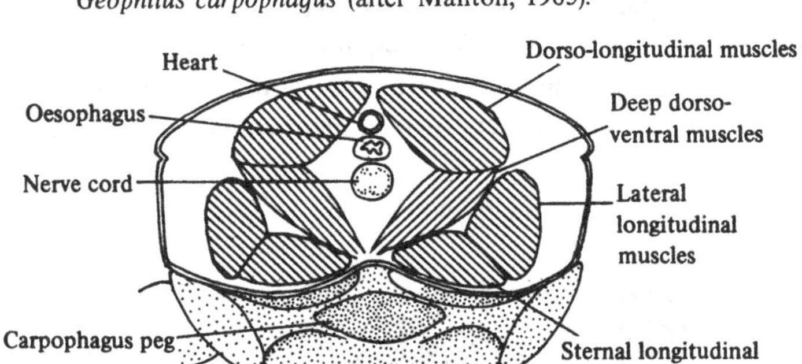

suffering minimum deformity during burrowing activities (Fig. 46) (Manton, 1965).

Manton suggested that the carpophagus structures form peg and socket articulations when the body is strongly flexed. Pushing against the substratum is possible in this position and the ventral intake of the body can be great enough to need no dorsal outlet; this is essential if the animal is to exert a thrust on the soil by a shortening of the flexed segments. If the cuticle of the segments used in this manner was not very strongly constructed, uncontrollable buckling might occur with internal damage to the animal. The major function of the carpophagus structures is, Manton concluded, to enable the animal to burrow at an angle to the surface.

The pits of *Nesoporogaster* may act as carpophagus pits but it is unlikely that the virguliform fossae of *Haplophilus* do so, for in order to allow strong downward flexion of the body the articulations must be near the mid-line to allow the free lateral bending which is essential for this type of activity. Clearly the parasternital clefts of *Nothobius* serve some quite different function. Verhoeff (1902–25) suggested that such pits in geophilomorphs were external storage chambers for the secretions of the sternal glands and that these are used to overcome large prey such as earthworms but there is currently little evidence to support this suggestion.

The great flexibility of the trunk of geophilomorphs can produce a flexion of the body of 180° so that it becomes folded back on itself. Further, many species seem to be able to move backwards as well as forwards and the last pair of legs are adapted to form 'posterior antennae' (Eason, 1964). They do not, however, show the variety of sensilla seen on the antennae. In long species the anterior end can be seen to move forward at the same time as the posterior end moves backwards.

Scolopendromorpha

In Manton's opinion, the scolopendromorphs unlike the geophilomorphs do not show morphological adaptations for burrowing. They are relatively fleet animals, capable of running fairly fast when required to do so. Speed is facilitated by having longer legs than geophilomorphs and by reducing the relative duration of the backstroke. As the backstroke decreases, however, the points of support for the body become further apart and there is an increasing tendency for the propulsive legs to throw the body into lateral undulations. Energy expended in such lateral movement is wasted and since the propulsive legs lie in the concavities of the undulating body and not at the convexities (Fig. 44*a*) and since the limb tips can only cross over slightly, the effective angle of swing is reduced. Anti-undulatory mechanisms associated with fast running are best developed in the Anamorpha but scolopendromorphs show some development to this end in that there is some degree of variation in tergite length so that the joints of the tergites no longer lie above the joints of the sternites.

The fact that the posterior legs in scolopendromorphs are longer than the anterior ones allows some degree of crossing by the limb tips and thus increases the angle of swing but the fact that they have 20 pairs of walking legs imposes limits on their length. As each leg must be capable of the same length of stride, the anterior legs cannot be very much shorter than the posterior ones. Leg 20 is often a little over twice as long as leg 1.

Not all scolopendromorphs are, however, relatively fleet: *Asanada sokotrana* Pocock frequently found in deserted termite mounds in Nigeria (Lewis, 1973) is a sluggish and slow-moving animal as is *Arrhabdotus octosulcatus* (Tömösváry) a species from the rain forests of Borneo that may be arboreal (Lewis, unpublished data). Lateral undulations of the body may not always be without benefit: *Scolopendra subspinipes* Leach can swim using lateral undulations (Lewis, unpublished data) and *Rhysida nuda togoensis* Kraepelin and *Otostigmus longicornis* (Tömösváry) escape when disturbed by violent squirming movements like those of some worms and millipedes (Lewis, unpublished data). It may be necessary to review our ideas on the Scolopendomorpha when more living material has been studied. Scolopendromorphs may not burrow as geophilomorphs do but they are better able to penetrate the soil than most Anamorpha. *Cryptops hyalinus* Say when disturbed in the leaf litter or lower horizons buries itself very quickly utilising all crevices and openings available, rather than running over the substratum as lithobiids are prone to do. The anterior appendages push away the dirt to permit penetration into the soil but the main progress is made by forcing and wedging the slender body through loose soil. Particles of soil generally fall in behind the retreating last pair of legs closing off the temporary subterranean burrow (Johnson, 1952). *Scolopendra cingutata* also burrows actively (Klingel, 1960*a*).

Lithobiomorpha and Scutigeromorpha

The Anamorpha have developed the 'top gear' gait. In association with this type of locomotion there has been an increase in leg length which results in a longer stride and hence an increase in speed. The maximum stride lengths of *Cryptops* (a small scolopendromorph), *Lithobius* and *Scutigera* of comparable sizes are 15, 21 and 38 mm respectively, the stride lengths of *Scutigera* greatly exceeding the length of the body.

When, as is the case in Anamorpha, the legs are long relative to the length of the segment, crossing over by consecutive legs is virtually unavoidable and in these animals takes place during the recovery stroke. An upper limit to the phase difference between the legs is set by the desirability of avoiding not only crossing over of successive legs in the propulsive phase, but also any crossing of the posterior legs of one group of successive propulsive legs by the anterior leg of the next group behind it. Thus the phase difference is less than 0.5 and therefore successive propulsive legs diverge from one another (Fig. 44*b*). In both Lithobiomorpha and Scutigeromorpha the number of walking legs has been reduced to 14, the last pair of legs is not used in locomotion, and each leg carries an approximately equal load. The coxae in the Anamorpha are enlarged to house the musculature which operates the long legs.

Lateral undulations of the body appear in the faster gaits of the Anamorpha just as they do in Scolopendromorpha. This tendency is countered, as in the scolopendromorphs by the development of tergite heteronomy (see above): in scutigeromorphs seven large tergites cover 15 sternites.

The Scutigeromorpha are the fastest moving myriapods: Manton (1952*b*) recorded a speed of 420 mm/s for a specimen 22 mm long whose stride length was 33 mm. The members of the order show morphological adaptations in connection with rapid running not seen in lithobiomorphs or scolopendromorphs. The subdivided tarsus consists of many joints (also seen in some lithobiomorphs and cryptopids) which bear spines. These grip the substratum so counteracting the leverage exerted by such long legs. The Scutigeromorpha alone amongst the Chilopoda possess compound eyes and dorsal spiracles opening into lung-like structures. The compound eye may be necessitated by the visual requirements of fast running: the blood and not the tracheae may transport respiratory gases to the tissues. The greater complexity of the maxillary gland, which is probably excretory (see Chapter 16), is suggestive of an ability for more rapid removal of waste in *Scutigera* than in other Chilopoda (Manton, 1952*b*).

Craterostigmus

Craterostigmus is a slow moving animal. The relative durations of the forward and backstrokes of the legs is 5.5:4.5 and the phase difference between successive legs is 0.75 so that the propulsive limb tips converge as in Epimorpha. When a fast gait is elicited by using intense illumination, the body is thrown into uncontrollable lateral undulations due to the presence of extra tergite and sternite joints and the absence of a skeletomuscular anti-undulatory mechanism. The extra tergital joints allow a flexibility and compressibility which permit the animal to make sharp ventro-dorsal body flexures which Manton believed are useful in hunting in shallow spaces. *Craterostigmus* is able to squeeze into narrow crevices in search of prey (probably termites) and can burrow through decayed wood using the poison claws in a spade and shovel action.

3

The integument, moulting and regeneration

The integument

The integument is the outer covering of arthropods and consists of a single layer of epidermal cells (often called the hypodermis) which rests on a basement membrane and secretes the cuticle. The cuticle covers the outer surface of the animal and lines the invaginations that arise from it such as the fore- and hind-guts, the tracheae, the lower parts of the genital ducts and the ducts of the epidermal glands.

The centipede cuticle appears to consist of three main layers: an outer, thin, refractile membrane usually about 1 μm in thickness called the epicuticle; a rigid, usually amber-coloured exocuticle and an inner thick elastic layer, the endocuticle, which is colourless and lamellated. The varying terminologies that have been used for these layers by different authors are shown in Table 1. The layers below the epicuticle are sometimes termed the procuticle. The outer part of the procuticle becomes tanned and sclerotised to form the exocuticle, the remaining undifferentiated part being the endocuticle. Between the exocuticle and the endocuticle there may be a region of hardened but not fully darkened cuticle which is fuchsinophil and lamellate like the endocuticle. This layer is termed the mesocuticle. The exocuticle and endocuticle show a variety of staining reactions and Blower (1951) regarded the optical appearance of the two layers as the only criterion which constantly differentiated them.

In surface view the cuticle of the sclerites often shows polygonal fields separated by clear boundaries and thought to correspond to the limits of the epidermal cells. In geophilomorphs pore canals are visible in the cuticle; these contain filamentous processes from the epidermal cells around which the cuticle is secreted.

Table 1. *The terminology used by various authorities for the layers of the centipede cuticle*

Füller (1963*a*)	Blower (1951)	Fahlander (1938)	Verhoeff (1902–25)	Duboscq (1898)
Epicuticle	Diffraction effect	—	—	—
Exocuticle	Outer non-staining exocuticle	Grenzlamella	Oberflachenschicht	Chitine achromatique
Mesocuticle	Exocuticle	Pigmentschicht	Farbschicht	Couche chromophile basophile
Endocuticle	Endocuticle	Lamellenschicht	Lamellenschicht	Couche lamelleuse acidophile

Geophilomorpha

Duboscq (1898) described non-staining *lentilles* (exocuticle) and acidophilic cones (mesocuticle) descending into a lamellate endocuticle in *Chaetechelyne vesuviana* (Newport) (Fig. 47). A

Fig. 47. Transverse section of the sclerite cuticle of *Chaetechelyne vesuviana* (after Duboscq, 1898).

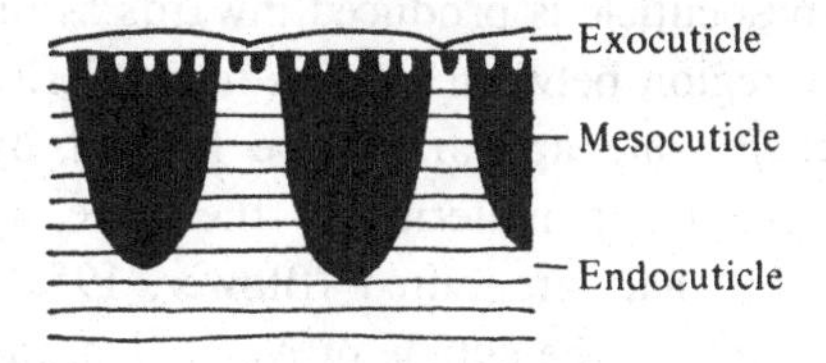

Fig. 48. The cuticle of *Haplophilus subterraneus*. *a*, transverse section of the sclerite cuticle; *b*, transverse section of an arthrodial membrane between the tergite and a pleurite (after Blower, 1951).

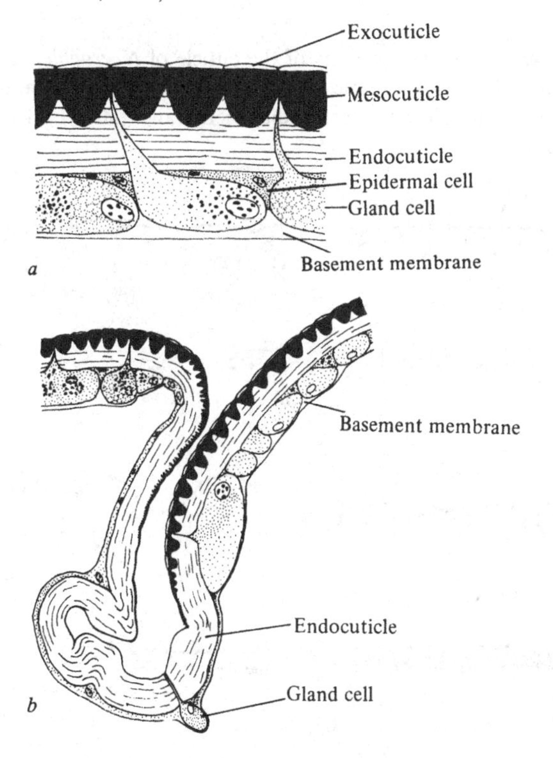

similar arrangement has been described in *Haplophilus subterraneus* (Fig. 48*a*) by Blower (1951) and in *Necrophloeophagus longicornis* (Leach) by Füller (1963*a*). Where the sclerite cuticle grades into the arthrodial membrane the exocuticle and mesocuticle gradually decrease in thickness. They are not developed at all in the arthrodial membrane (Fig. 48*b*). In the region of transition from the sclerite to the arthrodial membrane, which Blower termed 'intermediate sclerite', the mesocuticle is produced inwards as minute papillae (Fig. 49*c*). In the region between intermediate sclerite and arthrodial membrane, each cone appears to be formed by the merging together of exocuticular material in the pore canals: possibly the original pore canals are tufted (Blower, 1951). The detailed arrangement of layers in the cuticle of *Necrophloeophagus* is shown in Fig. 49. In this species mesocuticular cones are present only at the edges of the sclerites (Fig. 49*b*). Pore canals are visible in sections of tergite and intermediate sclerite (Fig. 49*a*, *b*, *c*) (Füller, 1963*a*).

Fig. 49. Transverse sections of the cuticle of *Necrophloeophagus longicornis*. *a*, The middle of a tergite; *b*, the edge of a tergite; *c*, intermediate sclerite; *d*, arthrodial membrane. (after Füller, 1963*a*).

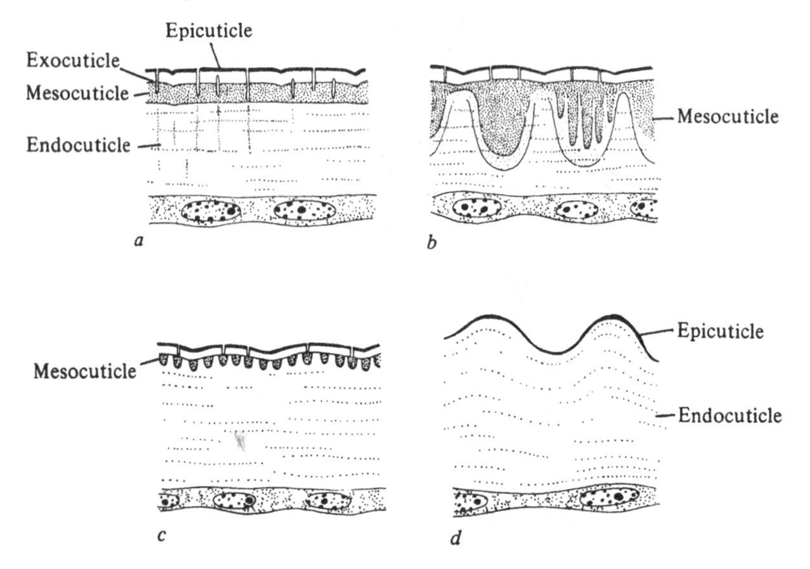

In *Haplophilus* there is a virtually continuous layer of glandular cells beneath the tergite cuticle but such glands are rare under the arthrodial membrane (Figs. 48*a*, *b*).

In *Geophilus carpophagus* the anterior sternites have far more mesocuticular cones than the tergites. In the region of the carpophagus pits and pegs the mesocuticle extends downwards to the epidermis. The sternite cuticle of the segments behind the carpophagus region is thinner, less sclerotised with virtually no cones (Manton, 1965).

Scolopendromorpha

Duboscq (1898) described an achromic exocuticle and mesocuticle with a combined thickness of 5 μm in *Scolopendra cingulata*. The underlying endocuticle is about 20 μm thick and, unusually, is acidophilic (Fig. 50*a*). The cuticle of *S. morsitans* is similar (Fuhrmann, 1922). Krishnan (1956) described an epicuticle 5 μm thick and exocuticle, mesocuticle, and outer and inner

Fig. 50. Diagrammatic representation of the structure of the cuticle in *Scolopendra* and the terminology used by various authors. *a*, *Scolopendra cingulata* (after Duboscq, 1898); *b*, *Scolopendra subspinipes* (after Krishman, 1956); *c*, *Scolopendra morsitans* (after Jangi, 1966); *d*, *Scolopendra morsitans* (after Shrivastava, 1971).

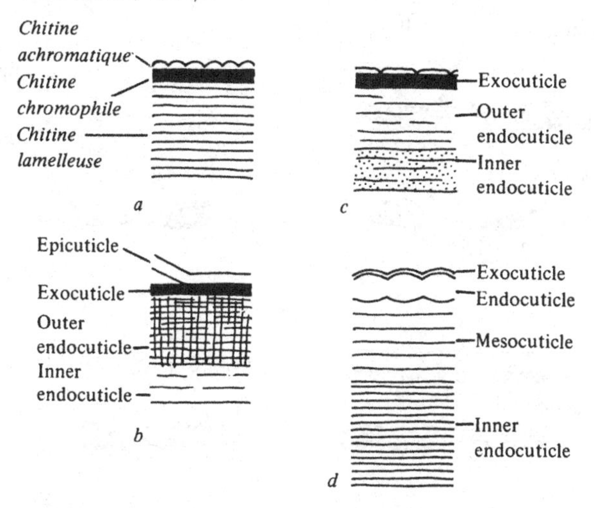

endocuticles in *Scolopendra subspinipes* (Fig. 50*b*). Jangi (1966) distinguished a basophilic exocuticle and lamellar acidophilic endocuticle, in *Scolopendra morsitans* (Fig. 50*c*). Following Blower (1951) he regarded the thin outermost colourless layer of the cuticle as a diffraction effect. The outer layer of the endocuticle would now be termed mesocuticle.

A very different account of the cuticle of *S. morsitans* was given by Shrivastava (1971). He regarded the cuticle as five-layered consisting of (1) an epicuticle of condensed lipoprotein, not demonstrated histologically, above which there is a mobile lipoid film; (2) an amber-coloured exocuticle 0.5 μm thick; (3) an outer colourless homogeneous and refractile endocuticle 5 μm thick; (4) a striated mesocuticle 30 μm thick and (5) an inner endocuticle 20 μm thick (Fig. 50*d*).

The terminology used by Shrivastava is confusing: his 'outer endocuticle' would appear to correspond to Jangi's exocuticle, the mesocuticle and inner endocuticle to the endocuticle.

Beneath the cuticle of *Scolopendra morsitans* there is a hypodermis 15 μm thick containing gland cells each with a chitinous

Fig. 51. Transverse sections of the cuticle of *Cryptops hortensis*. *a*, tergite in the region of a tergital suture; *b*, arthrodial membrane; *c*, weakly sclerotised middle region of sclerite; *d*, intermediate tergite (after Füller, 1963*a*).

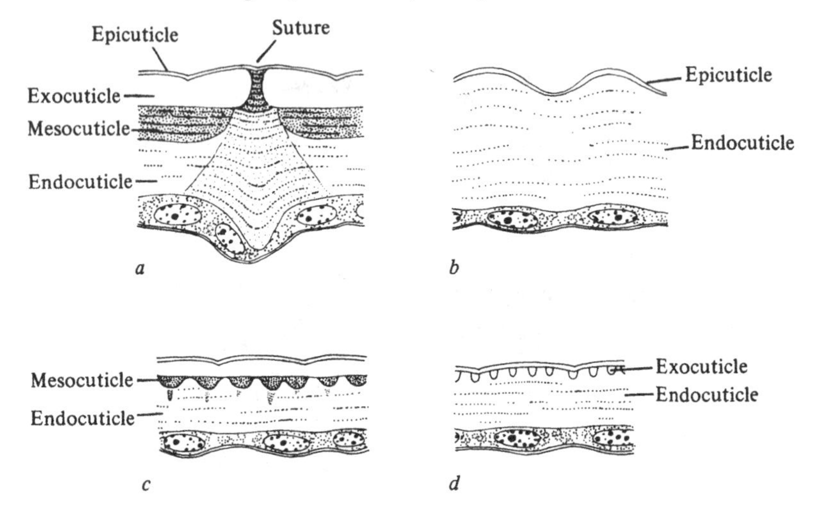

duct which commences within the cell as a flask-shaped annexe. The contents stain intensely with Sudan IV as does the surface film of lipid on the cuticle (Shrivastava, 1971). Similar glands have been figured for *S. morsitans* by Jangi (1966), for *S. cingulata* (Duboscq, 1898), *S. morsitans* (Fuhrmann, 1922) and for *Heterostoma australicum* (Grenacher, 1880). The gland cells resemble the small epidermal glands described in *Lithobius* by Rilling (1968) and Keil (1977). Further research may show that each consists of more than one cell.

In *Cryptops hortensis* the cuticle in the mid-part of the tergite is 13 μm thick. The epicuticle is 0.4–0.5 μm thick, the exocuticle 2–3 μm and the mesocuticle up to 7 μm. In less strongly sclerotised regions the mesocuticle forms small islands 1–1.5 μm thick (Füller, 1963*a*). In the region of a tergital suture the exocuticle is absent and the endocuticular lamellae wider than normal (Fig. 51*a*). Arthrodial membrane and intermediate sclerite cuticle are shown in Fig. 51*b*, *d*. No pore canals have been detected in *Cryptops*. Mesocuticular cones are present on the anterior and posterior borders of the sternites in *Cryptops* (Manton, 1965).

Lithobiomorpha

Duboscq (1898) distinguished three layers of cuticle in *Lithobius* the inner layer (endocuticle) being acidophilic. Detailed descriptions of the cuticle of *Lithobius forficatus* have been given by Blower (1951) and Füller (1963*a*). In sections of the tergite the cuticle reaches 50 μm in thickness. The epicuticle (regarded as a diffraction effect by Blower) is 0.4–0.8 μm thick and clearly two layered, the inner layer being pale yellowish. The exocuticle is 5–6 μm thick, the remaining 45 μm consists of the lamellate mesocuticle and endocuticle (Fig. 52*a*). In the region of transition from tergite to arthrodial membrane the exocuticle is reduced to small papillae (Fig. 52*b*) covered, according to Blower, by a thin layer of exocuticle but regarded as epicuticle by Füller. The papillate arthrodial membrane is almost entirely endocuticle. Blower described a thin layer of exocuticle, Füller a thin epicuticle with a small cone of mesocuticle under each papilla (Fig. 52*c*).

Treatment with silver has revealed filaments beneath the sclerite exocuticle of *Lithobius* that may represent the distal ends of the

Fig. 52. Transverse sections of the cuticle of *Lithobius forficatus*. *a*, central region of a tergite; *b*, procoxa; *c*, pleural membrane (after Füller, 1963*a*).

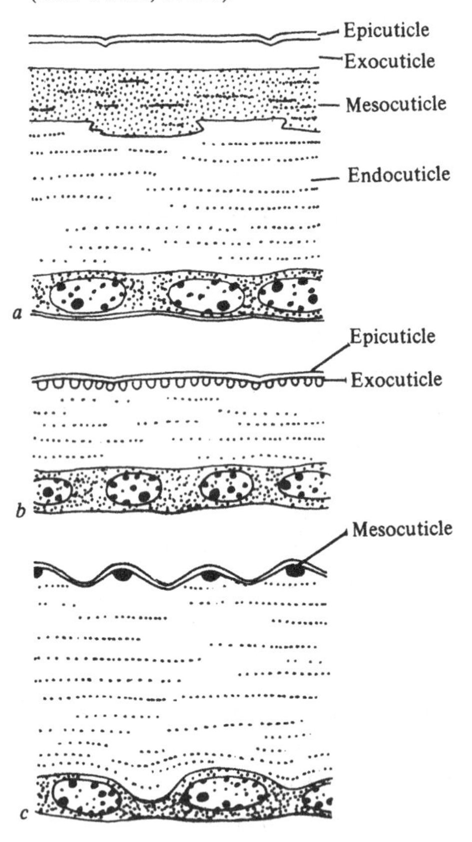

Fig. 53. Transverse section of the arthrodial membrane of *Lithobius forficatus* (after Blower, 1951).

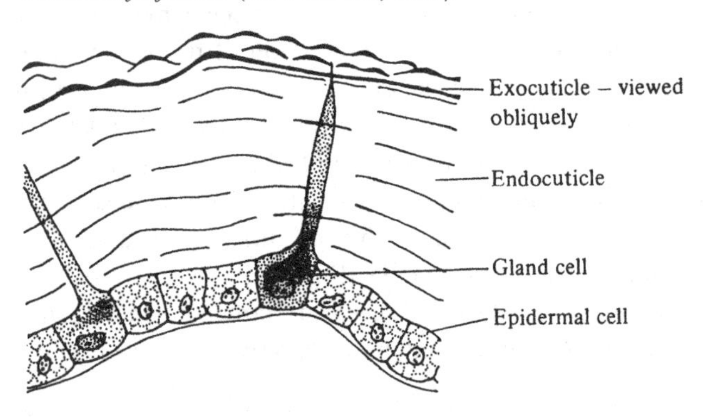

original pore canals. The minute papillae seen in the intermediate sclerite exocuticle may represent the distal ends of the original pore canals into which exocuticular material has penetrated (Blower, 1951). Scattered gland cells are present beneath all types of cuticle in *Lithobius* (Blower, 1951) (Fig. 53). Their ultrastructure was described by Keil (1977) (see Chapter 19).

Scutigeromorpha

In *Scutigera coleoptrata* the sclerite epicuticle is very thin. In the middle of the tergites the exocuticle is about 5 μm thick. The cuticle is about 25 μm thick and the lamellate region is entirely mesocuticle (Fig. 54*a*). The cuticular spines are formed of exocuticle containing fine canals, which may represent pore canals. The pleural membranes bear fine papillae (Fig. 54*b*) Füller (1963*a*).

Chemical composition of the cuticle

The centipede cuticle can be considered as a chitinous matrix impregnated to varying extends with a protein rich in

Fig. 54. Transverse sections of the cuticle of *Scutigera coleoptrata*. *a*, section through the tergite showing a spine; *b*, pleural membrane (after Füller, 1963*a*).

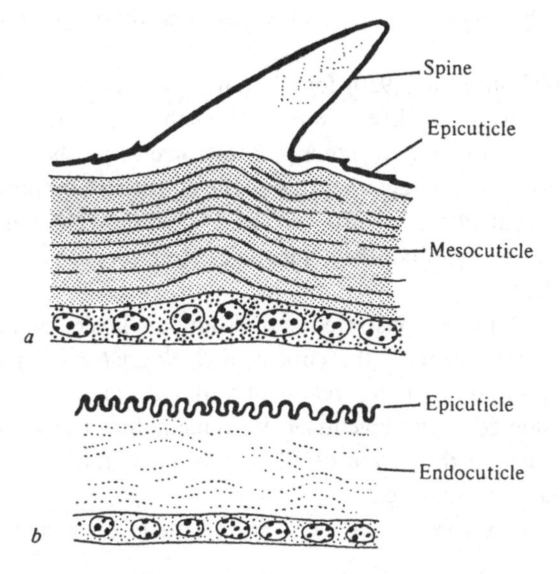

phenolic groups (Blower, 1951). Chitin is a nitrogenous polysaccharide probably made up of long chains of acetylated glucosamine residues; the phenolic-rich protein probably represents the precursor of typical sclerotin and was termed prosclerotin by Blower. Prosclerotin impregnates the exocuticle of *Haplophilus* and both the exocuticle and endocuticle of *Lithobius* but not the arthrodial membrane. It is resistant to acids and renders the impregnated region homogeneous and refractile. At a later stage an amber colour develops and the cuticle loses its staining reaction. At this stage the cuticle has the characteristics associated with complete sclerotin and is said to be tanned.

The procuticle of *L. forficatus* consists of alternating layers of chitinous lamellae and chitin-free protein layers. The chitin molecules run parallel to the body surface but are orientated in different directions. There are eight anisotropic chitin layers in *Geophilus*, 10 in *Cryptops*, 14 in *L. forficatus* and 17 is *Scutigera coleoptrata* (Füller, 1965).

Resilin, a protein in which the chains are bound together in a uniform three-dimensional network to form perfect rubber is present in small amounts between the chitin layers in ordinary insect cuticle (Wigglesworth, 1972). In centipedes it has been shown to be present in the arthrodial membranes of the last pair of legs of *Scolopendra morsitans* but not in other parts of the body (Rajulu, 1971*b*).

Cloudsley-Thompson (1950) found that a thin colourless layer less than 1 μm in thickness was left when the exocuticle of *Lithobius*, *Geophilus* and *Cryptops* was dissolved in concentrated chlorated nitric acid: he regarded this as the epicuticle composed of cuticulin like that of insects. Blower (1951) believed that this was not proven and regarded the 'epicuticle' seen in sections as a diffraction effect.

Krishnan (1956) demonstrated a continuous layer remaining unaffected by treatment of the cuticle of *Scolopendra subspinipes* with diaphanol and surviving treatment with chlorated nitric acid. This he considered to be epicuticle. Its outstanding feature is its ability to withstand the action of alkalis. X-ray diffraction studies suggested that chitin is present.

According to Shrivastava (1971) the epicuticle of *S. morsitans* consists of condensed lipoprotein above which is a mobile lipoid film, but his so-called chitinous exocuticle may well be part of the epicuticle. Electron micrographs of the sensory plate of the Tömösváry organ of *Lithobius* show a three-layered epicuticle 150 nm thick (Tichy, 1973).

It is clear that there is still confusion over the chemical composition of the epicuticle of centipedes.

The presence of lipoid material in the exocuticle is of considerable importance in centipedes. The lipoid appears to be secreted by the gland cells of the epidermis through ducts which open onto the surface of the cuticle. At the surface it appears to form a thin film. The exocuticle appears to be impregnated with lipoid and there seems to be an intimate association of the lipoid and sclerotin or prosclerotin. A *Lithobius* allowed to crawl in carborundum powder (an abrasive) for several hours and then plunged into silver nitrate solution showed a scratch pattern of brown lines. Blower suggested that the lipoid film covering the exocuticle undergoes a process similar to the drying of a coat of oil-bound paint and that the abrasive agent might remove this inert layer revealing the reactive lipoid which reduces the silver nitrate.

Functional properties of the cuticle

The nature of the cuticle is associated with the habits of centipedes. The elastic properties of the cuticle are utilised in promoting and controlling shape changes of the body in contrast to the millipedes where the strongly calcified sclerites are rigid (Manton, 1965). The apparent conflict of the needs for momentary shortening and elongation of the body on the one hand and the maintenance of axial rigidity on the other is solved by cuticular morphology associated with an endoskeletal-muscular system (see Chapter 4).

Centipedes are essentially hydrostatic systems in which the body movements are accompanied by alterations in shape of the cuticle. The major trunk sclerites do not articulate with one another, they are usually well separated and their margins furl and unfurl as each sclerite changes shape; muscular forces acting on the sclerites do so indirectly largely via the haemocoelic pressure.

The cuticle situated between the major trunk sclerites of

Chilopoda usually differs considerably from typical arthropodan arthrodial membranes (Manton, 1965). Typical arthrodial membranes are present at the leg joints but on the trunk occur only at the intertergite joints of the Anamorpha and at the intersternite joints of the Scutigeromorpha. The tergites and sternites of these joints alone have constant margins. Usually the joints on the trunk are formed by infoldings of the sclerites themselves, and the cuticle is thick and tough, gradually transforming so that there is no definite site of origin of the arthrodial membrane (Fig. 48*b*).

The apparent margins of the pleurites and sternites, the posterior margins of the tergites and, in the Geophilomorpha, the intercalary tergites also, are not fixed in position. The shape of a sclerite at any one moment depends upon the amount of intucked cuticle near the edges. By this means considerable shape changes of the body can easily be accomplished and a continuity of sclerite armour can be maintained during burrowing. The cones of sclerotised exocuticle give cuticular strength and flexibility in all directions. Flexibility along any plane between the cones is facilitated by the slight doming of the stiffer, but more elastic, surface layer of non-staining cuticle. When the exocuticular cones are present principally on one side only, as at the intertergite joint of *Geophilus*, *Cryptops* and *Cormocephalus* flexures between the cones result in a rolling movement of the hinder part of the tergite.

Füller (1963*a*) was the first worker to observe exocuticular cones in scolopendromorphs. No exocuticular cones are present in *Lithobius* where shape changes are less than in the Epimorpha but some degree of the rolling movement of the hinder part of the sternites can take place.

Moulting

Early accounts of the moulting process in *Lithobius forficatus* (Verhoeff, 1905; Attems, 1926) describe the first split in the cuticle as occurring between the head capsule and the forcipular tergite. Demange (1944), however, observed that the split takes place along the transverse suture which is situated one-third of the distance behind the front of the head capsule (Fig. 55*a*). He suggested that this indicated variation in the manner of moulting. Matic (1961) confirmed Demange's observation.

A detailed description of the moult in *L. forficatus* was given by Joly (1966*a*). The head capsule appears to swell before the moult and splits along the H-shaped exuvial lines which in most cases correspond with the sutures on the head capsule dividing it into four regions. During the moult the anterior region is turned forward and the two lateral regions rotate forward while the posterior region remains in place (Fig. 55). The antennae move backwards under the body. In some cases the split is of the simpler type described by previous authors. The mouthparts and forcipules are first removed from the old cuticle: the antennae are not normally freed until after the first three or four pairs of legs.

Moulting takes place by a series of jerks, the anterior end of the body being retracted and then moved forwards and upwards causing a telescoping of the exuvium. The movements initially take place at 10 to 20 s intervals, later at 20 s intervals. The complete process takes 30–40 min.

The animal remains motionless for a short period after the moult while the posterior appendages, initially stuck to the body by the moulting fluid, attain their normal position.

In *Scolopendra cingulata* the first split in the cuticle occurs behind the head capsule (Joly, 1966*a*) as it does in *Asanada sokotrana* (Lewis, unpublished data): the scolopendrids lack a frontal suture. In the geophilomorph *Strigamia maritima* (Leach) the split follows the frontal suture (Lewis, 1961). The line of rupture is also along the frontal suture in an Indian *Scutigera* sp. but there

Fig. 55. Moulting in *Lithobius forficatus*. *a*, dorsal view of head capsule showing exuvial lines; *b*, the splitting of the cuticle of the head capsule at the beginning of the moult (after Joly, 1966*a*).

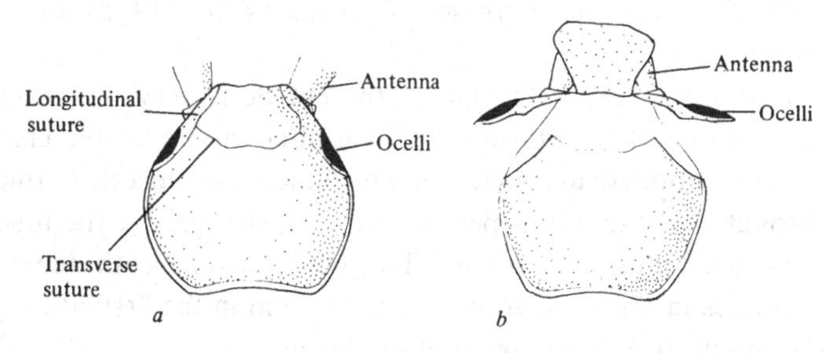

may be another behind the head capsule (Demange, 1948). Cameron (1926) described the moult of *Scutigera* as follows 'the shell splits along the median dorsal line and down across the head'.

Lithobius forficatus and *Bothropolys multidentatus* (Newport) eat the moulted exoskeleton leaving the remnants of the tibiae and tarsi in a small heap. *Cryptops hyalinus* leaves the tarsi and *Scutigera coleoptrata* the extremities of the appendages (Johnson, 1952). The exuviae of *L. forficatus* and *B. multidentatus* have the first 10 or 11 segments telescoped behind the head, while the remaining segments are normal. The exuvium is not eaten in *Ethmostigmus platycephalus spinosus* (Newport) (Rajulu, 1973*b*).

Newly moulted *L. forficatus* can be recognised by their pale colour and the presence of ovoid, brown faeces resulting from the consumption of the exuvium (Joly, 1966*a*). Newly moulted *Strigamia maritima* are white (Lewis, unpublished data).

Immediately after the moult a *Lithobius* is blue/grey-white; only the ends of the poison claws and the mouthparts are sclerotised and yellowish. After 24 h it is the normal pale brown but the cuticle is still soft, not regaining its rigidity for some 48 h. The tergites have a posterior light coloured band for some time after the moult.

Joly divided the intermoult period into three phases: the 'first fasting period' which lasts for about 48 h but during which, about 24 h after the moult, the exuvium is consumed: it is preferred to all other food. In the early anamorphic larval stages this period lasts only 8–12 h. There follows a long 'feeding period' followed by a 'second fasting period' of 2–4 days before the next moult.

The separation of the epidermis begins in the region of the extremities of the legs 16 to 18 days before the moult. The retraction of the appendages follows progressively before the moult. The new setae appear 3–4 days before the moult. The cuticle of the claw continues to increase in thickness for 12–13 days after moulting.

Immediately after the moult the cuticle is very thin consisting only of exocuticle. During the 24 h after the moult the epidermis secretes endocuticular material which increases steadily in thickness through the intermoult period but is most rapid for the first three days succeeding the moult. The epidermis beneath the tergites decreases in thickness from 15 μm to 7 μm in the first 30 days after the moult, increasing again after this period.

Five stages have been distinguished in the moulting cycle of *Ethmostigmus platycephalus spinosus* by Rajulu (1973*b*). They are:

Stage A: Duration 2–5 h. The new cuticle is wet and supple: the animals cannot move or feed and the legs are non-functional. The general colour of the animals is pale orange.

Stage B: Duration 9–12 h. The cuticle begins to harden. The legs are functional and the animal can crawl when disturbed but does not feed. The colour is bright orange.

Stage C: Duration 17–35 days. The intermoult period: the cuticle is hardened and bright orange in colour with black stripes.

Stage D: Duration 10–11 days. The premoult period: the animal ceases to feed and seeks refuge in a crevice. This stage is divided into

D_1: Duration 2–3 days. The cuticle becomes dull orange and the new claws start to form.

D_2: Duration 3–4 days. The epidermis simultaneously secretes a 2–3 μm thickness of new cuticle and moulting fluid.

D_3: Duration 2–3 days. The moulting fluid digests about half of the old endocuticle and an ecdysial membrane is seen to be present between the old and new cuticles. The latter increases to a thickness of about 8–10 μm. The ecdysial membrane perhaps represents the innermost sheet of the old endocuticle modified by leakage of sclerotizing compounds from the new cuticle. (Wigglesworth, 1972).

D_4: Duration 6–9 h. The new claws are completely retracted, the fresh cuticle increases in width from 12 to 16 μm and becomes pale orange. All the old endocuticle is digested.

Stage E: The moult. This takes 70–100 min and occurs at night.

The newly moulted animals are vulnerable to predators. Newly moulted *Lithobius forficatus* are found isolated deep within debris and recently moulted *Nadabius jowensis* (Meinert) are frequently found deep in the humus layer and Johnson (1952) concluded that it was probable that individuals about to moult moved to some protected spot. The littoral centipede *Strigamia maritima*, common at the edge of a salt-marsh in Sussex migrates up into a shingle bank or to the terrestrial habitat just behind it in order to moult (Lewis, 1961).

The hormonal control of moulting is described in Chapter 7.

Regeneration

In centipedes, as in other arthropods, damaged appendages are replaced at a subsequent moult and can often be recognised by their smaller size or abnormal characters such as the atypical arrangement of spines on the prefemur of the last pair of legs or atypical number of antennal segments in Scolopendridae (see Chapter 22). Verhoeff (1940) described two cases of regeneration of the poison claw. In a specimen of *Lithobius latro sellanus* Verhoeff the claw of the left prehensor was very small and the femur, tibia and tarsus not fully demarcated. The poison gland and its duct were lacking and the setae reduced in number (Fig. 56*a*). In a specimen of the South African geophilomorph *Eurytion dudichii* Verhoeff the entire telopodite had been lost: the regenerated telopodite consisted of the claw and a narrow segment representing the femur and tibia. The rudiments of the poison gland and its duct were present (Fig. 56*b*). Lewis (1968) described a specimen of *Scolopendra morsitans* in which the last left leg was bifurcated (Fig. 57).

Verhoeff (1902–25) distinguished two patterns of leg regeneration in centipedes: indirect in *Lithobius* and direct in *Scutigera.* Using mainly agenitalis II, immaturus, praematurus and pseudomaturus stadia of *Lithobius erythrocephalus* C. Koch he distinguished three developmental stages in the regeneration of a limb:

(1) Legs with stump-like telopodite probably formed at the first moult after the injury. The limb consists of a prefemur, femur, tibia and simple tarsus. The trochanter and second tarsus are absent as are setae, musculature and the claw tendon. The claw is a simple peg.
(2) Legs of half the normal length but with all the segments and the terminal claw well-developed. The muscles are developed but

weak and the telopodal glands and some setae are present. The leg spines are not differentiated.

(3) Legs normal with the exception of the size and number of the telopodal glands.

It would seem that Verhoeff regarded these stages as following each other after successive moults. He failed to discuss the effect of injury early or late in the intermoult period.

A case of partial telopodite regeneration was observed in which

Fig. 56. Regenerating forcipules. *a*, left forcipule of *Lithobius latro sellanus*; *b*, right (regenerating) and left (normal) forcipules of *Eurytion dudichii* (after Verhoeff, 1940).

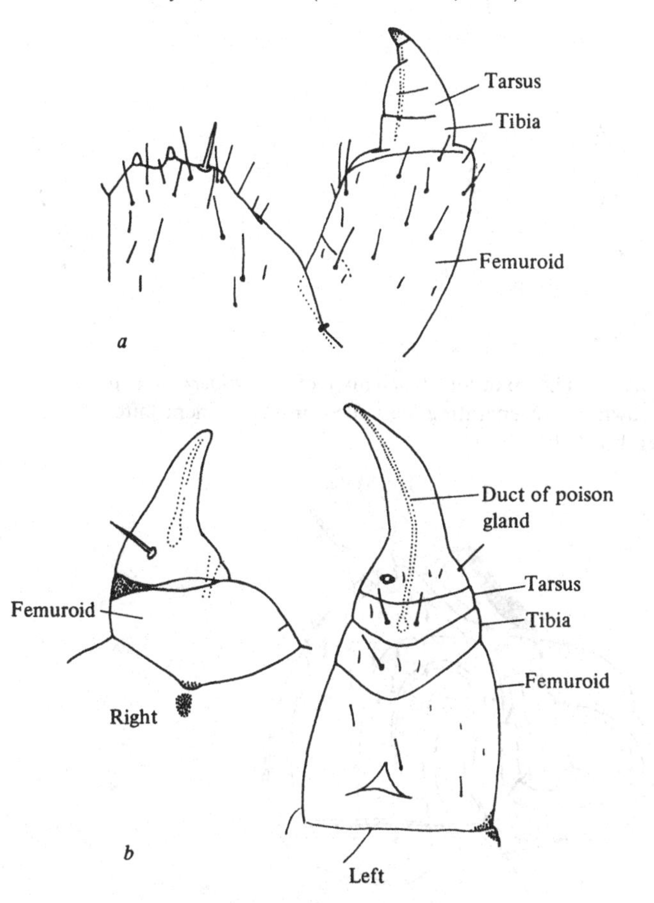

the segments distal to the prefemur had been lost and regenerated as indicated by their small size and sparse setae. This regeneration had effected the trochanter and prefemur which were somewhat smaller than those of other legs.

Fig. 57. Ventral view of a regenerated terminal leg of *Scolopendra morsitans* (after Lewis, 1968).

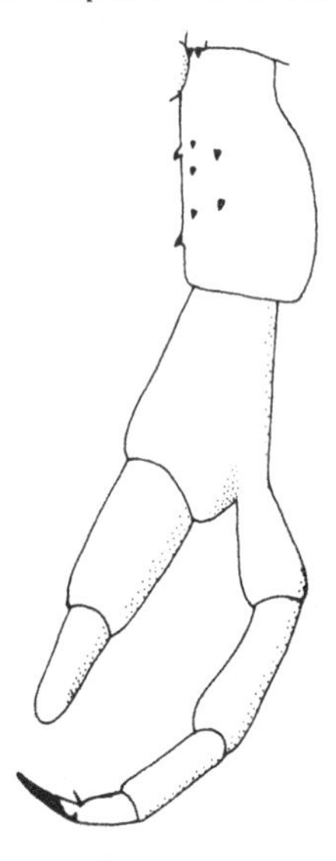

Fig. 58. The coxa and trochanter of a *Scutigera coleoptrata* showing a regenerating leg coiled up inside them (after Verhoeff, 1902–25).

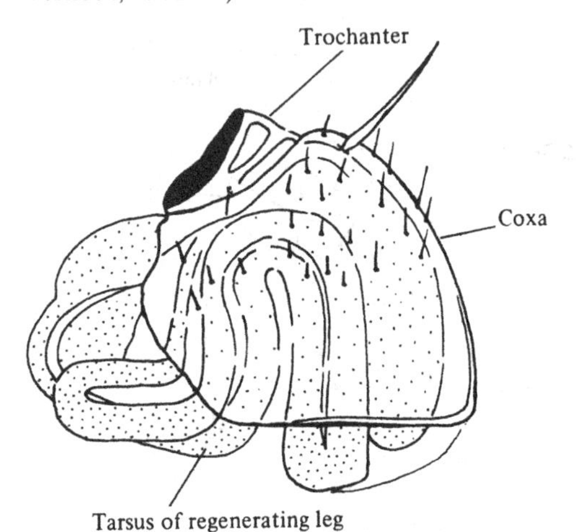

In contrast to the progressive regeneration in *Lithobius*, leg regeneration in *Scutigera* is sudden. In an agenitalis stadium *Scutigera* which had lost several legs, the replacement telopodite could be seen coiled up within the coxa, its tarsus projecting into the trochanter (Fig. 58): the setae were complete. In normal moulting the new telopodite develops inside the old leg, not in its coxa (Verhoeff, 1902–25). The author thought that partial leg regeneration was questionable in *Scutigera* but later (Verhoeff, 1938) described a specimen in which the newly replaced 7, 9, 10 and 13th legs were one-quarter to one-fifth shorter than the other legs and a second specimen in which almost all the legs were regenerated and of normal length except for the fifteenth pair of which one was shorter than the other. Cameron (1926) reported that small, incomplete or partially developed legs are never found in *S. coleoptrata*. In a few cases missing legs are not replaced until the next succeeding moult. In an Indian *Scutigera* sp. the regenerating legs are also coiled up in the coxae of the damaged limbs (Demange, 1948) as they are in the limb buds of scutigeromorph larval stadia (Fig. 199).

In *Bothropolys asperatus* Koch regeneration in the four larval stadia is both indirect and direct, and in the seven or more immaturus and maturus stadia is direct as in *Scutigera* (Murakami, 1958*b*).

Herbst (1891) described a strong diaphragm between the trochanter and prefemur in *Scutigera*. It contains bundles of elastic threads inserted into the hypodermis consisting in this region of several layers. The diaphragm is perforated by three holes, one for the leg nerve, the other two for blood vessels. The fibres are arranged in a circular manner around these holes. On the ventral side of the diaphragm there is a large cell surrounded by several smaller cells, thought by Herbst to be a sensory cell connected with the spine on the coxa. Herbst regarded the diaphragm as an adaptation to promote autotomy of the limbs. He pointed out that *Scutigera* is protected from attack by its long legs: potential predators find it difficult to reach the body so grasp the legs which are readily autotomised. When a leg is autotomised a single drop of yellowish viscous fluid is exuded which hardens almost immediately so that the loss of body fluid is very small (Cameron, 1926). When part of a leg is cut off distal to the point of autotomy the remainder is always dropped within a few hours of injury.

Using a group of *Scutigera coleoptrata* kept at 30°C and which moulted regularly every 30 days, Cameron (1926) showed that legs could be completely replaced at the next moult except when removed during the 6–7 day period before that moult. At 20°C they moulted every 60 days, at 25°C every 45 days and at 30, 35 and 40°C every 30 days. Recently captured animals with four legs or less dropped these within 24 h. Specimens that had sufficient legs to enable them to walk (4–30 legs) moulted at an interval of 45 days at 25°C but legless individuals moulted in 23 days. *Scutigera* has sufficient food reserves to sustain it through one complete moult cycle without feeding. The precise timing of the moulting cycle is remarkable: all the 127 specimens with legs kept at 25°C moulted after 45 days.

Joly (1966*a*) showed that leg removal accelerates the timing of the next moult in *L. forficatus* as does the loss of antennal segments (Joly & Lehouelleur, 1972).

4

The musculature and endoskeleton

The Chilopoda possess a bewildering array of muscles and no attempt will be made here to catalogue them. Meinert (1883) described the head muscles of *Scolopendra* and Bekker (1926, 1949) the trunk muscles of a geophilomorph and a scolopendromorph and compared the dorsal musculature of *Cryptops* and *Lithobius*. Bücherl (1940) described the musculature of *Scolopendra viridicornis* Newport, and Jangi (1966) that of *Scolopendra morsitans* though the trunk musculature was dealt with only briefly. The extrinsic and intrinsic musculature of the antennae of *S. morsitans* and the musculature of its terminal legs were described by Jangi (1960, 1961). Applegarth (1952) described the musculature of the head of *Pseudolithobius megaloporus* (Stuxberg), Rilling (1960, 1968) the head and trunk musculature of *Lithobius forficatus* in great detail and Füller (1963*a*) the musculature of *Scutigera*, *Necrophloesphagus* and *Cryptops*. Manton (1965) investigated the musculature of members of all five chilopod orders: she found Bekker's descriptions of the musculature of geophilomorphs and scolopendromorphs accurate but she could not entirely confirm his findings on *Cryptops*. She confirmed Rilling's (1960) observations on *Lithobius*.

Trunk muscles

The arrangement of the trunk musculature in myriapods has been summarised by Manton (1973). She distinguished six groups of trunk muscles: superficial, dorsal, lateral and sternal longitudinals, deep oblique and deep dorso-ventral muscles (Fig. 46). The superficial muscles control the alignment between successive segments but some of the superficial pleural muscles (Fig. 59; tcx) form part of the leg-rocking mechanism. The dorsal superficial muscles were listed among the dorsal longitudinal

Fig. 59. Dorsal view of the main muscles of a segment of *Haplophilus subterraneus* (from Manton, 1965). For further details see text.

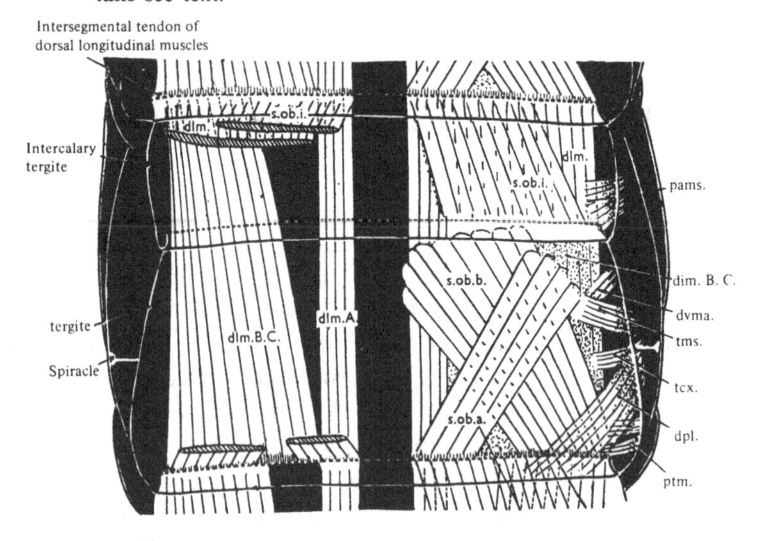

Fig. 60. Sagittal half of two segments of the middle part of the body of *Orya barbarica* with viscera and fat body removed to show the muscles and their tendons (from Manton, 1965). For further details see text.

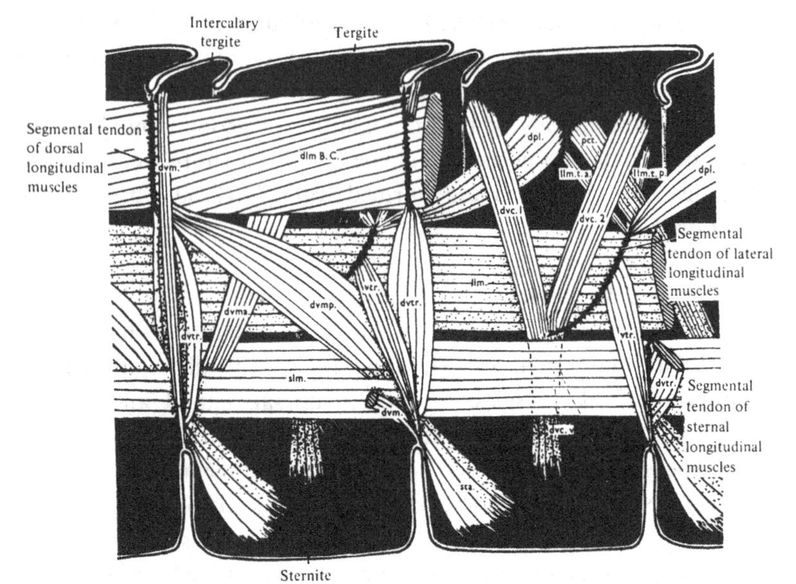

muscles (Manton, 1965) where they were termed 'superficial oblique' (Fig. 59; s.ob.) and 'deep oblique' (Fig. 61; dom) muscles. Dorsally and spreading laterally, the superficial muscles arise from the anterior strengthened margins of the tergites and insert on the faces of the sclerites on the preceding segment. Their course is slightly or decidedly oblique and they may form antagonistic pairs crossing each other as in the *Haplophilus* metatergite (Fig. 59; s.ob.a, s.ob.b) or one pair only may be present and set obliquely as in the *Haplophilus* pretergite (Fig. 59; s.ob.i) and *Cormocephalus*. Laterally the superficial muscles are roughly dorso-ventral in position: in epimorphic chilopods they are very well represented, extending between tergites, pleurites and sternites (Fig. 59; ptm, pams, tms).

The dorsal, lateral and sternal longitudinal muscles (Figs. 60, 61; dlm, llm, slm, stp, sta) give stability to the whole body. In Geophilomorpha, and to a lesser extent in Scolopendromorpha, they shorten the body so that a burrowing thrust is exerted upon the surrounding medium. In fast-running centipedes the musculature is complex and may lie in as many as three superimposed layers as in *Cryptops* and *Cormocephalus* (Fig. 61). The sectors are usually 'short', that is, cross one intersegmental joint or 'long', crossing more than one joint. In *Craterostigmus* the dorsal muscles extend between tendinous junctions situated in the middle of the tergites. This makes it easy for the tergites, which are subdivided, to slide over each other. In the Geophilomorpha short dorsal longitudinal muscles cause very strong shortening of the segments and dorsal buckling and overlapping of the sclerites during a burrowing heave. The normal short muscles are very thick and suspended from tendinous junctions sunk deeply into the body. The presence of separate intercalary tergites brings with it separate sectors of the dorsal longitudinal muscles, but long sectors are absent. These are present where there is need for controlling lateral flexures during running and for preventing sagging of the body when successive propulsive legs are far apart as in Scolopendromorpha and Anamorpha.

Lateral longitudinal muscles are present as separate entities in the epimorphic chilopods but are fused with the sternal longitudinals in the Anamorpha. In Geophilomorpha these muscles are massive: they extend between lateral segmental tendinous junctions

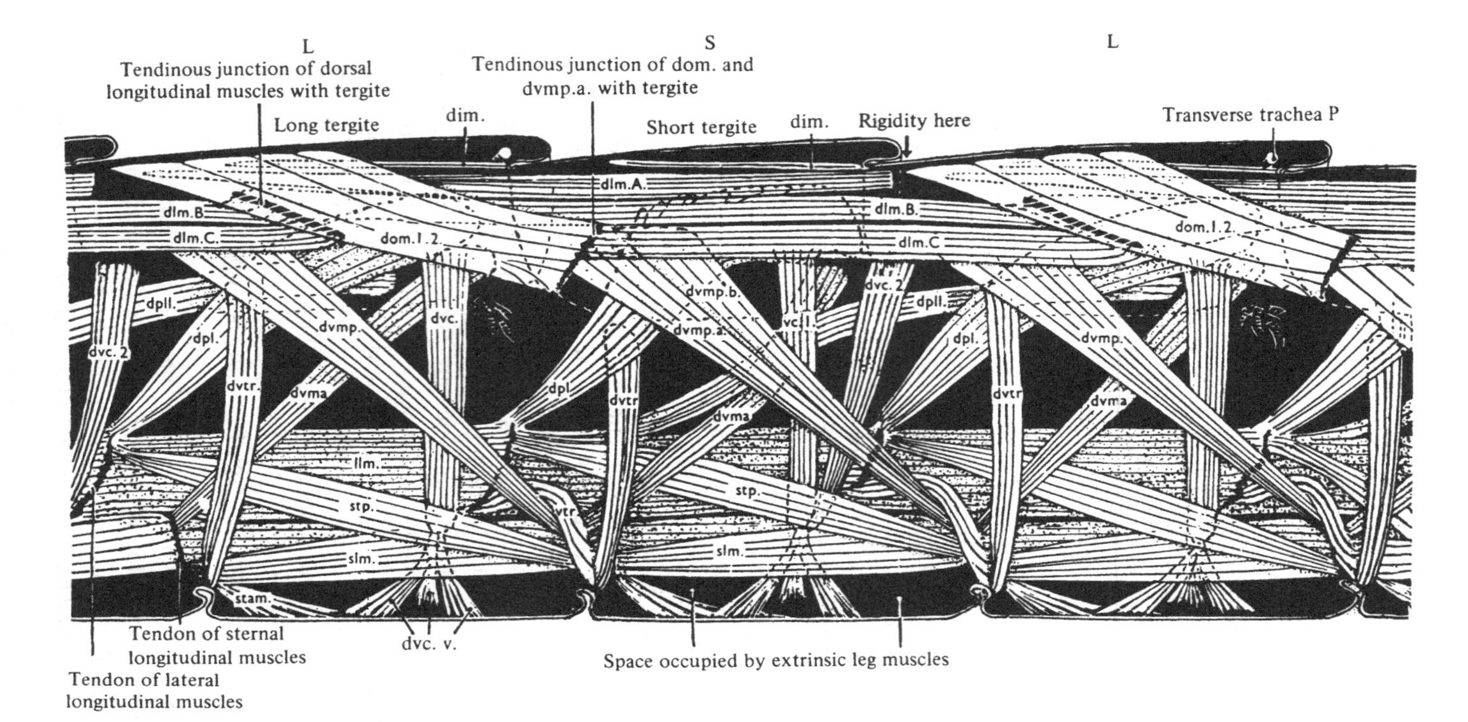

Fig. 61. Sagittal half of typical L- and S-segments of *Cormocephalus calcaratus* with viscera and fat body removed to show the muscles and tendons (from Manton, 1965). For further details see text.

situated in the posterior part of each segment and anterior to the tendons of the sternal longitudinal which are intersegmental in position (Fig. 60; llm). The tendons of the lateral longitudinal muscles are of great importance for the attachment of other muscles.

Sternal longitudinal muscles (Figs. 60, 61; slm, stp, sta) have the same function as the dorsal and lateral longitudinals. Variations in their arrangement were illustrated by Manton (1965).

Deep oblique muscles (Figs. 60, 61; dpl, drm, dvma, dvmp, vtr), whose tension strongly shortens and stiffens the body and provides an effective anti-undulatory mechanism, are antagonised by the deep dorso-ventral muscles (dvc, dvtr). Muscle vtr normally has a stabilising function but in the Geophilomorpha may provide an antagonistic force to the longitudinals, causing a narrower body shape. The deep obliques arise from the tendon of the sternal or ventral longitudinal muscles and extend obliquely upwards both anteriorly and posteriorly to insert on the tergites of the successive segments in Epimorpha. Secondary changes in the insertion of these muscles take place in association with tergite heteronomy.

Füller (1963*a*) regarded the basic arrangement of muscles as an outer circular layer beneath which is an outer diagonal layer, longitudinal muscles and inner diagonal muscles plus several transverse muscles from the body wall to the appendages. The circular layer is best developed in *Cryptops* and strongly reduced in *Lithobius* and *Scutigera.*

Manton (1965) pointed out that the hydrostatic pressure of the haemocoel in the trunk of Chilopoda is as important in mediating the action of the antagonistic muscles as it is in the annelid worms. The contractions of the massive sternal and lateral longitudinal muscles which shorten and thicken the body are antagonised by much weaker dorso-ventral and superficial pleural muscles by virtue of the haemocoel.

Trunk musculature and segmentation

A consideration of the trunk musculature led Demange (1963, 1967) to a number of conclusions regarding segmentation in centipedes. He regarded the juxtaposition of two long tergites on segments 7 and 8 in lithobiomorphs and scolopendromorphs, and in scutigeromorphs where they appear to be fused to form a single sclerite, as indicating the presence of a *région perturbée*. He suggested that the peculiarities of this region were due to the disappearance of a segment with a short tergite, the true segment 8, and, in addition to the arrangement of some trunk muscles, cited as evidence the arrangement of the extrinsic leg muscles, the distribution of spiracles and the pattern of development in *Lithobius* (see Chapter 18).

According to Demange's interpretation, tergite 8 in *Scolopendra* consists of a large tergite with the preceding small tergite fused to it and has thus acquired the muscles of that segment, although the other structures of the segment have totally disappeared. The homologies in *Ethmostigmus*, which unlike *Scolopendra*, possesses a pair of spiracles on segment 7, are slightly different.

The author considered that in *Lithobius* tergite 7 belongs to an incomplete segment lacking the part corresponding to the spiracles. He believed that the presence of narrow posterior projections on segments 9, 11 and 13 and rarely on 6 and 7 (Fig. 218 *f–j*) indicated the original length of the tergites, the central portions of which have become reduced. Eason (1970*a*) showed that in *Bothropolys grossipes* (C. L. Koch) these projections are relatively longer in the immature post-larval stadia than in the adults which suggests that Demange's explanation of their significance may be incorrect.

In *Scutigera* signs of perturbation are absent in the region of segments 7 and 8 due, Demange maintained, to the fusion of the homologues of segments 7 and 8 in lithobiomorphs and scolopendromorphs together with the original short tergite separating them. In the geophilomorphs there are no peculiarities in the region of segments 7 and 8.

It was concluded that the segments of chilopods are grouped in pairs or blocks, at least dorsally, the block consisting of a macrotergal metamere with spiracles and an incomplete microtergal metamere without spiracles. The dorso-ventral musculature of the bisegmental groups of chilopods and diplopods present 'large analogies which perhaps suggest the existence of true homology'. Demange considered that there is evidence of a tagma of eight metameres in both millipedes and centipedes. It is not clear how the geophilomorphs fit in with this theory.

Manton (1965) commented on Demange's (1963) assertion that a segment had disappeared between tergites 7 and 8 in Lithobiomorpha and Scolopendromorpha: 'Demange finds confirmation for his view in the dorsal musculature, but this complex system is neither figured nor adequately described by him. No evidence for such supposed exterpolations has been found in examples from all orders studied here'. Demange (1967) figured the musculature but Manton in later publications (1973, 1977) made no reference to this paper.

Leg muscles

There are two coxal movements in Chilopoda (i) the normal arthropodan promotor–remotor swing and (ii) the rotation of the leg on its long axis, resulting from the parasagittal rock of the coxa about a more or less ventral fulcrum.

Proximally there are pivot joints between the leg segments worked by pairs of antagonistic muscles, while distally the hinge joints are provided with flexor muscles only. Leg extension is accomplished indirectly by proximal depressor muscles aided by leg rocking and by hydrostatic pressure. With increase in ability to run faster the number of extrinsic leg muscles increases; 13 are present in Geophilomorpha, 19 in Scolopendromorpha, 20 in Lithobiomorpha and 34 in Scutigeromorpha. The intrinsic muscles also change where speed is achieved and many long sectors are present crossing more than one joint in most Chilopoda (Manton, 1973).

Histology of muscles

There are no smooth muscle fibres in arthropods but the striated fibres show variations in the arrangements of their myofibrils (Camatini & Saita, 1972). As in other groups the muscles consist of fibres subdivided into fibrils which are themselves composed of myofilaments composed of the proteins actin and myosin. The fibrils show alternating isotropic (I) and anisotropic (A) regions. A partition, the Z-disc divides each I-region separating the fibrils into sarcomeres. The A-band is sometimes traversed by a lighter H-band. The banding is the consequence of the arrangement of thick myosin filaments and the surrounding thinner actin filaments (Fig. 62). Both actin and myosin filaments are present in the darker portions of the A-band. Only actin filaments are present in the I-band and only myosin filaments in the H-band. The fibres are surrounded by an outer membrane, the sarcolemma enclosing the nucleated sarcoplasm which contains the fibrils. Transverse tubular invaginations of the sarcolemma form the so-called T-system. The endoplasmic reticulum of the muscle cell is termed the sarcoplasmic reticulum.

The histology of the skeletal muscle of *Necrophloeophagus longicornis*, *Cryptops hortensis*, *Lithobius forficatus* and *Scutigera coleoptrata* as seen with the light microscope was described by Füller (1963*a*). In *Necrophloeophagus* the Z-membranes are arranged in a spiral and do not form a disc. This spiral arrangement is also found in *Cryptops*.

There are generally two types of muscle fibres in arthropods, phasic fibres and tonic fibres. Phasic fibres have rather short myosin filaments offering reduced possibilities of filament sliding as seen in insect flight muscles. Here the myosin filaments are arranged in a hexagonal pattern and a 1:3 ratio between myosin and actin filaments. Tonic fibres have longer myosin filaments and a 1:6 ratio between the myosin and actin filaments (Camatini & Saita, 1972). *Scutigera coleoptrata*, *Lithobius forficatus*, *Scolopendra cingulata* and *Himantarium gabrielis* show tonic fibres with an orbital of up to 12 actin filaments giving a ratio of myosin filaments to actin of 1:6 (Camatini, 1970).

The sarcoplasmic reticulum forms a fenestrated envelope surrounding the myofibrils along their entire length. Mitochondria are

Fig. 62. Diagram of the arrangement of actin and myosin filaments and banding in striated muscle.

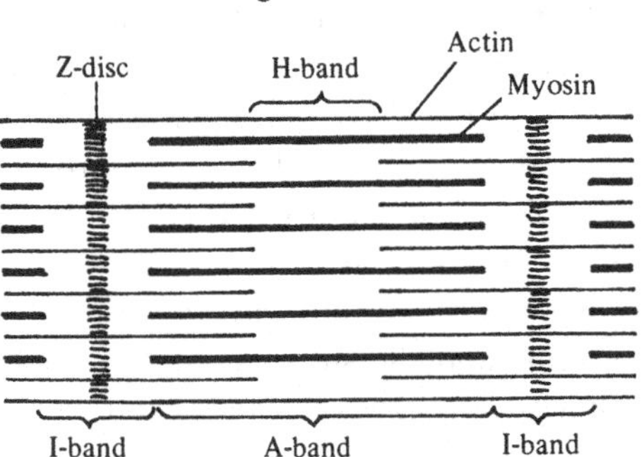

mainly in the area of the Z-lines in the dorsal and ventral longitudinal muscles but in the dorso-ventral and tergo-coxal muscles of *Lithobius* and *Scutigera* they are also present at the edges of the fibres among the myofibrils. At the Z-lines there is a network formed of mitochondrial expansions elongated to surround the myofibrils.

Fahlander (1938) described fibrils of the skeletal muscles of centipedes as being arranged radially around a central core of cytoplasm containing the nucleus, like the tubular muscles of insects. According to Camatini (1970) the nuclei of the ventral longitudinal muscles are centrally placed but those of the dorsal longitudinal muscles are located underneath the plasma membrane. The dorso-ventral and tergo-coxal muscles are 'subdivided into numerous myofibril groups with a greater number of nuclei'.

In *Lithobius forficatus* the mid-gut fibres measure 2.5×4.0 μm and have a peripheral nucleus. The T-system is well developed, the sarcoplasmic reticulum represented by small vesicles under the plasma membrane or inside the fibre. The ratio of myosin to actin filaments in the A-band is 1:6. The Z-lines are extremely irregular and sometimes the fibrils appear to run in different directions (Camatini & Castellani, 1978).

The muscle fibres of the testis of *L. forficatus* each correspond to a single myofibril. The external longitudinal fibre (2.6×3.5 μm in diameter) have a central nucleus; the circular fibres (1.9×4 μm in diameter) laterally situated nuclei. The fibres are clearly striated but the Z-bands are very irregular and the I- and H-bands are not well defined. The sarcoplasmic reticulum is not well developed and the T-system is apparently absent. The ratio between actin and myosin in the A-band region is very low for muscles of the tonic type and they are irregularly arranged. The fibres may be intermediate between the smooth and striated type (Camatini & Castellani, 1974, 1978).

Füller (1963*a*) regarded the muscle bundles of the Epimorpha as more primitive than those of the Anamorpha.

Muscle insertions

The muscles of centipedes are attached to the subectodermal basement membrane or to subectodermal layers of connective tissue. In scutigeromorphs and large scolopendromorphs the basement membrane consists of the thick *innere Cuticula* (3 μm thick in *Thereuopoda clunifera* Wood) and a thin *cutis*. The *innere Cuticula* may be enlarged to form endoskeletal structures (Fig. 63). It is often absent in lithobiomorphs and geophilomorphs (Fahlander, 1938). The muscles attach by tonofibrils extending into the cuticle.

Fig. 63. Muscle attachments to the cuticle in centipedes (after Fahlander, 1938).

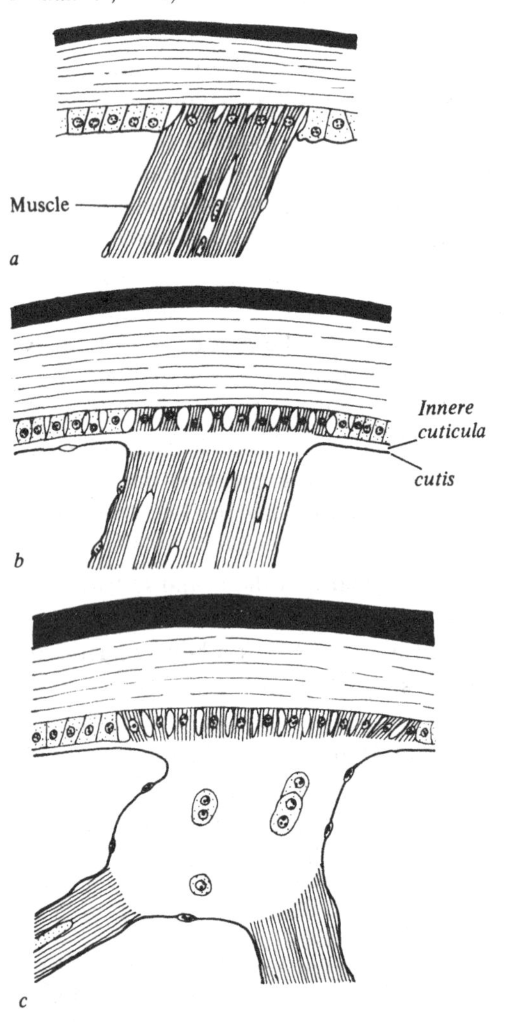

The method of muscle attachment to the cuticle is only now being elucidated in the insects (Richards & Davies, 1977). At the junction of muscles and epidermis the cells show regular interdigitations, the processes being lined with desmosomes. Within the epidermal cells microtubules connect the desmosomes with cone-like depressions of the outer epidermal plasma membrane. From each cone an electron-dense muscle attachment or tonofibrilla runs through the procuticle in a pore canal and finally inserts on the epicuticle.

Where muscles in centipedes are attached to endoskeletal structures these are either compacted connective tissue fibrils forming skeletal tendinous sheets of various thicknesses which directly carry the muscles or apodemes. Where the attachments are to apodemes, the covering apodemal epidermis is associated with the muscles as on the body wall, but the ectodermal cells and tonofibrils may be very attenuated. The Chilopoda need sufficient internal areas for the insertion of bulky intersegmental musculature. There is insufficient space on the flat surfaces of the sclerites, and hollow apodemes and rigid flanges of skeleton which project inwards are largely inappropriate where an ability to vary the shape of the segment is a primary requirement. Instead extensive use is made of flexible tendinous junctions anchored to the sub-epidermal layer of connective tissue fibres and carrying muscles on both sides (Manton, 1965). The sites of origin and insertion of the dorsal longitudinal muscles may be thin pads of fibres or, in Geophilomorpha, massive ingrowths at the anterior ends of the intercalary tergites (Fig. 60). Some of the principal tendinous junctions between muscles become detached from the subcutaneous layer or almost so, forming the tendons of the lateral and sternal longitudinal muscles. They also support a mass of other muscles.

Tentorium and endoskeleton of trunk

The tentorium is the endoskeleton of the head in arthropods and is formed by the fusion of apodemes and serves as a region of attachment for muscles and as a structure giving the head greater strength. The degree of development of the tentorium varies considerably between the different centipede orders being best developed in the Scutigeromorpha and least developed in the Geophilomorpha.

A single pair of head apodemes, called the *Komandibular Gerüste* by Attems (1926) and Verhoeff (1902–25) corresponds with the anterior tentorial apodemes of the insects and other myriapod classes (Manton, 1964). These apodemes arise as paired intuckings from the wall of the pre-oral cavity, lateral to the hypopharynx and dorsal to the mesial part of the mandible. The openings are usually wide in the transverse plane. The apodeme projects posteriorly into the head as the posterior process which usually carries the sternal longitudinal muscles (Fig. 64).

In the Scolopendromorpha and Lithobiomorpha, where the

Fig. 64. *Scutigera coleoptrata. a*, dorsal view of head capsule showing the tentorium and transverse mandibular tendon; *b*, transverse mandibular, maxilla 1 and maxilla 2 tendons (after Manton, 1965).

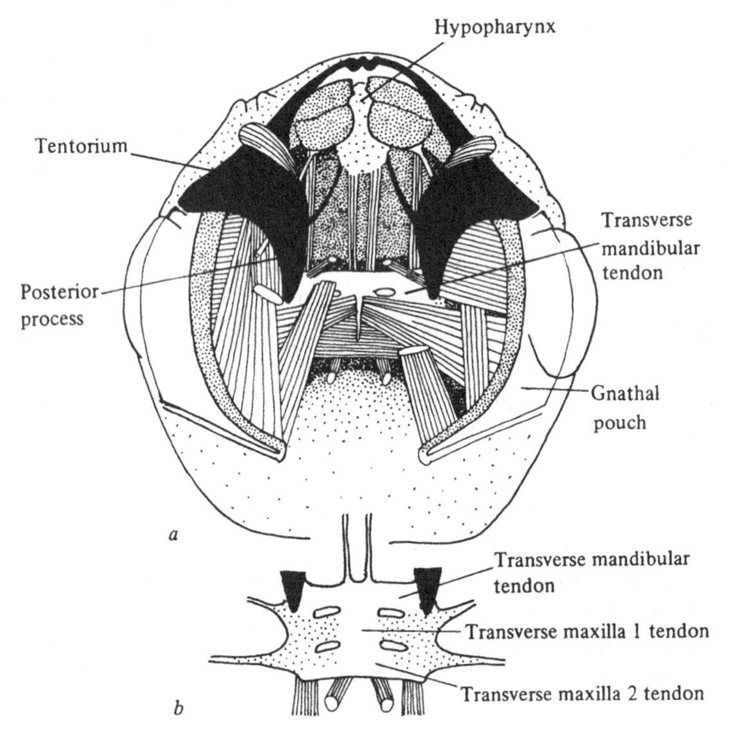

tentorium swings within the head and promotes mandibular movements, a stout mandibular process extends to the mandible from the ventral mesial edge of the tentorial orifice (Figs. 66, 67). The mandibular process also supports the hypopharynx either directly, or by a small separate branch. When the tentorium does not swing

Fig. 65. *Thereuonema tuberculata. a*, tentorium and transverse mandibular, maxilla 1 and maxilla 2 tendons; *b*, transverse forcipular tendon (after Fahlander, 1938).

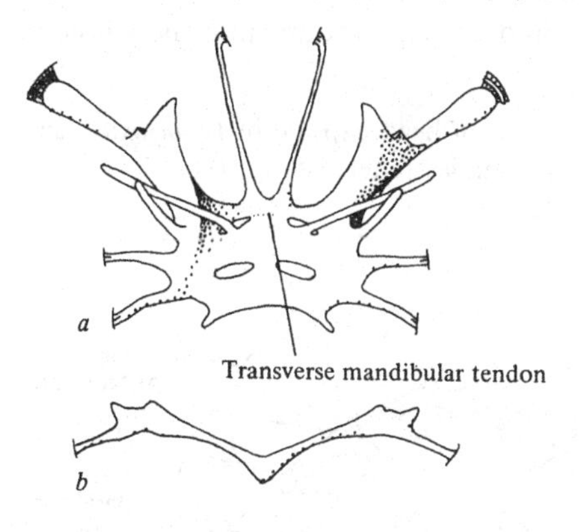

Fig. 66. Tentorium of *Cormocephalus calcaratus* (after Manton, 1964).

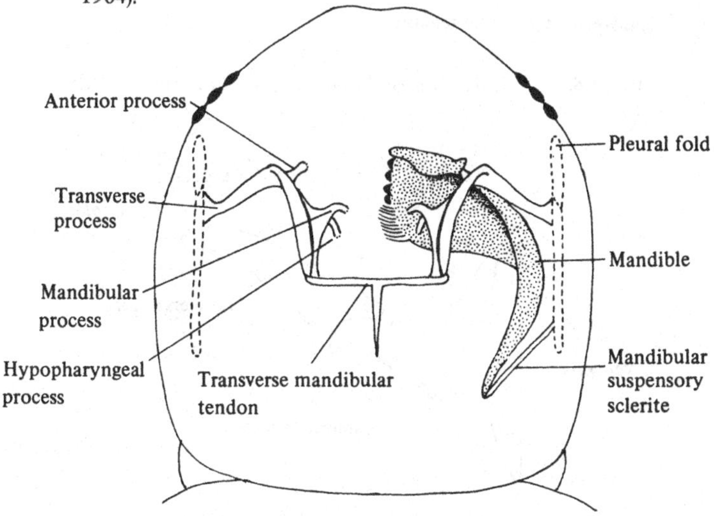

in the head (Geophilomorpha and Scutigeromorpha) this same branch supports the hypopharynx (Figs. 64, 69). It is not united with the mandible which is thus free to move independently (Manton, 1965).

In the Scolopendromorpha and Anamorpha, where the preoral cavity is wide, the lip of the tentorial apodeme is continued laterally as a surface sclerotization forming the transverse process lying just under the lateral part of the labrum. There is no transverse process in Geophilomorpha because the preoral cavity is not open laterally to the tentorial origin and in Scutigeromorpha the whole apodeme

Fig. 67. Dorsal view of head capsule of *L. forficatus* showing tentorium and mandibles (after Rilling, 1968).

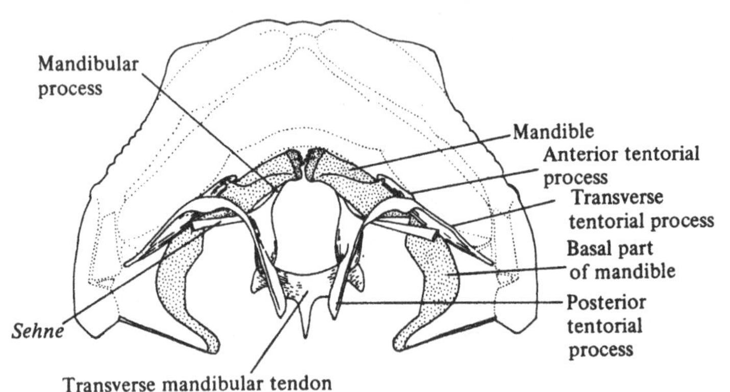

Fig. 68. Left tentorium of *Lithobius forficatus* (after Verhoeff, 1902–25).

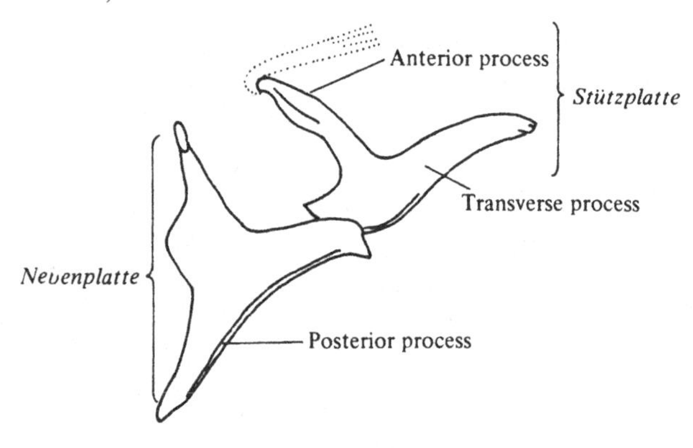

is so extensive that there is no clear distinction between the transverse and posterior processes.

The tentorium is largest in the Scutigeromorpha (Figs. 64, 65). The posterior tentorial processes are linked by the transverse mandibular tendon (Fahlander's *Tentorial Körper*) which represents the intersegmental tendons of the sternal longitudinal muscles continued into the head. The tendons of the mandibles and of the first and second maxillae are fused to form a fenestrated plate which bears three pairs of processes in *Scutigera coleoptrata* (Fig. 64*b*) and five in *Thereuonema tuberculata* (Wood) (Fig. 65*a*). Fahlander (1938) may have overlooked the anterior part of the tentorium in *Thereuonema* (compare Figs. 64*a* and 65*a*).

Verhoeff (1902–25) described the tentorium of *Lithobius* as a jointed structure, an anterior *Stützplatte* articulating with a posterior *Nebenplatte* (Fig. 68). Manton (1965) stated that the joint doubtless facilitates the swinging movement of the large tentorium in the confined space of the head capsule. Rilling (1968) could not confirm that this articulation was present.

The tentorium of the Scolopendromorpha (hypopharyngeal suspensoria, Jangi, 1966) resembles that of the Lithobiomorpha: its ventral part is horseshoe-shaped in cross section. In Geophilomorpha the tentorium is small and immobile (Fig. 69) and

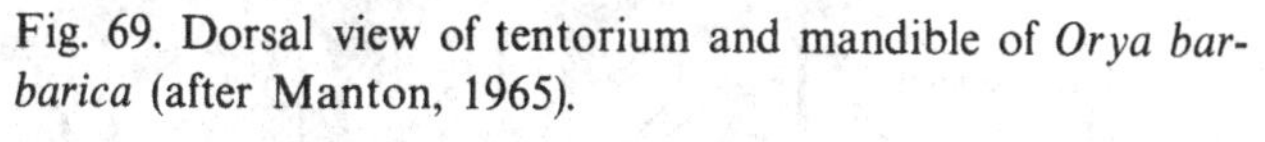

Fig. 69. Dorsal view of tentorium and mandible of *Orya barbarica* (after Manton, 1965).

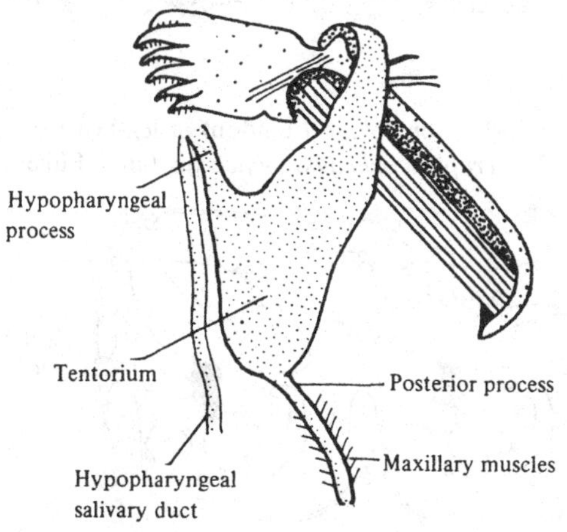

the transverse mandibular tendon is reduced to its connective tissue investment. Manton (1965) pointed out that Verhoeff used the terms 'tentorium' and 'nebententorium' in a confused manner: the nebententorium of *Lithobius* is not the equivalent of the narrow terminal portion of the posterior process in the Geophilomorpha.

The tentorium of *Craterstigmus* is small but typically scolopendromorph in form (Manton, 1965).

In each leg-bearing segment and in the forcipular segment of scutigeromorphs there is an endoskeletal bridge (Figs. 65, 70) to which the leg muscles attach and from which a small process runs dorso-medially (Fahlander, 1938). The bridge lies ventral to the gut. In the lithobiomorphs, scolopendromorphs and geophilomorphs

Fig. 70. Endoskeleton of leg-bearing segments 3–6 of *Scutigera coleoptrata* (after Füller, 1963*a*).

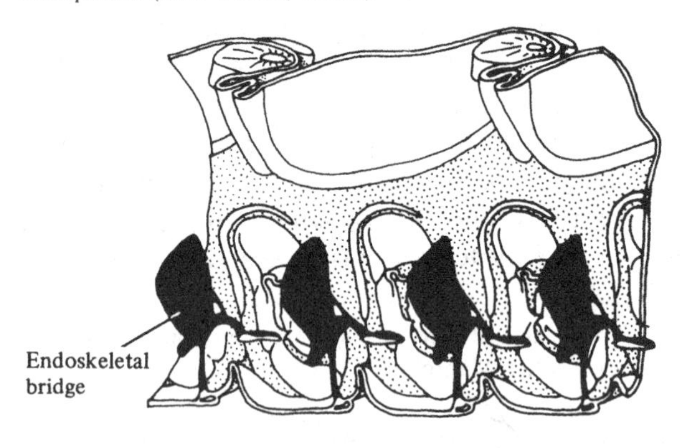

Fig. 71. Endoskeleton (pleural tendon) of leg-bearing segments 5 and 6 of *Necrophloeophagus longicornis* (after Füller, 1963*a*).

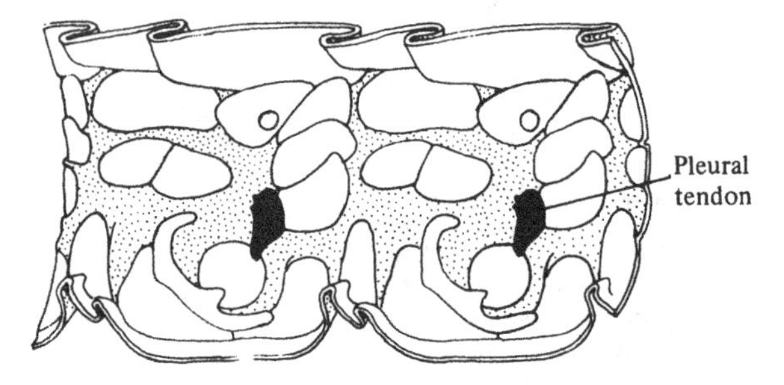

(Fig. 71) the endoskeleton is represented by a pair of separate lateral plates in each segment (Rilling, 1960, 1968; Füller 1963*a*). The endoskeleton consists of a fibrillar connective tissue whose intercellular substance consists of collagen bundles and a few elastic fibres (Füller, 1963*b*). In *Scutigera* the connective tissue is cartilage-like but contains no chondro-mucoid.

Manton (1965) described a number of segmental tendons in the trunk region. Segmental tendons of the dorsal longitudinal muscles are present in the Epimorpha (Fig. 60) but in Lithobiomorpha there are no conspicuous dorsal tendons. A dorsal tendon is present near the anterior end of the spiracle bearing segments in *Scutigera*. In Epimorpha there are segmental tendons of the sternal longitudinal muscles and lateral longitudinal muscles (Fig. 60), the latter appear to correspond to the endoskeleton of other authors. In the Anamorpha the sternal and lateral longitudinal muscles are confluent as are their tendons ('endoskeleton' of Füller and Rilling).

Fig. 72. Ventral view of head capsule and first two leg-bearing segments of *Orya barbarica* to show coxal apodeme. The cavity of the apodeme is stippled (after Manton, 1965).

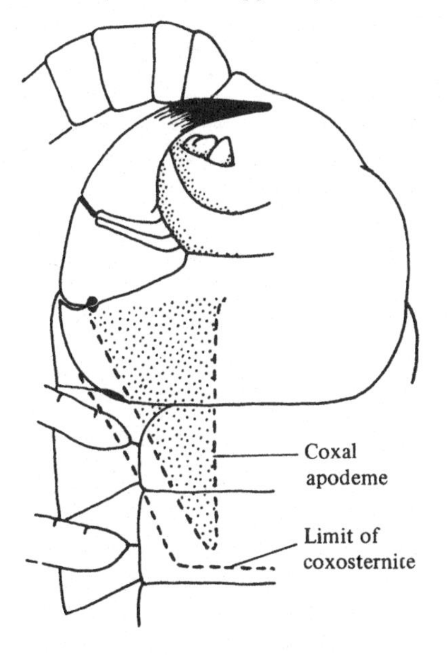

Attems (1926) referred to paired *Coxal-platten* in the forcipules of centipedes. These are extremely large paired apodemes arising widely from the dorsal face of the distal part of the syncoxite (Fig. 72). These coxal apodemes are not present in *Scutigera* (Manton, 1965).

5

The nervous system and sense organs

The nervous system

The arthropod nervous system may be conveniently considered under three headings:

(*a*) the central nervous sytem, consisting of a dorsal brain connected by circum-oesophageal connectives to a double ventral nerve cord along which there are ganglia, one pair per segment,
(*b*) the visceral nervous system supplying the gut and
(*c*) the peripheral nervous system including all the nerves radiating from the ganglia of the central and sympathetic systems.

Data on the anatomy of the nervous system were reviewed by Verhoeff (1902–25) and Hilton (1930). Subsequently, detailed studies have been carried out on *Scolopendra cingulata, Lithobius forficatus* and *Thereuopoda clunifera* by Fahlander (1938), on *Pseudolithobius megaloporus* by Henry (1948), on *Scolopendra morsitans* by Jangi (1966) and on *L. forficatus* by Rilling (1968).

In the Lithobiomorpha the brain lies transversely above the stomodaeum. It consists of the cerebral ganglia each composed of a lateral and an antero-lateral lobe. The lateral lobe receives nerves from the ocelli and Tömösváry organ and from the protocerebral, or cerebral glands. These lobes represent the protocerebrum – the ganglion of the pre-antennary segment. The deutocerebrum, the ganglia of the antennary segment, is represented by the antero-lateral lobes which receive nerves from the antennae. The ganglia of the third head segment, the tritocerebrum are flattened lobes closely joined to the ventral side of the cerebral ganglia (Fig. 73). They are connected anteriorly by a commissure, the stomodaeal bridge or frontal ganglion from which arises a recurrent nerve which runs along the stomodaeum (oesophagus) forming the stomogastric system (see below).

The tritocerebral ganglia also give rise to the circumoesophageal connectives which pass postero-ventrally around the oesophagus to fuse below it to form the suboesophageal ganglion which supplies the mandibular and first and second maxillary segments. Immediately behind the suboesophageal ganglion is the ganglion of the forcipular segment and behind that the ganglion of the first leg-bearing segment. Posterior to this the ganglia are of uniform size: each ganglion giving rise to four pairs of nerves. The last ganglion of the chain, sometimes termed the genital ganglion, is smaller than the rest.

In the Geophilomorpha the protocerebrum is reduced; the sub-

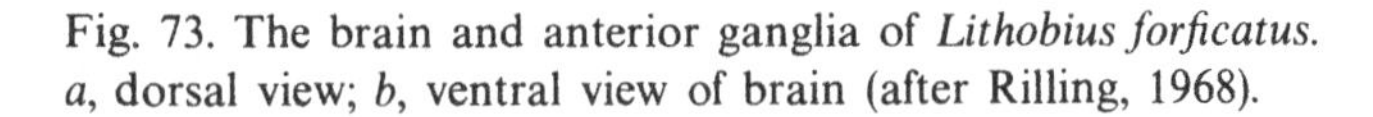

Fig. 73. The brain and anterior ganglia of *Lithobius forficatus*. *a*, dorsal view; *b*, ventral view of brain (after Rilling, 1968).

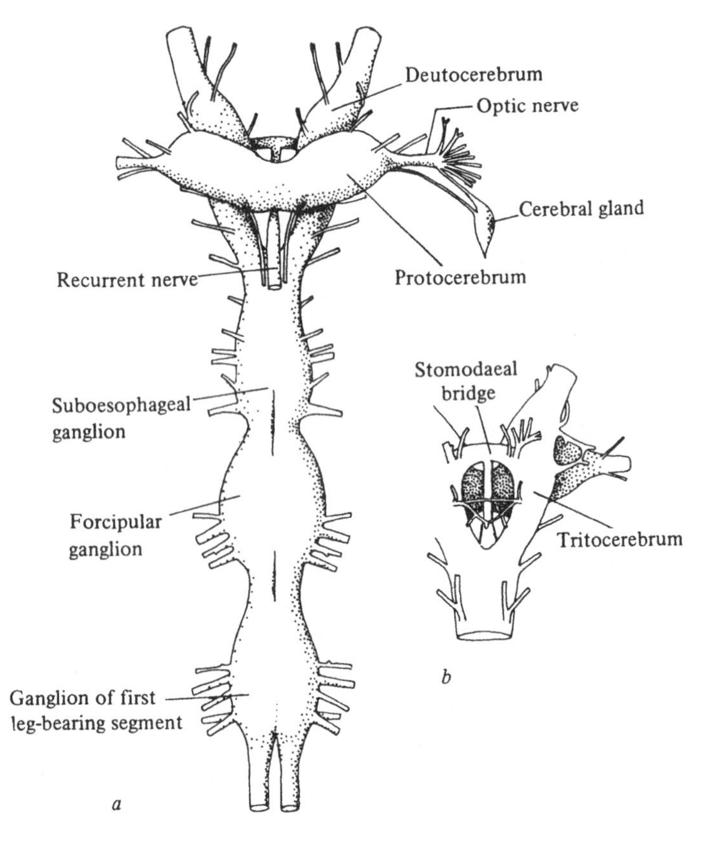

oesophageal and forcipular ganglia are clearly separated by connectives (Fig. 74*a*) and there are only three pairs of nerves in each segment. The arrangement in Scolopendromorpha resembles that in *Lithobius* (Fig. 74*b*). In the Scutigeromorpha the frontal ganglion is free and not obscured in the stomodaeal bridge as in other groups (Fig. 74*c*). The connectives between the trunk ganglia are obscured by a thin layer of ganglion cells.

Fahlander (1938) suggested that the cerebral gland of centipedes was not to be homologised with the frontal organ of Crustacea as Holmgren (1916) had suggested, but was of endocrine function and analogous with the corpora allata of insects, the glands lying along the oesophagus and producing a juvenile hormone. Gabe (1952) showed that neurosecretory cells in the brain of *Lithobius forficatus* were joined by axons to the cerebral glands where the secretions were stored.

Details of the nerve tracts and cell groups within the brain of centipedes have been given by Fahlander (1938).

The fast withdrawal responses of centipedes, most easily observed in long geophilomorphs, suggest that the ventral nerve cord contains fast through-conducting pathways (giant fibre systems) characteristic of many other invertebrates. That this is the case has been shown by Babu (1964) for *Otocryptops* sp., *Scolopendra viridis* Say and *Scolopendra heros* Girard. In these species there are three to six groups of 'AP' fibres responding to stimulation in the anterior half of the animal and there to six groups of 'PA' fibres which respond to mechanical stimulation in the posterior half of the body. The fastest group conduct at 3–4.5 m/s. The largest fibres measure 10–15 μm in diameter. 'Startle times' for centipedes (24 m/s) compare with those of other arthropods, for example cockroaches (24 m/s), flies (45 m/s) and moths (75 m/s). In *Scolopendra morsitans* (Varma, 1973) the giant fibres are 28–58 μm in diameter and arranged in dorso-lateral, dorso-intermediate and dorso-medial groups. The dorso-median and dorso-intermediate groups correspond to the ascending pathway of Crustacea and insects, the dorso-lateral groups to the descending pathway. The fibres are syncytial.

Seifert (1967*b*) investigated the stomogastric system of *Cryptops hortensis*, *Scolopendra cingulata*, *Rhysida afra* (Peters),

Fig. 74. Ventral view of the brain and anterior ganglia of various centipedes. *a*, *Strigamia hirtipes*; *b*, *Scolopendra cingulata*; *c*, *Thereuopoda clunifera* (after Fahlander, 1938).

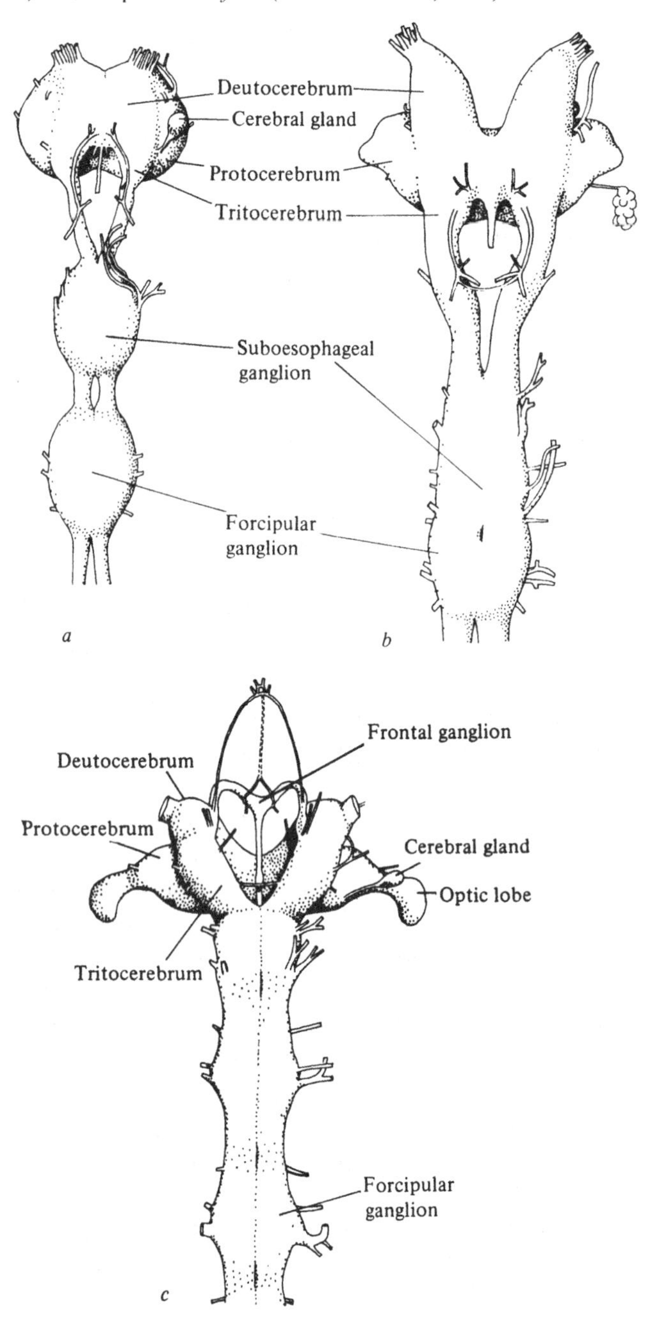

Necrophloeophagus longicornis, Strigamia acuminata (Leach), *Lithobius piceus, L. forficatus* and *Scutigera coleoptrata.* The recurrent nerve originates from the ventro-caudal region of the stomodaeal bridge, or the ganglion in *Scutigera*, supplying the muscles in the pharynx and running posteriorly between the outer circular and inner longitudinal muscles of the gut. A hypocerebral ganglion is situated on the nerve. Labral nerves originate from the tritocerebrum and connective nerves from the circumoesophageal connectives supply the hypopharyngeal muscles. In *Lithobius piceus* the connective nerve joining the stomodaeal bridge to the protocerebrum gives off a fine dorsal cardiac nerve which supplies the dorsal aorta and heart.

Rosenberg & Seifert (1978) examined the structure of the 'outer neural lamella' in *Clinopodes linearis* and *Necrophloeophagus longicornis* and found it to consist of myelinised cells supporting the neural lamella. Ganglia, connectives and thick nerve trunks are totally enclosed by the sheath. It is formed by several cells, the myelinised parts of them arranged spirally. The number of their 'windings' and therefore the number of cells taking part in their construction depends on the region and its diameter: ganglia have few windings: they may even be missing, sporadically. The windings are numerous in the connectives and nerves.

Mechanoreceptors

The sensilla for mechanoreception in arthropods may be classified according to their structure or their function, e.g., auditory, tactile, stretch and position receptors. One structural type may have more than one function; some mechanoreceptors may also function as chemoreceptors.

Two general types of mechanoreceptors are known. Type I in which the dendrite of the bipolar neuron is associated with the cuticle or its invagination and Type II (commonly called stretch receptors), sensilla with multipolar neurons which are associated with the walls of the alimentary canal, the inner surface of the body wall muscles and connective tissue but not the cuticle. Type I may be subdivided into (*a*) sensilla which although associated with the inner aspect of the cuticle lack an external cuticular component, the scolopidial or chordotonal sensilla and (*b*) those with an external cuticular portion, the cuticular mechanoreceptors (McIver, 1975).

Trichoid sensilla

Hair sensilla in arthropods are hair-shaped or clearly derived from a hair. They have been variously termed hairs, setae, sensilla chaetica, trichoid sensilla or sensilla trichodia. Typically, the mechanoreceptive hair bears no pores or openings in the hair wall and is innervated by one neuron. The hair is attached to the socket by an articulating membrane which may consist of the rubber-like protein resilin.

The sensory neuron is composed of axon, cell body and dendrite. The last attaches to the cuticular sheath. Typically the dendrite is divided into an inner (proximal) *Sinnesfortsatz* and an outer (distal) *Sinnescilium* (Fig. 75). These two segments are separated by the narrower ciliary region which is usually composed of nine sets of peripherally arranged double microtubules. It lacks the central pair characteristic of motile cilia. The characteristic feature of cuticular mechanoreceptors is an accumulation of microtubules, the tubular body, in the distal region of the outer segment. The tubular body is the most likely site of sensory transduction (for details see McIver,

Fig. 75. Structure of an arthropod sensillum and associated cells (after various authors).

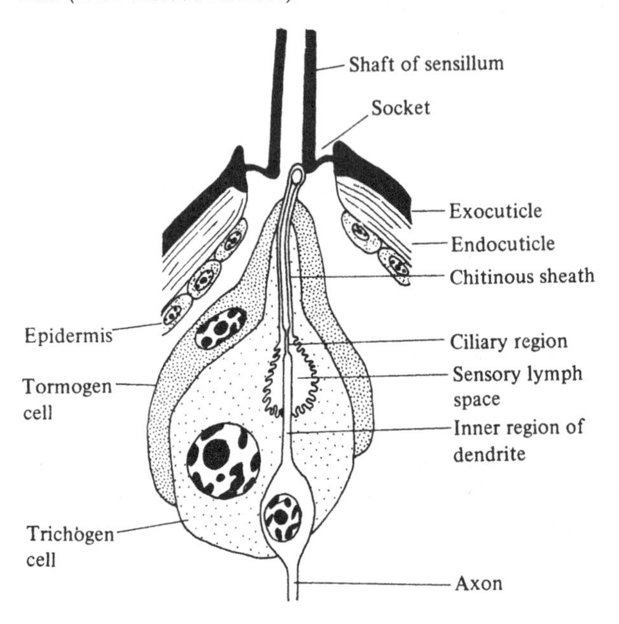

1975). Distally the sensory process or dendrite may be enclosed in a sheath of cuticular material which continues to the surface of the cuticle at the base of the hair and is shed with the cuticle at moulting (Chapman, 1971). This tube is called the scolopale or cuticular sheath. The dendrite may end in a scolopale cap or it may leave the scolopale and extend into the lumen of the hair.

Two or more specialised cells known as sheath cells (*Hüllzellen*) surround the neuron. In sensilla with two sheath cells the names trichogen and tormogen are commonly used. Where there are three they are either numbered or referred to as neurilemma, trichogen and tormogen cells. Variations in the terms used and in the number of sheath cells make it difficult to establish the identity of corresponding sheath cells in the different sensilla.

In development the trichogen cell forms the hair while the tormogen cell forms the socket region. Upon completion the cells withdraw leaving a fluid-filled cavity the receptor-lymph cavity. The surfaces of the sheath cells contacting the extracellular space bear numerous lamellae or microvillae which show signs of secretory activity.

Some trichoid sensilla function both as mechanoreceptors and chemoreceptors. In this case the hair typically has a blunt tip perforated by a single large pore. The chemosensitive dendrites extend the length of the hair to the pore and there may, in addition, be a mechanoreceptive neuron which attaches to the base of the hair.

The trichoid sensilla of centipedes are found on most parts of the body but are particularly abundant on the antennae. There are five types of sensilla on the antennae of the geophilomorph *Necrophloeophagus longicornis*. Each antenna carries about 1200 in all. There are about 600 trichoid sensilla which measure 50–75 μm in length. Some 150 of these are on the last antennal segment. Segments 9–13 have about 70 each, segments 1–5 two to eight, the number gradually increasing on the intermediate segments (Ernst, 1976). In addition there are short thick hairs which are probably thick-walled basiconic sensilla (see below), thin-walled basiconic sensilla, long acuminate hairs and small sense knobs (*kleiner Sinneshügel*) of unknown nature. The distribution of the sensilla is shown in Fig. 76.

Dense fields of small trichoid sensilla are frequently present in

Geophilomorpha particularly on the ventral surface of the last pair of legs in males. The clypeus of *Dicellophilus carniolensis* (C. L. Koch) has a group of about five large setae at each anterior lateral extremity and almost the whole of the posterior portion is occupied by an extensive group of numerous smaller setae interspersed with pores (Fig. 77). The South African oryid *Diphtherogaster flava* Attems has a dense covering of short setae on the sternites of segments 32 to 41 in the male and segments 36 to 47 in the female. Males have from 111 to 115 pairs of leg-bearing segments, females 127 to 133.

In *Scolopendra morsitans* the basal six segments of the antenna do not carry typical trichoid sensilla. There are a few basiconic sensilla (*Hohle Borsten*) and short cones (*Herzförmiger Sinneskegel*)

Fig. 76. Diagrammatic representation of the distribution of sensilla on the terminal antennal segment of *Necrophloeophagus longicornis*. *a*, dorsal side; *b*, ventral side. △ short, thick hairs; ○, trichoid sensilla; ●, basiconic sensilla; ▲, small sense knobs; ■ long acuminate hairs (from Ernst, 1976).

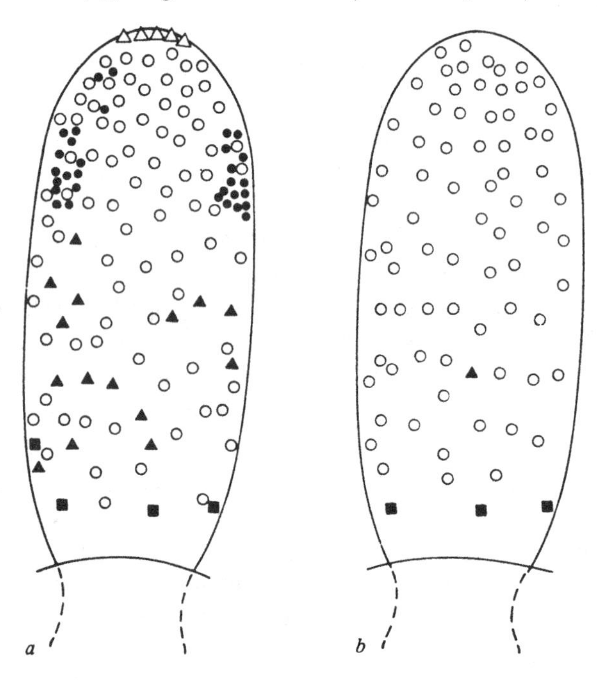

which appear to be modified trichoid sensilla (Fig. 78*a*). The distal part of the antenna bears trichoid sensilla both with a solid hair (*massiven Borsten*) in which the shaft is bent and tapers to a point and with the socket in the form of a tubercle (Fig. 78*b*), and with hollow hairs (Fuhrmann, 1922). The short cones are set in pits. Each contains a canal opening at the apex. The sensory fibres originating from a group of 10–20 neurons end in a small swelling at the base of the canal. Similar but more elongated structures (Fig. 78*c*) are revealed by scanning electron microscopy on the heavily

Fig. 77. Ventral view of the clypeus and labrum of *Dicellophilus carniolensis* (after Eason, 1964).

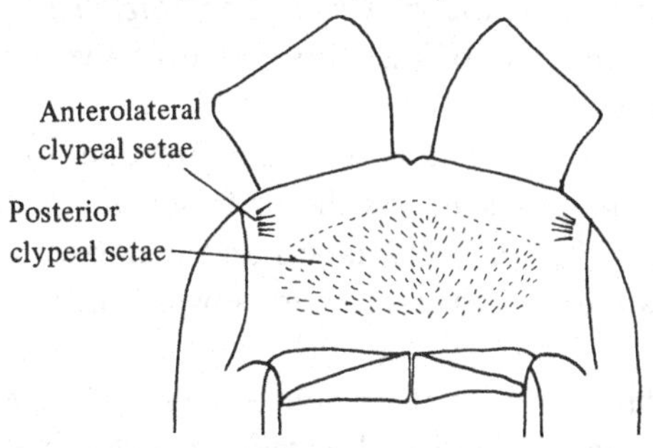

Fig. 78. Modified trichoid sensilla from *Scolopendra morsitans*. *a*, short cone; *b*, sensillum with solid hair; *c*, cone from the poison claw (*a* and *b* after Fuhrmann, 1922; *c* after Jangi & Dass, 1977).

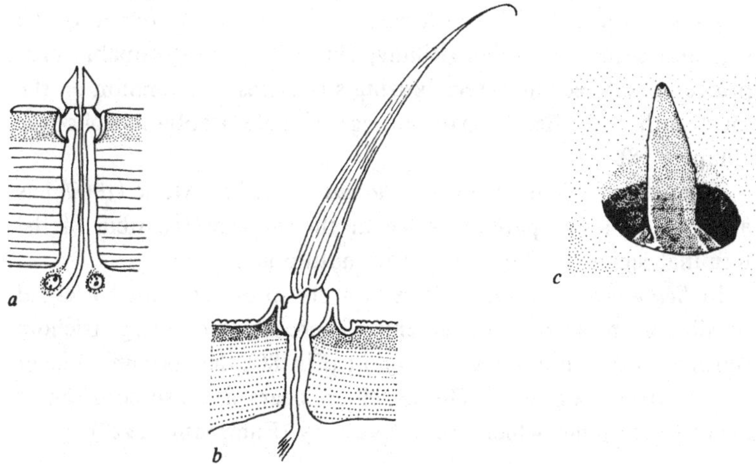

sclerotised region of the poison claw (Jangi & Dass, 1977). When stimulated with saturated glucose solution they evoked a feeding reflex with movement of the maxillae and mandibles in a decapitated animal.

In *Scolopendra* the trichoid sensilla are distributed profusely on the antennae, labrum, hypopharynx and the maxillae, and rather inconspicuously on the legs and the general body surface (Jangi, 1966).

Jangi (1964) showed that there was an increased rate of discharge from the crural nerve when the last legs of *Scolopocryptops sexspinosus* (Say) were raised and attributed this to the stimulation of very small setae, apparently the same as Rilling's *Stellungshaare* (see below) situated at the base of the dorsal side of the femur and succeeding segments.

In some species of *Cryptops* fine setae are abundant on the last pair of legs. In *Cryptops anomalans* the prefemur and femur bear short spine-like setae but the tibia, tarsus and pretarsus bear very fine setae. There is a dense brush of setae between the tibial tooth comb (Fig. 28*a*).

In *Lithobius forficatus*, normal trichoid sensilla are arranged in about three whorls on each antennal segment. Smaller setae (*blasse Borsten*) at the end of the last antennal segment are basiconic sensilla. Very small setae in groups of two to four are present on the basal third of each segment (Fuhrmann, 1922). These *Stellungshaare* are also arranged in rows or fields on the anterior borders of the sternites and tergites and in the coxal region of the legs and second maxillae (Rilling, 1968). They are probably proprioceptors. Specialised seta-bearing structures are common on the fourteenth and fifteenth pairs of legs in male lithobiomorphs (see Chapter 18).

The cavernicolous Spanish *Lithobius san-valerii* Matic (1960) has a group of long spatulate setae at the apex of the tibia of the terminal legs (Fig. 79). Their exact nature is unknown.

In *Scutigera coleoptrata* there is a group of very small trichoid sensilla on the scape of the antenna. There are several large trichoid sensilla on the base of the flagellum I (Fig. 80); they become smaller on the distal segments. The flagellum bears a dense covering of small solid spines which are not sensory (Fuhrmann, 1922).

Fig. 79. Setae of the apex of the tibia of the terminal leg of *Lithobius san-valerii* (after Matic, 1960).

Fig. 80. Basal antennal segments of *Scutigera coleoptrata* (after Fuhrmann, 1922).

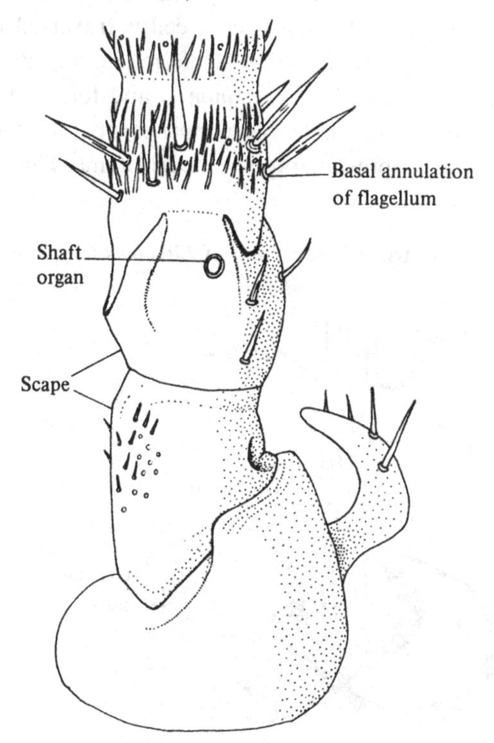

In *Necrophloeophagus longicornis*, *Lithobius* and *Scutigera* and probably other centipedes, one side of the polygonal area that surrounds the hair socket may be thickened to form a stop (*Kragen*) limiting the movement of the hair (Fig. 81).

Fine structure and innervation of the trichoid sensillum. In *Necrophloeophagus longicornis* the trichoid sensillum receives two mechanoreceptive dendrites each with a tubular body. They terminate at the base of the hair. Sixteen chemoreceptive dendrites enter the hair lumen at the tip of which there is a small pore. There are two sheath cells and a larger inner and outer sensory lymph space (Ernst, 1976).

The antenna of *Lithobius forficatus* bears about 2000 trichoid sensilla each consisting of 18 sensory cells and at least three sheath cells. Dendrites of 17 of these sensory cells extend into the shaft which is ridged externally and has a terminal pore. The remaining dendrite which contains a tubular body ends in a complicated structure at the hair base. The first sheath cell is traversed by the dendrites. It has a rich granular endoplasmic reticulum, surrounds a sensory lymph space (*Sensillenliquorraum I*) and forms an inner dendritic sheath which surrounds the dendrites up into the hair lumen (Fig. 82). This is presumably the neurilemma. The second

Fig. 81. Base of trichoid sensillum of *Lithobius forficatus* (after Keil, 1976).

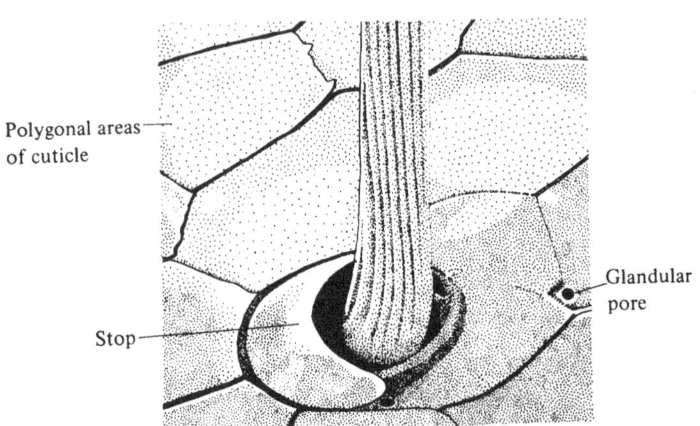

sheath cell is the tricogen cell. It contains numerous microtubules and sends a narrow process into the hair lumen. The third sheath cell, the tormogen cell, surrounds a second sensory lymph space. A second cuticular sheath lies between the trichogen and tormogen cell. It is connected to the body wall (Keil, 1976). It is possible that Keil's terminology does not correspond with that of other workers.

The very small sensilla (probably *Stellungshaare*) at the base of each antennal segment and arranged, according to Keil (1977) in three groups around the periphery each have a single sensory cell. They are typical trichoid sensilla.

Before the advent of electronmicroscopy there was no report of a terminal pore in centipede trichoid sensilla. How many sensilla have this structure, associated with mechanoreception and chemoreception is not known. The sensilla of the tibia of *Scutigera coleoptrata* and *Necrophloeophagus longicornis*, of the first antennal segment of *N. loricornis* and the sensilla of the poison claw of *Lithobius martini* Brolemann have only one sensory neuron. The trichoid sensilla of the tibia of *Lithobius piceus* may have one or several neurons; those on the coxa several (Duboscq, 1898). This suggests the former may be simply mechanoreceptors, the latter mechanoreceptors and chemoreceptors.

Campaniform sensilla

Structures similar to the campaniform sensilla have been described from the epipharyngeal wall of *Scolopendra* (Jangi, 1966).

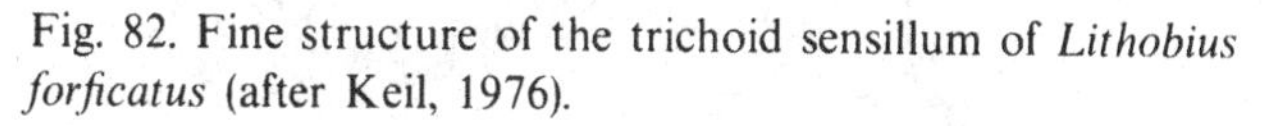

Fig. 82. Fine structure of the trichoid sensillum of *Lithobius forficatus* (after Keil, 1976).

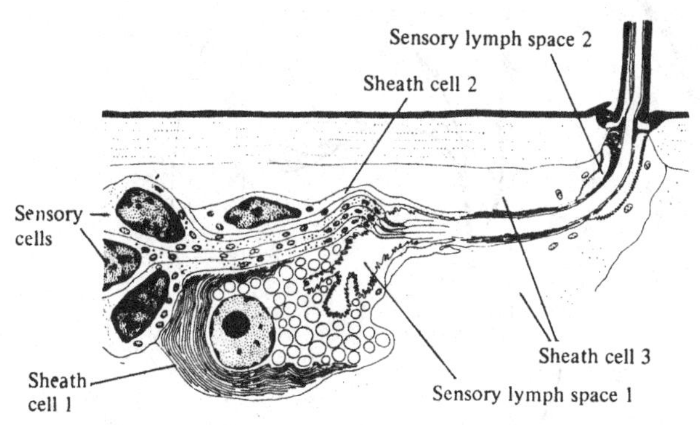

These epipharyngeal papillae resemble vacant hair follicles but are more elliptical in outline and consist of a sclerotised dome of cuticle to which is attached a sensory process (Fig. 83). Campaniform sensilla are mechanoreceptors.

Scolopidia

In insects the scolopidia consist of three cells: the neuron, an enveloping scolopale call, and an attachment, or cap cell. They

Fig. 83. Campaniform sensilla from the epipharyngeal wall of *Scolopendra morsitans* (after Jangi, 1966).

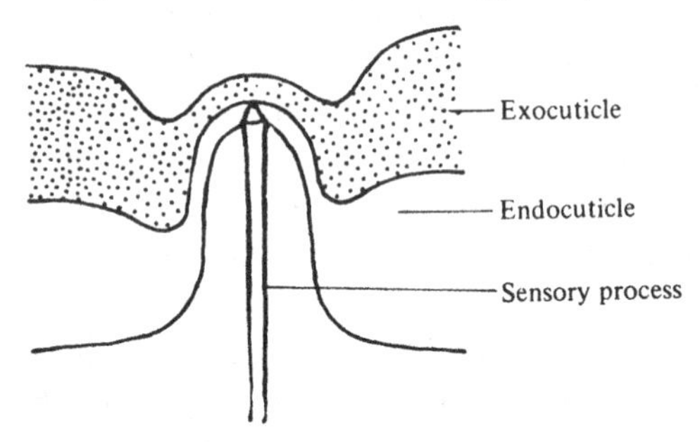

Fig. 84. Left mandible of *Lithobius forficatus* showing scolopidia (after Rilling, 1968).

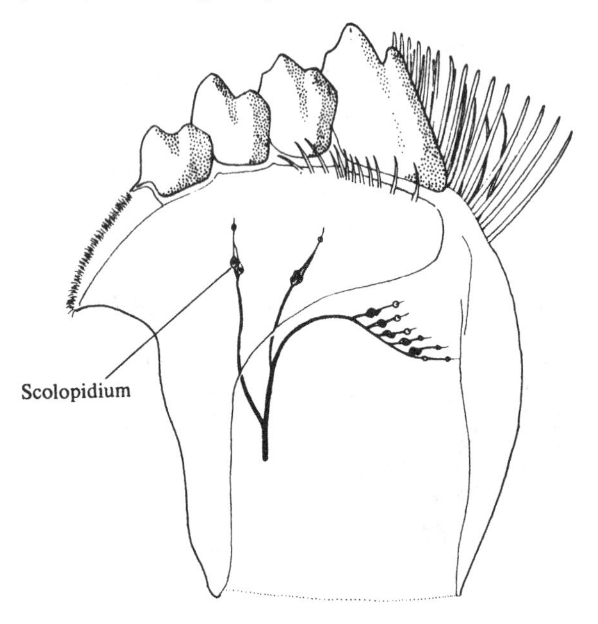

are probably stimulated by changes in tension but 'are not to be regarded as structures of constant function, but as nervous organs ... which may be adapted to diverse physiological functions' (Wigglesworth, 1972).

Rilling (1968) described receptors in the mandible of *Lithobius forficatus* consisting of two or three cells forming a string with a distal sensory spike (*Sinnesstift*) identical to that of the *Stellungshaaren* (Fig. 84). He suggested that they were sense organs that had lost their setae. His figures of these structures lack detail, however, and his identification of them as scolopidia should, perhaps, be treated with caution.

Stretch receptors

Stretch, or muscle, receptors consist of neurons associated with strands of muscle and have a proprioceptive function. In *Lithobius forficatus* there are two rows of tergal organs consisting of branched muscle bands between succeeding long tergites (Fig. 85*a*). Further stretch receptors, *Pedale Organe*, consisting of a simple

Fig. 85. Stretch receptors in *Lithobius forficatus*. *a*, left sense organ between tergite 5 and tergite 7; *b*, pedal muscle stretch receptor of coxa of twelfth leg; *c*, the sense cells of this organ; *d*, dorsal view of antennal muscle sense organ (*a* and *c* after Rilling, 1960; *b* and *d* from Rilling, 1968).

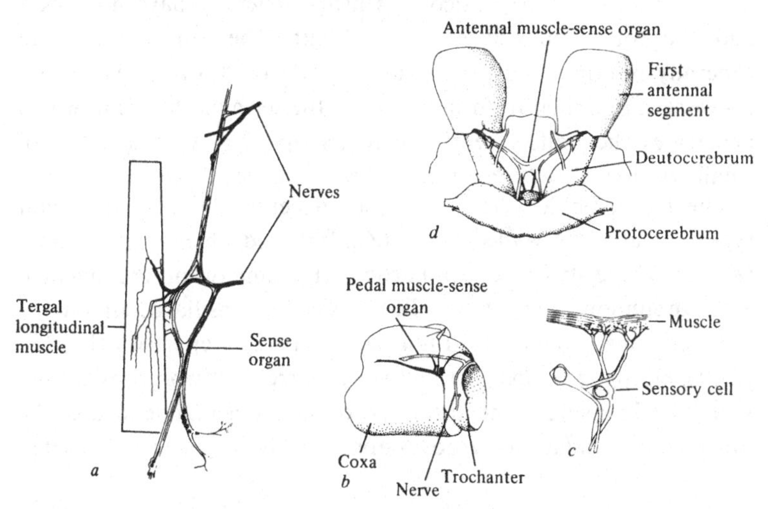

band of muscle and a pair of sensory cells link the coxa and trochanter of each leg (Fig. 85*b*, *c*). A similar structure is found in each half of the second maxilla. The X-shaped antennal muscle-sense organ is associated with the basal segments of the antennae (Fig. 85*d*) (Rilling, 1960, 1968).

Stretch receptors have also been reported in *Scolopendra morsitans*. Here they consist of pairs of thin muscles surrounded by connective tissue (Varma, 1972). These are attached at one end to the tergite and at the other to the intersegmental arthrodial membrane. There are 10 receptors on each side of the body and they follow a zig-zag course from the head to the last segment. They are innervated by 13 nerve cells that are both bipolar and multipolar.

Spines

Bipolar sensory cells supply many of the spine and tubercle-like structures in centipedes such as the mandibular teeth and the teeth of the forcipular coxosternite of *Lithobius*. They are found within the leg spines of *Lithobius* and the long spines of the forcipular coxites of *Scutigera* (Duboscq, 1898). Rilling (1968) confirmed that the teeth of the forcipular coxosternite of *L. forficatus* are each innervated by three or four bipolar nerve cells.

Chemoreceptors

Basiconic sensilla

In insects the basiconic sensilla generally have no socket and the tormogen cell is often absent. The sensory cells are generally multiple and may form a cluster of 20 or 30. The distal processes are united to form a bundle, the terminal filament, which penetrates the hollow sensillum to its tip often with a group of small rod-like structures on its course (Wigglesworth, 1972).

The basiconic sensilla are similar in centipedes and two main types occur: thin-walled pegs (*Zapfen*) and thick-walled cones (*Kegel*). These structures are particularly common on the antenna.

In geophilomorphs thin-walled basiconic sensilla occur on the last antennal segment. Fuhrmann (1922) reported that in *Necrophloeophagus longicornis* they occurred in two grooves each with 20–25 sensilla at the distal part of the segment, one median the other lateral. In fact the grooves appear to be artefacts as indicated by the disposition of the trichoid sensilla in his figure. The arrangement of these sensilla varies considerably from one geophilomorph genus to another. The distribution of sensilla on the last segment of the antenna of *N. longicornis* is shown in Fig. 76.

Demange (1943) gave the term *microchètes* to thick-walled sensilla which occur on the ventral surface of segments 5, 9 and 13 of the antennae of *Hydroschendyla submarina* (Grube). These are also present on the same segments of *Strigamia maritima* (Lewis, 1961) and are probably thick-walled basiconic sensilla. In *Pectiniunguis pampeanus* Pereira & Coscaron, small groups of trifid setae (Fig. 86) are present on the end of the apical antennal segment and on segments 2, 5, 9 and 13 (Pereira & Coscaron, 1976).

In *Scolopendra morsitans* hair-like basiconic sensilla are present on the basal six segments of the antenna (Fig. 87), termed *hohle Borsten* by Fuhrmann (1922). They are relatively short, about 0.02 mm long, and unique in that a strongly tanned ring of cuticle is found around the pore canal between the exocuticle and endocuticle. Intermediates between these and thin-walled sensilla are found but the thin-walled basiconic sensilla are very rare.

In *Lithobius forficatus* thin-walled finger-shaped basiconic sensilla are found in groups of one to three on the dorsal side of the

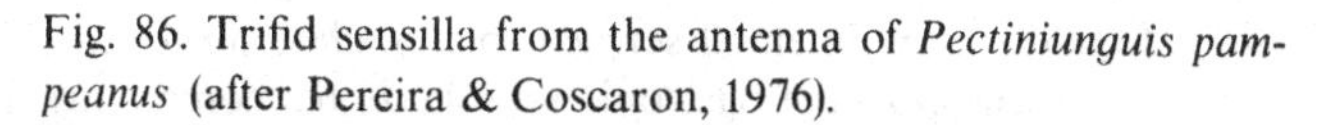

Fig. 86. Trifid sensilla from the antenna of *Pectiniunguis pampeanus* (after Pereira & Coscaron, 1976).

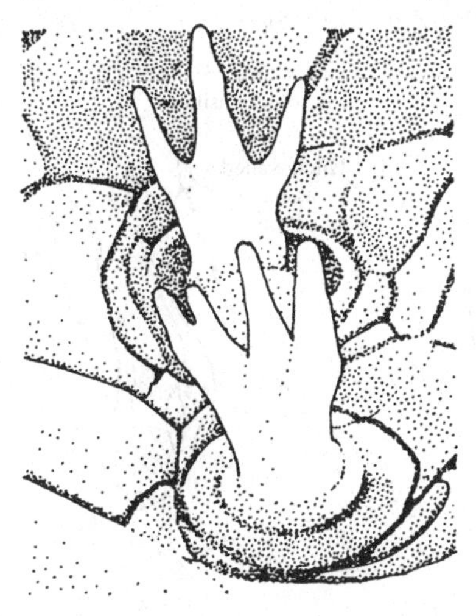

Fig. 87. Basiconic sensilla from the antenna of *Scolopendra morsitans* (after Fuhrmann, 1922).

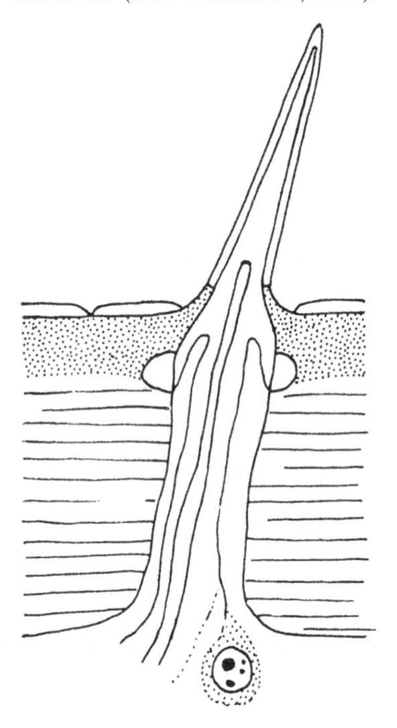

Fig. 88. *a*, an antennal segment of *Lithobius forficatus* showing the position of the basiconic sensilla; *b*, basiconic sensilla showing innervation (after Rilling, 1968).

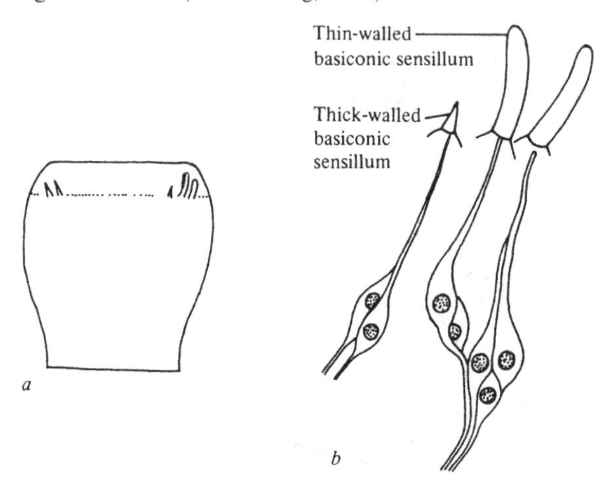

front edge of each antennal segment at the point where normal cuticle and arthrodial membrane meet (Fuhrmann, 1922; Rilling, 1968; Keil, 1977) (Fig. 88). Thick-walled basiconic sensilla (*blasse Borsten* of Fuhrmann) are also found on the front of each segment (Fig. 88) where they are about 10 μm long, but the eight at the end of the last antennal segment are 50 μm long.

Basiconic sensilla do not appear to have been described from Scutigeromorpha with the exception of those in the antennal *Schaftorgan*. The latter is a nearly spherical pit with an opening measuring about 0.02 mm in diameter. From the floor of the organ there arise about 20 short, spine-like *Zapfen* which are 5 μm long (Fig. 89). The function of the organ is unknown (Fuhrmann, 1922). It is possible that the tarsal papillae (Fig. 219*g*) are sensilla.

The fine structure of the basiconic sensilla has been studied only in *Lithobius* (Keil, 1977). The thin-walled *Zapfen* is supplied by three sensory cells from each of which two outer segments run to the apex of the sensilla (Fig. 90*a*). The neurons are surrounded by several sheath cells which run into the sensillum. The wall of the peg is penetrated by cavities along its entire length and these open to the exterior by numerous widely ramified slits. The terminal opening of the sensillum is about 0.4 μm wide. Beneath it is a drop of unknown composition. Keil suggested that these structures might be humidity receptors.

The thick-walled basiconic sensilla (*Kegel*) are supplied by from

Fig. 89. *Schaftorgan* of *Scutigera coleoptrata* seen in vertical section (after Fuhrmann, 1922).

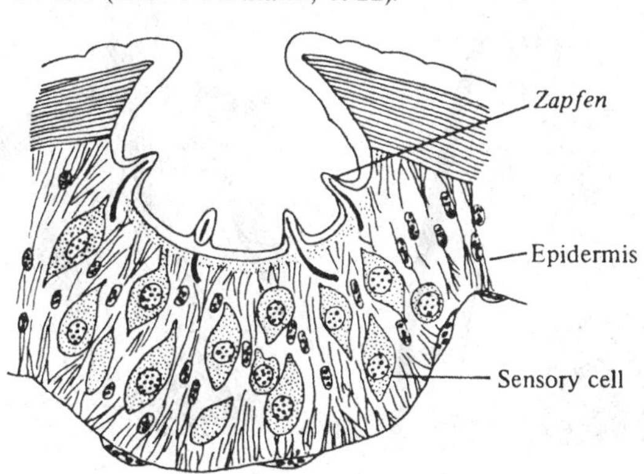

four to six, usually five, sense cells which are surrounded by several sheath cells (Fig. 90*b*). There may be fine canals in the wall of the seta. Keil suggested that these sensilla might be thermoreceptors. Rilling considered that they were olfactory organs.

Plate organs

A sensillum that may be a plate organ similar to that of insects occurs on the end segment of the last pair of legs of the geophilomorph *Himantarium samuelraji* Rajulu. Each consists of a circular or oval depression 25–30 μm long. The sensillum is supplied by a large number of sensory cells whose distal processes are said to coalesce to form a terminal strand which is inserted into, or touches, the cuticular plate (Rajulu, 1970*e*). The terminal strand is surrounded by a pair of cap cells (Fig. 91). The fine structure of the sensillum has not been investigated.

When the end segment of the last pair of legs of decapitated animals is removed or coated with celloidin animals ceased to

Fig. 90. *a*, thin-walled basiconic sensillum of *Lithobius forficatus*; *b*, thick-walled basiconic sensillum of *Lithobius forficatus*. Only one sensory cell is shown in each case but in *b* the dendrite of a second cell is shown (after Keil, 1977).

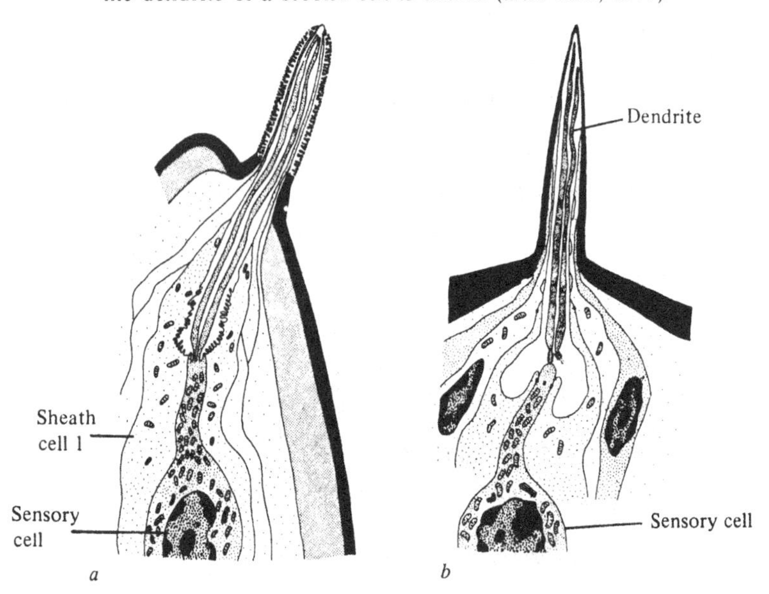

Fig. 91. Vertical section of a chemoreceptor from the last leg of *Himantarium samuelraji* (after Rajulu, 1970*e*).

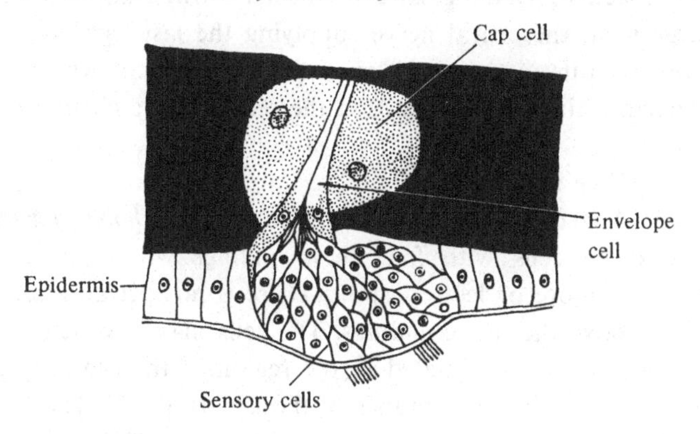

Fig. 92. Cells of *Lithobius forficatus* with free nerve endings. *a*, from the pleural membrane posterior to the coxa; *b*, sensory cells from the trochanteral sense organ of the second maxilla (after Rilling, 1960).

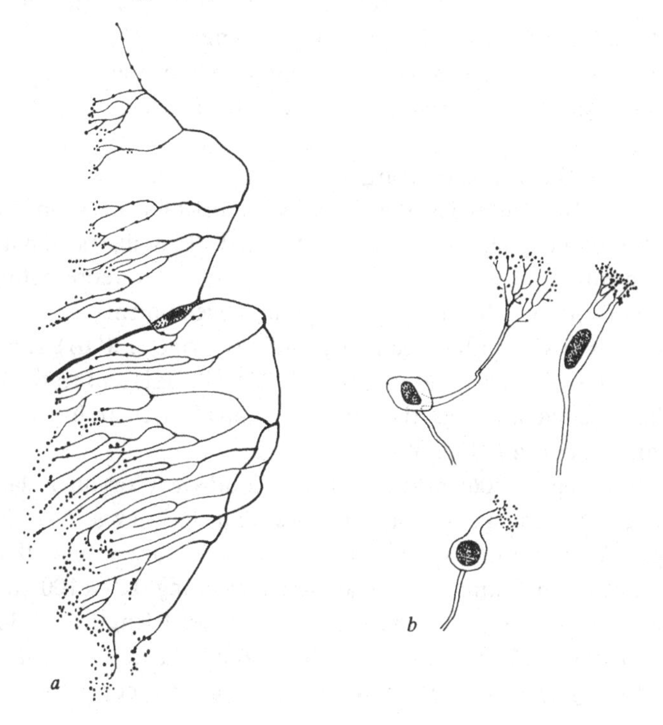

respond to volatile chemicals such as ethyl alcohol, turpentine and ether. Electro-physiological experiments showed an increased discharge from the crural nerve supplying the last legs upon stimulation but this disappeared when the last segment was coated or removed. Rajulu concluded that the sensilla were chemoreceptors.

Free nerve endings

Apart from the sensilla, the cuticle of *Lithobius forficatus* is supplied by cells with free nerve endings whose position and number is more or less constant. There is no specialisation of the cuticle above the nerve endings. The cells may have few or many branches. They are situated in the region of the coxae (Fig. 92*a*) and the arthrodial membranes of the legs (Fig. 93). They are also found on the forcipules, second maxillae (Fig. 92*b*) and the basal three antennal segments. They are presumably proprioceptors (Rilling, 1960).

At the end of most segments of the legs there is a narrow transverse line of unsclerotised cuticle (Fig. 93), the *Sinnesnäht* (Rilling, 1960). Free nerve endings ramify in this suture. These structures are also found on the gonopods of the female and the telopodites of the forcipules and second maxillae.

The Tömösváry organ

In centipedes the Tömösváry organ occurs only in the Lithobiomorpha, Scutigeromorpha and (Crabill, personal communication) *Craterostigmus*. A group of cells representing this organ appears during the embryonic development of *Scolopendra* but it is absent in the free-living stadia (Heymons, 1901). It is found immediately below and in front of the ocelli, adjacent to the base of the antenna and has the form of a small circular fossa with an annular margin (Fig. 30).

Hennings (1906) noted that the Tömösváry organ is larger in cavernicolous species of *Lithobius* that lack eyes. In *Lithobius matulicii* Verhoeff the organ is three times larger than that of a *Lithobius forficatus* of similar size measuring 187 × 100 μm. The ratio of the length of the organ to the head length is 1:36 in *L. forficatus*, 1:17 in *L. reiseri* Verhoeff and 1:11 in *L. matulicii*.

In *L. forficatus* the organ has the form of a cup with a central circular chitinous disc. The groove between the disc and the wall of the cup is perforated by the pores of the ducts of unicellular glands (Fig. 94*a*). The central disc has a central perforation through which the terminations of fusiform sensory cells pass to spread out on its

Fig. 93. A left walking leg of *Lithobius forficatus* (anterior aspect) showing the position of sense cells with free nerve endings. ●, cells on the anterior surface of the leg; ○, cells on the posterior surface. *Sinnesnähte* and the pedal muscle-sense organ are also shown (after Rilling, 1968).

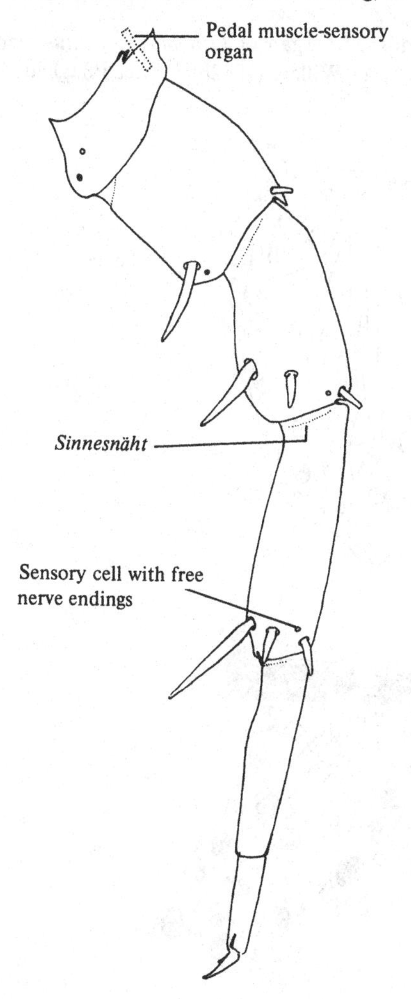

outer surface. This region is covered by a fine cuticular membrane (Willem, 1892*b*). A second description of the organ in *L. forficatus* was given by Pflugfelder (1933). He described a small opening in the centre of the organ where terminate a number of sensory cells. The distal region of each cell contained a thin fibre which appears to project into a delicate chitinous cap (Fig. 94*b*).

The ultrastructure of the organ was described by Tichy (1973). He found that the sensory fibres run through the pore which passes through the exocuticle and the endocuticle, spreading out and

Fig. 94. The Tömösváry organ of *Lithobius forficatus* seen in vertical section. *a*, after Willem (1892*b*); *b*, after Pflugfelder (1933).

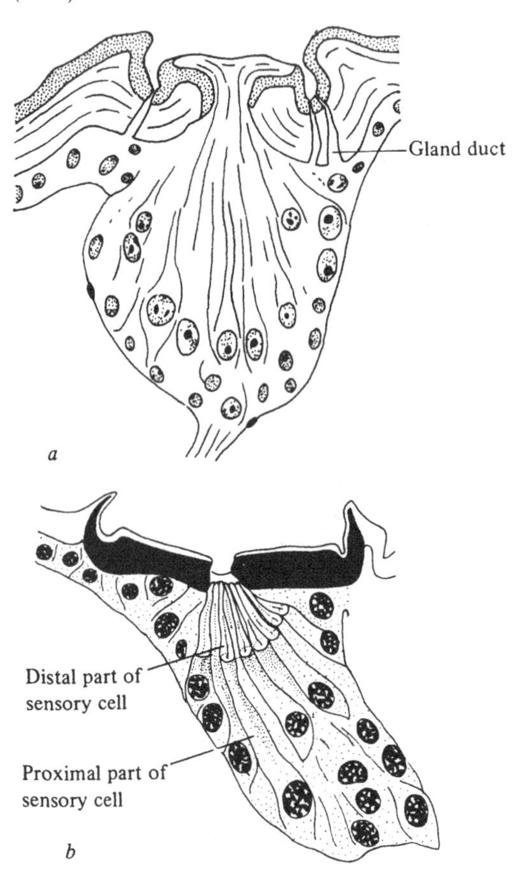

branching in a cavity covered by epicuticle (Fig. 95). There are from two to four sensory cells, each with the normal ciliary region containing two cilia. Each sensory cell is surrounded by two sheath cells whose distal walls are produced into microvilli. Sheath cell 2 reaches into the cuticular pore in which is found the receptorlymph cavity. The numerous gland cells opening into the groove around the sensory plate consist of a gland cell proper containing secretory granules and a canal cell with a cuticle-lined duct.

It is difficult to reconcile Pflugfelder's figure with the findings of Willem and Tichy. It would appear that the epicuticle and sensory plate were lost in his sections. His 'sensory cells' are clearly not bipolar neurons. They may be gland cells, in which case the 'fibres' in the distal parts of the cells are probably the gland ducts.

The Tömösváry organ of an unidentified species of *Scutigera* is formed of a conical wall with a very small opening beneath which are found the sensory cells. The ends of these are capped with a cuticle-

Fig. 95. A diagramatic reconstruction of the Tömösváry organ of *Lithobius forficatus* based on Tichy (1973).

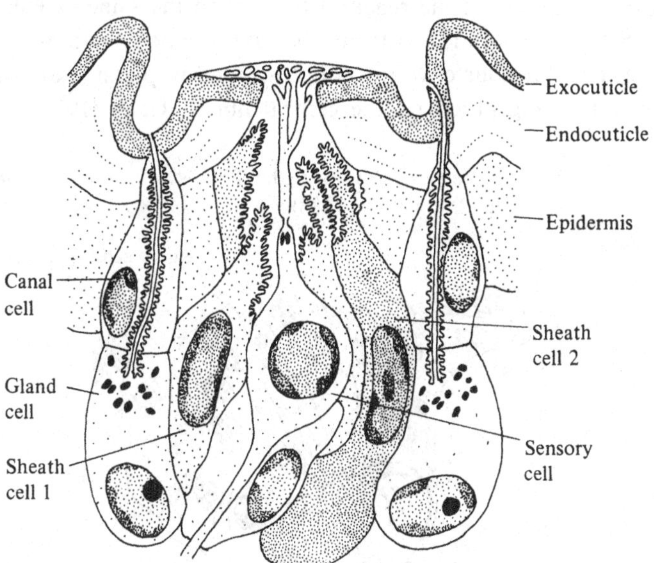

like substance and they are grouped into the form of a spinning-top (Fig. 96) (Pflugfelder, 1933). Knoll (1974) described the development of the organ in the embryonic *Scutigera coleoptrata.*

Function of the Tömösváry organ

Hennings (1904*a*) thought it unlikely that the Tömösváry organ in the millipede *Glomeris* was either an auditory organ or a chemoreceptor. He quoted (Hennings, 1906) Tömösváry's suggestion that it might be a humidity receptor or register changes in atmospheric electricity, pressure or temperature. It has also been regarded as a light receptor. Pflugfelder (1933) considered the organ to be a pressure receptor concerned with the perception of concussion, vibration and perhaps sound and tone. Verhoeff (1902–25) also considered it to be a tone receptor.

Meske (1960) investigated the function of the organ experimentally. Having shown that *Lithobius forficatus* responds to sounds between 50 and 5000 Hz he coated the Tömösváry organs with nail varnish and showed that the reaction was diminished. When the organ was damaged mechanically, there was a similar diminution in the response to sound. Application of nail varnish next to the organ diminished the reaction but not to the same extent.

Rilling (1968) argued from the structure as known at that time, that the chitinous covering could respond to pressure and that this would be registered by the central fibrils. Tichy (1973), however,

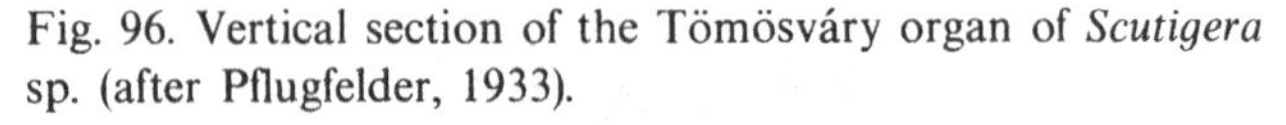
Fig. 96. Vertical section of the Tömösváry organ of *Scutigera* sp. (after Pflugfelder, 1933).

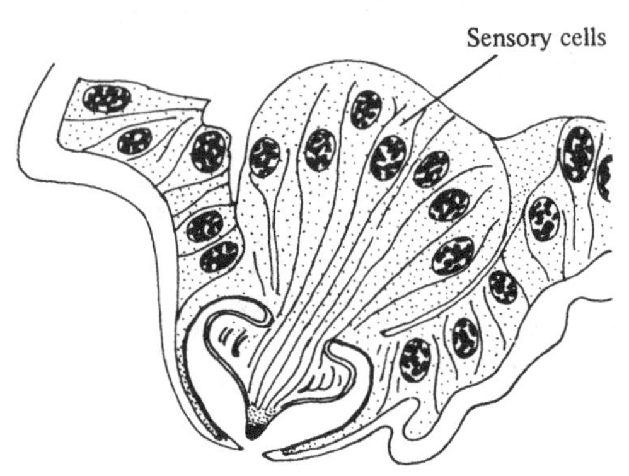

drew attention to the similarity, as shown by an electron-microscopical investigation, to the olfactory sensilla of insects: these have a thin cuticle pieced by numerous minute pores and the distal tips of the dendrites branch repeatedly in the lumen of the sensillum and end in the pores. The Tömösváry organ appears, in fact, to resemble a large placoid sensillum. He investigated the function of the organ experimentally using animals with either the antennae removed and the Tömösváry organ intact or the antennae intact and the Tömösváry organ coated with varnish. In a humidity gradient, 76 per cent of animals came to rest in the highest humidity region (95 per cent relative humidity), 64 per cent did so with the Tömösváry organ covered and 46 per cent with the antennae removed. Only 23 per cent of animals without antennae and with Tömösváry organ varnished came to rest in the highest humidity. It was concluded that the organ supplemented the humidity reaction.

Evidence as to the function of the Tömösváry organ is then, still conflicting and it is possible that it serves more than one function. Previous authors have not attempted to explain the significance of the increased size of the organ in cavernicolous lithobiomorphs where it appears to compensate for the loss of eyes. Since the Tömösváry organs are placed widely apart on the head they may be used in tropotactic responses, that is responses where the intensity of stimulation on the two sides is compared simultaneously by means of symmetrically placed receptor organs. The humidity in caves is likely to be high and unvarying so the organs may be used as chemoreceptors in the location of prey. Bedini & Mirolli (1967) drew attention to the similarity in the structure of the Tömösváry organ in the millipede *Glomeris romana* and the basiconic and coeloconic sense organs of some insects which are olfactory receptors.

Ocelli and compound eyes

The Geophilomorpha and Cryptopidae lack ocelli, the Scolopendridae have a group of four ocelli on each side of the head and in the Lithobiomorpha the number of ocelli varies from one (in *Lamyctes*) to 40 on each side in some species of *Lithobius*: some cavernicolous lithobiomorphs are blind. The Scutigeromorpha have well-developed faceted eyes. *Craterostigmus* has one ocellus on each side of the head capsule.

The eyes of lithobiids are better known than those of other centipedes. The ocellus consists of a biconvex cuticular lens overlying a layer of corneagenous cells beneath which is a cup-shaped retina (Fig. 97). The sensory cells have a fringe of processes on their inner borders (Hesse, 1901).

Grenacher (1880) was the first to note a difference between distal receptor or hair cells, and the basal cells (retina). Willem (1892*a*) and Hesse (1901) recognised the *Haarzellen* as sensory structures.

An electronmicroscopical study of *Eupolybothrus fasciatus* (Newport) has confirmed the general organisation as described by earlier workers (Bedini, 1968). The optic cup is surrounded by flattened sheath cells (satellite cells, Bedini). The lens has a thin epicuticular layer 1 μm thick and an underlying striated endocuticle. The inner surface of the lens rests directly on a thin layer of corneagenous cells about 2 μm thick. In direct contact with the corneagenous cells are long retinal cells, their arrangement is shown in Fig. 97*b*. Distally, their outer surface is covered with microvilli about 3.5 μm long and perpendicular to the longitudinal

Fig. 97. The ocellus of lithobiomorphs. *a*, *Lithobius forficatus* (after Bähr, 1974); *b*, *Eupolybothrus fasciatus* (after Bedini, 1968).

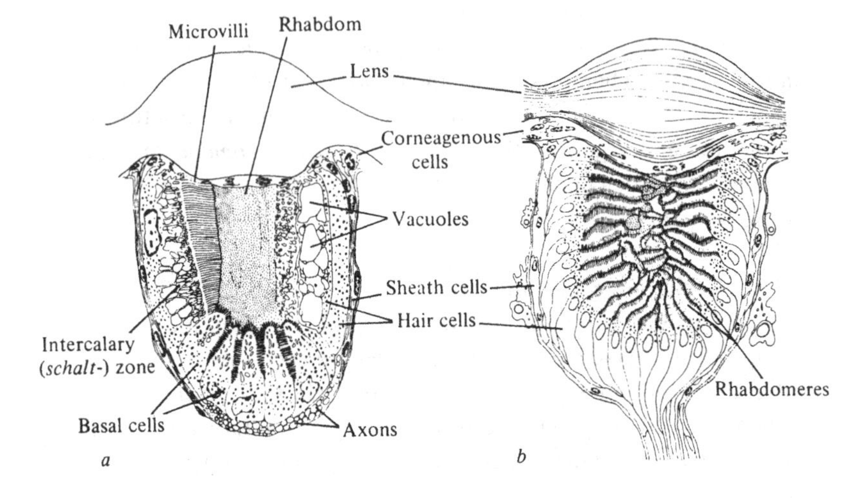

axis of the cell. Pigment granules are mostly scattered in the proximal part of the cell. The periphery of the optic cup is covered with a thin lamella of granular appearance about 1 μm thick which seems to be the direct continuation of the sub-epidermal lamella.

In *Lithobius forficatus* there are up to 110 visual cells which contribute either to the distal or to the proximal rhabdom (Fig. 97*a*). Small pigment cells which Grenacher (1880) believed to surround the internal cuticle between the lenses do not appear to be present (Bähr, 1974). They are in the same position as Bähr's corneagenous cells. The lens is biconvex, normally 50 μm in diameter and 35 μm thick. Externally an epicuticular layer 2–3 μm thick can be distinguished from the underlying endocuticle.

The corneagenous layer does not have the regular appearance of the *Eupolybothrus*. The cell bodies are usually located laterally to the internal vault of the lens. In the central part only three to nine dilations of the corneagenous cells are present. Grenacher's figure which showed nuclei situated in the midline resulted from tangential sectioning. The cells contain numerous free ribosomes and a high proportion of microtubules.

The sheath cells near the corneagenous cells are transitional in form between the latter and the cells covering the eye-cup where they form a flat epithelium. They may function (*a*) in the secretion of the endocuticular part of the lens, (*b*) to separate the ocelli from each other and (*c*) to isolate one or more axons. There is evidence that they serve as a physiological barrier between the haemolymph and the receptor layer (Bähr, 1971).

The distal and greater part of the eye-cup is occupied by 25 to 70 large, somewhat prismatic sensory cells (the hair cells). In the dark adapted state they show a succession of sectors from the axis outwards: (1) the rhabdomere, (2) the intercalary zone (*Schaltzone*), (3) the main cytoplasmic part with organelles and (4) the nerve fibre process. As in other photoreceptors the rhabdomeres are composed of microvilli which are large compared with those of other arthropod visual cells (1–17 μm in length and 750–2500 Å in diameter) and were therefore detected in light microscope studies. The *Schaltzone* consists of cytoplasmic bridges which surround small cisternae of smooth endoplasmic reticulum the 'palisades', and, further from the centre, of vacuoles of low electron density.

The less developed *Schaltzonen* in Hesse's illustration and those shown electronmicroscopically by Joly (1969) seem to be an intermediate stage in the process of dark-adaptation. Under conditions of dark-adaptation the *Schaltzone* extends over the whole vertical axis. From the level of the lowest microvilli the oblique inner membranes converge at the cell base to form the nerve fibre. The main cytoplasmic part of the cell contains numerous pigment granules. The axons occur at the periphery of the eye cup surrounded by sheath cells.

The basal cells (retina) are quite different from those of the distal part. The microvilli are inserted on all sides of the distal and apical membrane (Fig. 97*a*) and the microvilli of the neighbouring cells interdigitate. Beneath the microvilli there lies a *Schaltzone*. Pigment grains are abundant in the cytoplasm.

Grenacher (1880) investigated the structure of the ocelli of 10 species of scolopendrids. In all species there is a large biconvex lens whose inner bulge is surrounded by a layer of cells continuous with the hypodermis, the *Gläskorper*, which probably corresponds to the

Fig. 98. Vertical section of an ocellus of *Heterostoma australicum* (after Grenacher, 1880).

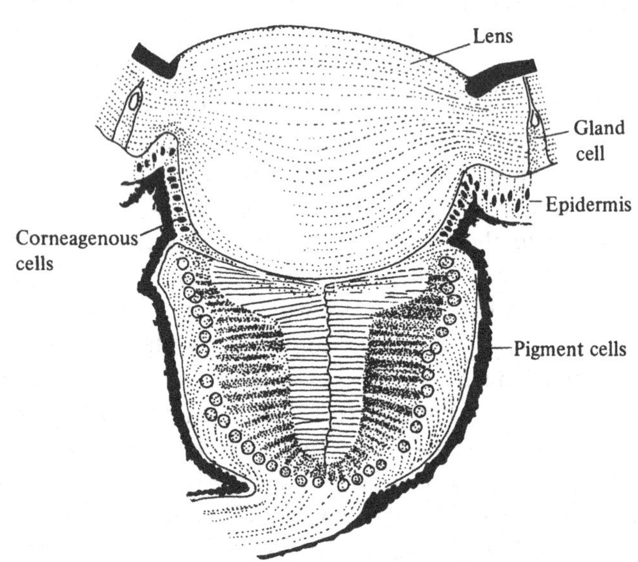

corneagenous cells of other authors. The retina is U-shaped and consists of about 100 pigmented cells with distal microvilli (*Stabchen*) which are unpigmented. In '*Heterostoma australicum*', presumably a species of *Ethmostigmus*, the retinal cells immediately below the lens are short and bear long microvilli. Those further below are longer with correspondingly short microvilli (Fig. 98). In '*Scolopendra tahitiana*' the retinal cells are of approximately equal proportions with the exception of those at the base of the cup, where there is a small U-shaped invagination of retinal cells. In *Cormocephalus foecundus* Newport and *C. aurantiipes* Newport (Grenacher's *C. gracilis*) the basal part of the retina forms a hump: the cells of this hump lack microvilli. Grenacher referred to an *Innere Cuticula* surrounding the retinal cells not seen in lithobiomorphs. In Grenacher's figure of *S. tahitiana* the pigment is concentrated at the bases of the retinal cells, in his figure of *H. australicum* it is evenly dispersed through the cells: these conditions probably represent different stages in dark-adaptation (see below).

In the Scutigeromorpha the eyes consists of numerous ommatidia: from 100 to 200 in *Scutigera coleoptrata*, 360 in *Thereuonema tuberculata* and 600 in *Thereupoda clunifera*. The structure of the individual ommatidium was described for *Scutigera coleoptrata* by Grenacher (1880), Adensamer (1894) and Hemenway (1900). Beneath the cuticular lens lies a central vitreous body surrounded by three tiers of cells (Fig. 99). The vitreous body is composed of six to twelve segments. Although without nuclei in the adult, nuclei were reported to be present in these cells in an individual 5 mm long (Adensamer, 1894). Knoll (1974) however could find no cells associated with the vitreous body in an embryological study of *S. coleoptrata*. Adensamer reported the presence of large yellowish bodies in the vitreous body of adults: they appeared to be fat droplets.

Distally the vitreous body of *S. coleoptrata* is surrounded by a ring of 8–10 large pigment cells, below which is a ring of 9–12 elongated sensory cells and a proximal group of 3–4 sensory cells. Both have distally placed nuclei and (Hesse, 1901) an inner border of microvilli separated by a clear *Schaltzone* from the cytoplasmic region that contains the nucleus. The whole ommatidium is surrounded by 16–18 elongated pigment cells with expanded distal and proximal ends in which the granules collect (Hemenway, 1900). In addition there are four delicate supplementary cells between the distal sensory cells which each send fine processes proximally and distally. The large nuclei found distally between the ommatidia may represent corneagenous cells. Snodgrass (1952) gave a composite figure based on the work of the above-mentioned authors and Hanström (1934) gave a detailed comparison of their findings.

Paulus (1979) has carried out an electronmicroscopical study of the ommatidium of *Scutigera coleoptrata* which sheds light on the nature of the vitreous body. Each segment of the vitreous body consists of a hyaline substance and along the cell membranes there

Fig. 99. The ommatidium of *Scutigera coleoptrata*. *a*, longitudinal section; *b*, transverse section through distal region; *c*, transverse section through central region; *d*, transverse section through proximal region (after Paulus, 1979).

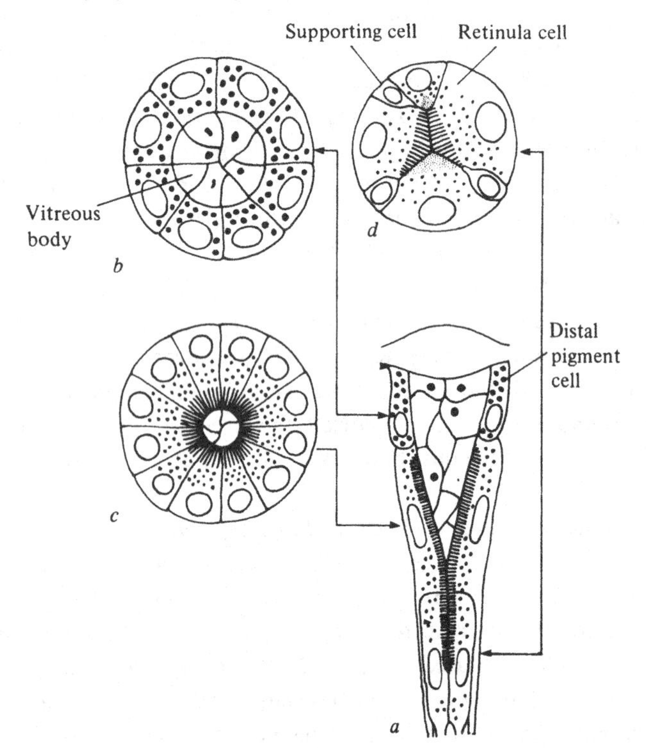

are dense cytoplasmic zones and sometimes small spheres (Fig. 99*b*). The cone-segments appear to be extracellular secretions of the distal pigment cells (Fig. 99*a*). Paulus believes that these cells secrete both the cone-segments and the cornea. Each cell produces more than one segment during its lifetime. The rhabdom is two-layered and the number of cells varies according to the region of the eye. In the centre of the eye there are about 7–10 retinula cells and in the lateral parts (dorsally) 9–23. The proximal layer consists in all cases of four retinula cells forming a triangular rhabdom (Fig. 99*d*). Between the four proximal retinula cells there are three or four smaller supporting cells. The author concludes that the faceted eye of *Scutigera* is not homologous with those of insects or Crustacea and seems to be derived from the eye of *Lithobius*.

In the late larval *S. coleoptrata* two pigmented spots, the *Stirnocellen* which are situated near the front of the head capsule (Fig. 109). These later develop into neurosecretory structures (Knoll, 1974).

Light induced changes of visual cells

After exposure for 5 min at 250 lux the distribution of *Schaltzone* elements in *Lithobius forficatus* which have been dark-adapted for a long period is reduced to less than half their original extent (Bähr, 1974). The cytoplasmic sector enlarges and the pigment grains and mitochondria move against the rhabdomere. After 25 min illumination the peripheral part of the *Schaltzone* is occupied by pigment grains which form a screening ring around the rhabdom which seems to be complete in 60 min. Under 250 lux the *Schaltzone* cross-sectional area is reduced from 35–42 per cent to 6–7 per cent of the entire ocellus area. The perirhabdomal areas of the palisade are diminished or disappear, while the connections of the larger vacuoles to the rhabdomeric evaginations become more numerous. Bähr was unable to explain the mechanism of these changes or the mechanism of pigment migration. Intense and prolonged illumination affects the arrangement and organisation of the microvilli.

The migration of pigment grains within the basal cells is not influenced in the same manner as that in the hair cells. The position of pigment in the dark-adapted cells and the slight movement of grains between the *Sehstäbchen* during illumination suggest that pigment translocation had already occurred while the animal was handled under dim red light.

On return to darkness after illumination the large vacuoles in the *Schaltzone* extend and are surrounded by small vesicles. After 60 min exposure, mitochondria and pigment granules have moved towards the ocellus periphery. The reconstitution of fine cytoplasmic bridges within the palisades is, however, a long-term process, more pronounced in the distal than in the basal cells.

It seems that the reversible changes of *Schaltzone* and pigment distribution resemble an adaptive mechanism regulating the rate of incoming light similar to those described for some insects.

The retinal action potentials of the ocelli of *Lithobius forficatus* were investigated by Bähr (1967).

6

Sensory responses and related behaviour

This chapter is concerned primarily with responses to such stimuli as light, sound, chemicals and touch. Other aspects of behaviour are discussed in the appropriate chapter: locomotion in Chapter 2, feeding behaviour in Chapter 10, reproductive behaviour in Chapter 17, defensive behaviour including meeting behaviour in Chapter 19 and rhythms of activity and social behaviour in Chapter 21.

Reactions to light

Plateau (1886) showed that *Lithobius forficatus*, *Necrophloeophagus longicornis*, *Cryptops hortensis* and *Cryptops punctatus* were negatively phototactic (taxes are directed responses dependent on discrimination of the direction of the stimulus). Chilopods in general appear to be negatively phototactic although Demange (1956) noted that in captivity *Lithobius piceus gracilitarsus* Bröl. does not appear to seek darkness: if the light is not too bright it will leave its hiding place and behaves normally. It hunts in the middle of the day and reproduces in the light. Demange maintained that it was forced underground in order to seek the required humidity.

Klein (1934) showed that *Lithobius forficatus* ran towards a black screen at the side of an arena (skototaxis). Unilateral blinding did not affect the result. When illuminated from one side it showed a negative phototaxis which overruled the skototaxis. Görner (1959) demonstrated that *Lithobius forficatus*, *Scutigera coleoptrata* and *Scolopendra cingulata* exhibit a skototactic response. Neither *Scutigera* nor *Scolopendra* modify their runs when illuminated from the side but *L. forficatus* reacts to lateral artificial light by a negative phototaxis. Some *Scolopendra* while reacting skototactically also exhibit a positive phototaxis, that is to say they sometimes move towards light.

Meske (1961) confirmed that *L. forficatus* comes to rest in the shaded regions of a Petri dish (84 per cent of 240 observations) and likewise shows a preference for a dark background (77 per cent of 240 observations).

Given a choice of colours in a T-maze, *L. forficatus* preferred the dark to pale red and deep violet but preferred dark red and infrared to the dark and avoided ultra-violet (Scharmer, 1935). In electrophysiological studies *L. forficatus* showed maximum sensitivity to green light (Bähr, 1965). *Lithobius* shows no reaction to polarised light.

Meske (1961) investigated the optomotor response of *Lithobius* to a slowly revolving drum of black and white stripes and found a slight tendency for the species to run towards the moving dark stripes. There is a very slight reaction to vertical red and grey stripes.

Although without eyes, geophilomorphs are light sensitive. Weil (1958) conducted a series of choice-chamber experiments using different light intensities. At high light intensities *Necrophloeophagus* can distinguish a 50 per cent difference in light intensity (between 500 and 1000 lux). At lower light intensities it distinguishes clearly between 44 and 100 but not between 60 and 100 lux. When suddenly illuminated the animals react either by drawing back or turning round. The entire body surface appears to be light sensitive: removal of the antennae has no effect on its responses to light.

Scutigera coleoptrata is able to distinguish optically between certain mutants of the fruit fly *Drosophila melanogaster* (Le Moli, 1970, 1972, 1975).

Sound perception

Scolopendra cingulata and *Scutigera coleoptrata* show no marked reaction to sound (Klingel, 1960*a*) but *Lithobius forficatus* reacts to stimuli ranging from 50 to 5000 Hz, the reaction being maximal between 500 and 2000 Hz. Moving animals respond to a sudden tone of 0.5 s duration by stopping and usually remain motionless for as long as the sound lasts and sometimes for several seconds longer, up to a maximum of one minute. In some cases the animal responds either by running faster for a few seconds or by changing direction. If the sound is particularly loud, the *Lithobius* may stop and then move off at great speed. Stationary *Lithobius* do not respond. There is no reaction to vibration or concussion of the experimental chamber (Meske, 1960).

The adaptive significance of this behaviour may be in its probable advantage in avoiding predators. *Lithobius* is protectively coloured and may escape detection by 'freezing'. A rapid increase in movement may also be effective in that it may lead the animal to a safe refuge.

Meske (1960) believed the Tömösváry organ to be the sound receptor (see Chapter 5) but this may be open to doubt.

Chemoreception

'Nerve endings of many kinds are doubtless sensitive to irritant chemical substances. From this has been evolved the common chemical sense of primitive invertebrates. In insects and vertebrates this sense has become further differentiated into the two senses of taste and smell. There is no absolutely satisfactory distinction between these two ... although the sensitivity is usually far greater in the case of smell, as estimated by the lowest molecular concentration that can be detected, yet there are very great differences between the species, ... for terrestrial organisms taste is a sense of solutes and smell a sense of vapours' (Wigglesworth, 1972).

The reaction of *Lithobius* to airborne chemicals was first investigated by Dugès (1838) who exposed a *Scolopendra* sp. to alcohol, turpentine and ether vapour and noted that the antennae were withdrawn and rolled spirally. Hennings (1904*b*) exposed *Lithobius forficatus, Cryptops hortensis, Clinopodes linearis* (C. L. Koch) to clove oil, turpentine, xylol, acetic acid, ammonia and chloroform and found that the intensity of reaction, measured as distance at which it was perceived, increased in the order given. *Lithobius* with the antennae removed did not react at all to clove oil, turpentine and xylol and their reaction to the other three chemicals was very much reduced. *Cryptops hortensis* reacted almost equally to clove oil, acetic acid and xylol, but to a much smaller degree to the other three chemicals. *Clinopodes linearis* responded most strongly to acetic acid and least to clove oil.

In a second set of experiments Hennings tested the ability of three normal and three antennaless animals which had been starved for three months to detect raw cow meat. The normal animals found the food source in a quarter of an hour, the other animals had not found it within a day.

Scharmer (1935) reported that both smoke and amyl alcohol released a shock reaction in *L. forficatus*. The animals reacted positively, however, to hydrogen sulphide, ammonia and musk. No obvious preference was shown for clove oil. After the removal of both antennae the animals were less active but the preferences shown were similar, as they were when the '*Taster*' (presumably the telopodites) of the second maxillae were removed. When the head and first segment were varnished, however, the animals became indifferent to hydrogen sulphide and musk, but still reacted negatively to amyl alcohol. Clove oil and ammonia were not tested.

Antennal cleaning is frequently observed in centipedes, the appendages being drawn through the mouthparts. The frequency of these cleaning movements has been shown to vary with the concentration of volatile chemicals (Meske, 1961). The average number of antennal cleanings per hour of eight animals was 14.7, but in the presence of a worm it increased to 49.2. In a circular chamber with 12 radially arranged peripheral chambers separated from the main arena by gauze, *Lithobius* almost invariably went to the single chamber containing horse meat. The animal made frequent unsuccessful attempts to reach the food, tearing at the gauze with its poison claws and mouthparts. Scharmer (1935) was unable to demonstrate that *Lithobius* located its food by smell.

Scharmer (1935) showed that *Lithobius* became active when the *Taster* on the second maxilla was touched with the juice of an earthworm. The reaction was not elicited by contact with the antennae or other parts of the body. No reaction was obtained with glycerine, quinine or sodium chloride solution. Rilling (1968) considered that chemical stimuli were probably perceived by the *Sinneskegel* (basiconic sensilla) of the first and second maxillae and perhaps other areas around the mouth.

Humidity reactions

Scharmer (1935) found that in a T-maze *Lithobius* showed no preference for high or low humidity but it preferred a damp to a dry surface. The Tömösváry organ did not seem to be important in humidity perception.

Bauer (1955) showed that *L. forficatus* preferred a high humidity and when offered a choice of 76 per cent and 90 per cent relative humidity (above distilled water) preferred the latter. On a surface of damp towelling providing relative humidities of 25, 40, 90, 76, 62 and 28 percent, normal animals and animals without antennae showed a shock reaction, stopping and turning round at the border of 90 and 76 percent relative humidity. With normal antennae, but varnished tarsi, discriminatory powers were reduced. Animals without antennae and with varnished tarsi no longer responded to relative humidity but with few exceptions came to rest after long searching in the dampest section of the arena.

Bauer concluded that the tarsi were the most important carriers of hygroreceptors but also the antennae and the body itself must be sensitive to relative humidity. Humidities were controlled mainly by using different concentrations of sulphuric acid and it is possible that the animals may have reacted to these.

Further experiments were carried out on *L. forficatus* by Tichy (1973). The highest humidity offered (95 per cent) was selected by the majority of animals. Removal of the antennae reduced the positive reaction as did occlusion of the Tömösváry organ.

When offered alternatives of 20 and 100, 78 and 100 and 97 and 100 per cent relative humidity in choice chambers, the geophilomorph *Pachymerium ferrugineum* (C. L. Koch) showed a response to the moist side in 6–12, 28–32 and 130–170 h respectively. The species showed clear differences in locomotor activity at less than 10 per cent and 100 per cent relative humidty. The average distance covered in 1 h by a single individual was 1605 cm in dry air and 525 cm in moist air (Palmen & Rantala, 1954).

In a series of choice chamber experiments, Weil (1958) showed that *Necrophloeophagus longicornis* preferred high humidities (Fig. 100). Sensitivity is greater at high humidities. A humidity of less than 85 per cent is eventually lethal: the humidity amongst leaves where the animals were found was 85 per cent. Observations on the behaviour of the animal suggested that the antennae are important in the humidity reaction, strong stimulation leading to quick forward and backward movements of the antennae. Removal of the antennae showed that the humidity receptors are localised on these organs.

The above data suggest that both lithobiomorphs and geophilomorphs prefer high humidities but the response may be much more rapid in some species than in others (*Pachymerium* is exceptional in its slow response). The humidity receptors are located on the antennae and possibly also on the tarsi in *Lithobius* but the nature of these tarsal receptors is unknown. The body surface and Tömösváry organ may also be involved in *Lithobius*. There are no data on the humidity preferences of scolopendromorphs and scutigeromorphs: information on the latter would be of particular interest. The tarsal papillae of *Scutigera* may well be basiconic sensilla.

Temperature reactions

The data on the temperature preference of *Lithobius forficatus* are conflicting. Scharmer (1935) showed that in a tempera-

Fig. 100. The track of an individual *Necrophloeophagus longicornis* over a 7 min period in a choice chamber with a 60 % relative humidity on the left and 10% relative humidity on the right (after Weil, 1958).

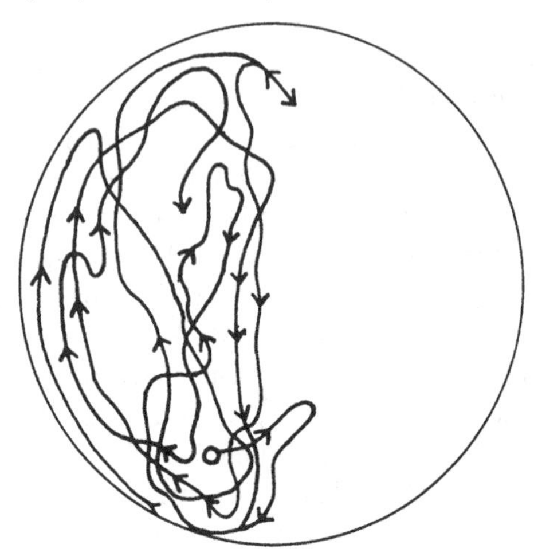

ture gradient the specimens selected 27 °C. Bauer (1955) showed that there were two distinct temperature preferences in *L. forficatus*. These were 7° and 27 °C in males and slightly higher temperatures in females (Fig. 101). Females prefer a higher ground temperature than males even when the antennae are removed or the tarsi lacquered. In the field, animals show a preference for 8–16 °C. This 'natural temperature' is avoided more and more as antennal segments are progressively removed. Bauer concluded that the temperature receptors were located in the basal half of the antenna.

Meske (1961) attempted an investigation of the causes of the major differences in the results of previous authors. *L. forficatus* showed a preferred average value of 21.9 °C but if the blotting paper on the floor of the arena was damp this fell to 17.3 °C. Specimens of *Necrophloeophagus longicornis* on a damp substratum showed a preference for 13 °C.

The previous temperature at which the animals were kept also affects preferences. *Lithobius* preconditioned for one and a half hours at 10 °C showed a preference for 18.4 °C under dry conditions and 16.3 °C under damp conditions. Conditioned at 25 °C the results were 22.6° and 21.8 °C respectively. *Necrophloeophagus longicornis* selects 12.5 °C after conditioning at 10 °C, and 17.3 °C after being kept at 25 °C.

The results are unexpected in that animals kept in dry conditions prefer higher temperatures than those kept in a damp atmosphere.

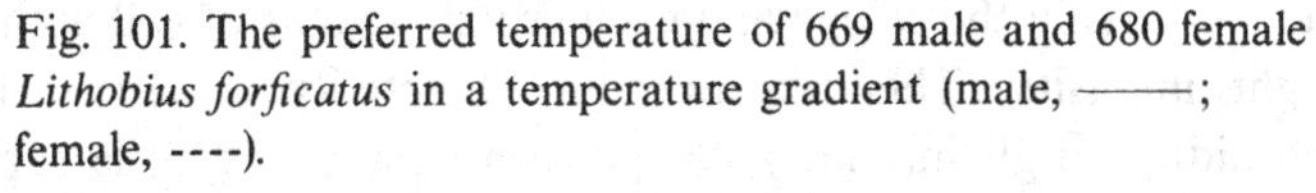
Fig. 101. The preferred temperature of 669 male and 680 female *Lithobius forficatus* in a temperature gradient (male, ——; female, ----).

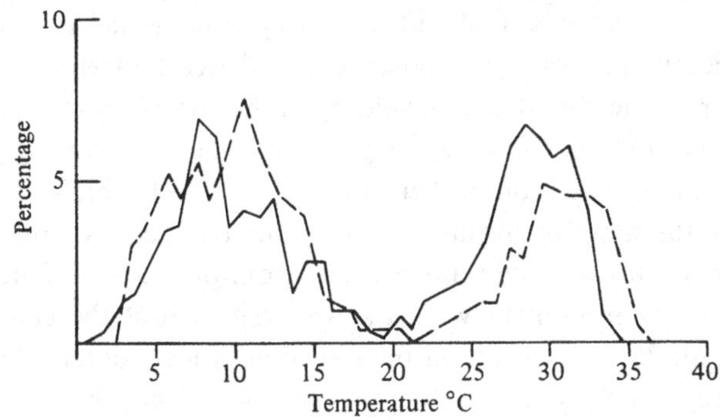

This presumably has some advantage. Perhaps the toleration of high temperatures enables the animal to move across unsuitable areas rather than being repelled by them and thus trapped in an unsuitable habitat.

Palmen & Rantala (1954) showed that *Pachymerium ferrugineum* has a 'preferred temperature range' of 10–35 °C in a linear temperature gradient with a moist substratum. On a dry substratum the animals aggregated at the cooler end, possibly due to the presence of a relative humidity gradient.

Richards & Davies (1977) pointed out that not all experiments with a temperature gradient are satisfactory since the method fails to distinguish between convective and conductive transfer and humidity variations within the apparatus are not always eliminated.

Thigmotaxis and related behaviour

Plateau (1886) showed that centipedes not only seek dark humid resting places but also regions where most of the body is in contact with a firm object (thigmotactic behaviour). In *Lithobius* the antennae are used to pick up information from dark environments such as passages in the the soil and under logs and stones. When the centipede's refuge such as a stone or a plank is removed it runs at high speed (28 cm/s has been measured) until a small stone or some leaves are reached. *L. forficatus* crawls under these and remains stationary. After a while the animal again becomes active leaving the refuge and even crossing areas of soil with a high light intensity. This second bout of activity is caused by low humidity. High humidity (90 per cent)(Bauer, 1955) is a primary influence on the behaviour of *Lithobius*: thigmotactic information as well as negative phototactic responses can lead to certain refuges but do not override the stimulus of increasing dryness (Schäfer, 1976*a*).

L. forficatus exhibits a 'forward going tendency' preferring to crawl over an obstruction rather than to turn. The higher the obstruction, the more often this forward going tendency is reduced and the animal turns to run along the bottom of the obstruction. Animals running in contact with a single wall turn at the end of this wall (with 25° as a mean) in the direction of lost contact. This value can be varied by amputating one or both antennae. If walls are used that bend away from the animals with 30°, 60° and 90° the angle of turning, 25°, remains constant. Although these walls offer leading lines for thigmotactic behaviour, the animals do not, in most cases, follow the bend of the walls exactly. The results show the interaction of the tendency to run straight ahead and asymmetrical tactile stimuli inducing turning responses.

Lithobius running in a corridor shows an asymmetrical carriage of antennae. One antenna is curved so that its anterior edge is in contact with the wall: this is the 'direction antenna'. The other antenna is held in front of the body and is straight or nearly so (Fig. 102). The direction antenna indicates the direction of turn after leaving the corridor: in 74 per cent of runs the animal turns towards the side of the direction antenna. Thus the animal has chosen the direction of turning on entry to the corridor: a fact that makes experiments on the learning capacity of *Lithobius* in T- and Y-mazes (Scharmer, 1935) very difficult. The antenna which makes first contact with a wall becomes the 'direction antenna'. In a series of runs in a corridor with a single animal the position of the 'direction antenna' normally changes randomly as it gets first wall-contact equally to left and right so the mean value of turnings is 0° (Schäfer, 1976*a*).

As is the case with many animals *Lithobius forficatus* exhibits 'reverse turning' otherwise known as correcting behaviour, alternation behaviour, reactive inhibition, spontaneous alternation and

Fig. 102. Examples of the assymetrical carriage of the antennae by a *Lithobius forficatus* running in a corridor. For further explanation see text (after Schäfer, 1976*a*).

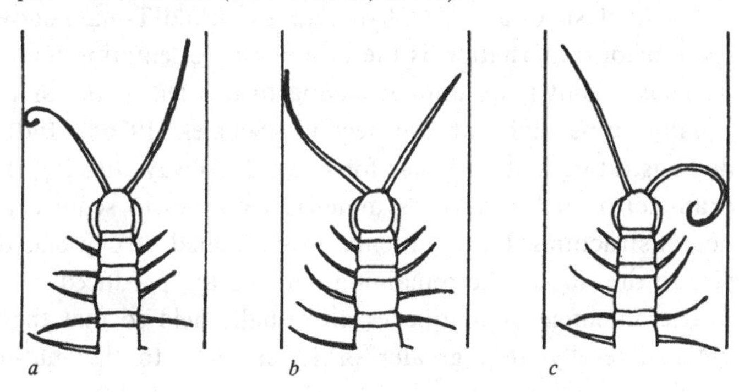

turn alternation. This is observed when an animal is introduced to a corridor with an abrupt turn. On emerging from the corridor the animal shows a turn in the opposite side (Schäfer, 1972). The resulting magnitudes of the reverse turning angles appeared to be almost a linear function of the forced turns. Although every animal demonstrated a correlation between its reactions and the forced turning angles, there were considerable individual differences between the angle of forced turn and the angle of reverse turn, that is, some turned through a greater angle than others. The amputation of one antenna leads to a bias to the side with the remaining antenna. Removal of both antennae reduced but did not eliminate the turning response.

Since a straight path had to be run from the deviation to the exit of the corridor the animals must retain information of direction and amount of the forced turn in their memory. When the distance from deviation to choice point increased, reverse turning decreased. Schäfer (1972) suggested that this behaviour was probably due to kinesthetic mechanisms, thigmotaxis could not control quantitative reverse turning. Kinesthetic orientation can be defined as endokinetic orientation based on the recording and storing of bodily movements and displacements in local and temporal sequence. The information about the forced turn would be obtained by measuring bodily displacements which appear during the execution of the turn. This information is then used to control the reverse turn. *Lithobius* has several organs suspected of having proprioceptive functions such as sensilla at the edges of tergites and sternites; free nerve endings and stretch receptors in the muscles (Rilling, 1968) which would be suitable to measure body deformations.

The analysis of a series of runs in a fourfold T-maze shows that spontaneous alternation is the dominating orientation tendency in *Lithobius*. That is an animal having turned left at the first choice usually turns right at the second (Schäfer, 1976*b*). Individuals demonstrating a strong bias for one side (always turning right for example) probably have asymmetrical injuries to sense organs or neural structures. If one antenna is amputated, strong bias tendencies to the side of the remaining antenna are produced.

The antennae of centipedes are usually held so that they point antero-laterally at a greater or lesser angle to the mid-line. In

Lithobius the antenna in contact with a surface is, however, coiled: the so-called 'direction antenna'. A similar disposition is shown for the antennae of *Scolopendra morsitans* in photographs by Manton (1965). When the terminal antennal segment of *Strigamia maritima* is removed the antenna is curved backwards (Lewis, unpublished data) (Fig. 103). The Bornean scolopendromorph *Otostigmus longicornis* has very long antennae which reach the ninth segment when reflexed. They are held in a spiral coil (Fig. 104) (Lewis, unpublished data).

Proprioception

There is as yet no experimental evidence to support the presence of proprioceptors in centipedes, but by analogy with other arthropods it seems probable that a number of structures might serve this function. In *Lithobius forficatus* these are the free nerve endings, the pedal, antennal and muscle sense organs (stretch receptors); the hair rows or fields, most of which have a spike-like

Fig. 103. Position of the antennae in *Strigamia maritima*. *a*, normal; *b*, when the terminal segment of each antenna is removed.

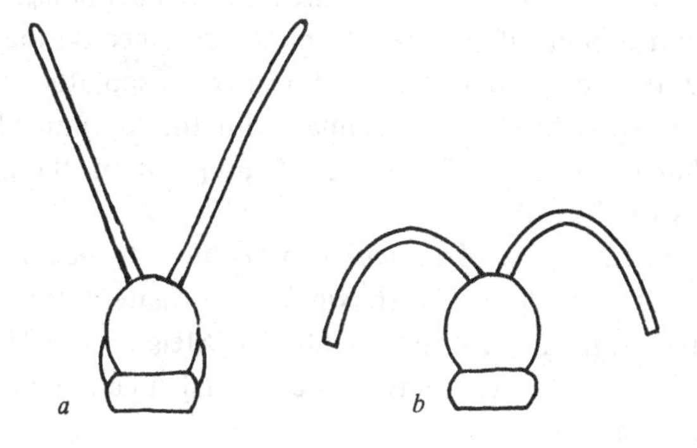

Fig. 104. The normal position of the antennae in *Otostigmus longicornis*.

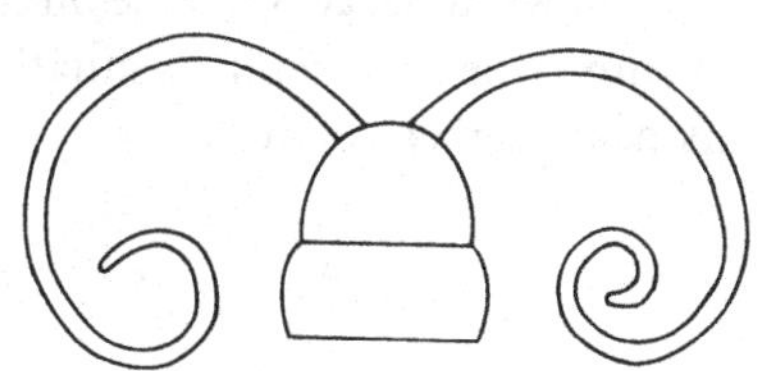

structure, and the scolopidia-like structures of the mandible (Rilling, 1968). These proprioceptive organs, Rilling suggested, must respond to environmental stimuli to some extent. It is possible that the stretch receptors at the base of the antennae are also gravity receptors.

Grooming behaviour

Cleaning or grooming behaviour occurs in all centipedes: a number of authors have described antennal cleaning. In *Scutigera coleoptrata* the antennae and legs are gripped by the poison claws, brought up to the mouthparts and drawn through them from base to tip: the broad-ended first maxillae are the most important cleaning organs (Verhoeff, 1938). Manton (1965) regarded the maxillary organs as structures for grooming. The tarsi of legs 1–6 are particularly thoroughly cleaned. The appendages may be lubricated by the secretions of the head glands (Verhoeff, 1938). Bennett & Manton (1963) suggested that the maxilla II and forcipular head glands might be responsible. The cleaning behaviour is performed on a simple and strictly fixed schedule (Le Moli, 1977). It commences with an antenna, followed by the first to fifteenth legs of the same side. Immediately afterwards cleaning is resumed on the other side in exactly the same order. If, for any reason, the animal temporarily ceases cleaning, it resumes with the leg immediately following that last cleaned. Even after its amputation, the animal still attempts to clean the leg.

In second instar larvae of *Lithobius forficatus* antennae and legs are groomed and a preference is shown for contralateral transitions in antennal grooming (left-right, right-left) (Meissner, 1978). The author concluded that grooming has a special function of information reception.

Displacement activity

Scutigera coleoptrata shows a large increase in the frequency of grooming movements when in conflict situations. Le Moli (1977) classified this as displacement activity.

7

Endocrinology

Protocerebral neurosecretory cells and cerebral gland

Holmgren (1916) described the cerebral gland of centipedes calling it the frontal organ. Fahlander (1938) realised that it was endocrine in nature and proposed for it the name cerebral gland. Gabe (1952) showed that axons from a group of lateral neurosecretory cells on each side of the protocerebrum carried secretory material to the cerebral glands. This was confirmed by Palm (1955).

In *Lithobius forficatus* the cerebral gland appears to be a hollow sac which is bathed by a stream of haemolymph and innervated by a nerve from the optic stalk, also receiving a pair of nerves originating from lateral neurosecretory cells in the protocerebrum (Palm, 1955) (Fig. 105*a*). The lateral neurosecretory cells are of two types, large and small, the former being far more numerous than the latter. In addition to these lateral cells there is a group of small neurosecretory cells in the posterior median region of the protocerebrum which corresponds to the *pars intercerebralis* of insects (Scheffel, 1961) (Fig. 105*a*). The nerve from the optic nerve to the cerebral gland in *L. forficatus* runs from the optic lobe to the gland in *Lithobius calcaratus* C. L. Koch. The nerve is absent in geophilomorphs and scolopendromorphs (Joly & Descamps, 1970). The large lateral neurosecretory cells (Type 1) are phloxinophil after Gomori staining, the smaller cells (Type 2) stain with haematoxylin. The neurosecretory cells of the *pars intercerebralis* are of Type 2. Three cell types (*a*, *b* and *c*) may be distinguished using the electron microscope. They differ in their size and number and electron opacity of their granules. Types *a* and *c* correspond to the optical Type 1 and some Type *b* to Type 2 (Fig. 105*b*) (Jamault-Navarro & Joly, 1977).

The histology of the cerebral gland of *Lithobius forficatus* has been described by Scheffel (1965*c*) and Joly (1966*b*). The glands are surrounded by a connective tissue capsule and contain intrinsic cells and the axons of the protocerebral neurosecretory cells. The cells are of two types, neurosecretory cells and less common interstitial cells without excretory granules. The glands contain both endogenous and exogenous secretions. The endogenous secretions are formed in the Golgi apparatus of the neurosecretory

Fig. 105. Diagrammatic representation of the distribution of neurosecretory cells in the protocerebrum of *Lithobius forficatus*. *a*, as seen with the light microscope with cell types 1 and 2; *b*, as seen with the electron microscope with cell types *a*, *b* and *c* (after Jamault-Navarro & Jolly, 1977). For further details see text.

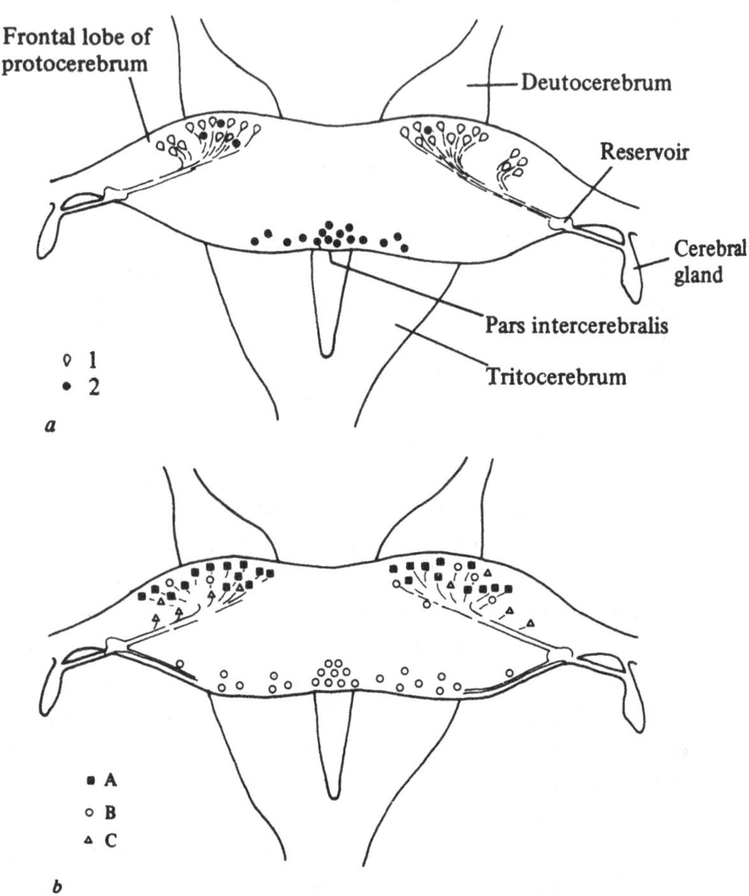

cells of the gland (Joly & Devauchelle, 1970). Both endogenous and exogenous secretions seem to be non-histone proteins containing cysteine and cystine; the endogenous secretions also contain arginine or lysine. It seems possible to differentiate aminergic and peptidergic fibres on the basis of cytochemical tests. Cyclical excretory activity is reflected in the changes in the Golgi apparatus of the neurosecretory cells (Joly, 1970). The first signs of secretory activity occur during the premoult period. Secretory material appears to be released shortly after ecdysis.

The exogenous and endogenous secretions of the gland differ in their staining reactions, the former are phloxinophiles, the latter Gomori-positive. In lithobiomorphs there is an abundance of phloxinophil areas; in geophilomorphs, Gomori-positive material predominates. In *Cryptops* the gland shows two distinct zones, a peripheral phloxinophil zone and a lighter staining central zone (Joly & Descamps, 1970).

Joly (1966*a*) investigated the function of the protocerebral neurosecretory cells using 'sub-mature' and 'young mature' *Lithobius forficatus*. Surgical removal of the cerebral glands increases the percentage of animals moulting at a given time as it does in *Scolopendra cingulata*. Sectioning the nerves to the cerebral glands also increases the number moulting. Implantation of cerebral glands decreases the number of animals moulting and repeated implantation virtually inhibits the moult. Repeated implantations of fragments of tergite, however, also decrease the number of moults, an effect ascribed by Joly to operative shock. Removal of the glands 10 or 18 days after the moult increases the number of animals moulting and decreases the time of the intermoult period as compared with controls. Removal of the glands 27 days after the moult has no effect on the time of the next moult. The author suggested that there is a critical period during which the hormone is effective, from the tenth or eleventh to the twenty-sixth to twenty-seventh day of the intermoult. The length of the intermoult period in controls is 51 days.

Destruction of the antero-lateral neurosecretory cells of the protocerebrum also stimulates moulting but destruction of the *pars intercerebralis* decreases the number of animals moulting suggesting that it either produces a moulting hormone or acts on a moulting gland. Destruction of the *pars intercerebralis* also causes a loss in weight as compared with controls (Joly, 1971; Herbaut & Joly, 1971).

If the *pars intercerebralis* is destroyed before the rapid increase in oocytes in mature female *L. forficatus* in the autumn, the growth of oocytes is blocked or curbed in 50 per cent of operated animals. Implantation of two *pars intercerebralis* restores the normal oogenetic cycle (Herbaut, 1975*a*).

Implantation of cerebral glands into stadium pseudomaturus I females may trigger vitellogenesis but not oocyte maturation (Herbaut, 1977*a*): oocytes mature only in maturus junior and maturus senior stadia. The destruction of the cerebral glands or the protocerebral gland cells causes the degeneration of most oocytes at the end of previtellogenesis and of numerous vitellogenic and mature oocytes. The cytological changes in the nucleoli of the oocytes resemble those induced by electrocoagulation of the *pars intercerebralis.* Implantation of cerebral glands allows normal growth of oocytes in animals without the *pars intercerebralis* but cannot restore the oogenetic cycle in starved females. The regulation of oogenesis seems to be a function of the neurosecretory cells of the protocerebral frontal lobes and the cerebral gland controlling nucleolar metabolism of previtellogenetic oocytes, stimulating vitellogenesis and maintaining ripe cells (Herbaut, 1976*a*).

Destruction of the *pars intercerebralis* increases the time required for spermatogenesis from 30 to 70–80 days in *L. forficatus* but the maturation divisions are not inhibited (Descamps & Joly, 1971). There is an increase in the number of cells undergoing spermatogenesis (Descamps, 1972*a*). Implantation of two *pars intercerebralis* returns the spermatogenetic cycle to normal (Descamps, 1974). Electrostimulation of the *pars intercerebralis* brings about the completion of spermatogenesis in the pseudomaturus stadium in which it does not normally occur (Descamps, 1977*a*).

Destruction of the lateral secretory cells of the protocerebrum or cerebral glands stimulates spermatogenesis (Descamps, 1975). Induction of moulting either by injections of ecdysone or by the removal of antennae or limbs causes a delay in the appearance of spermatids and spermatozoa. Injections of ecdysone also increase the number of degenerating cells (Descamps, 1977*b*).

Injection of the moulting hormone ecdysone stimulates moulting in *L. forficatus* although this is accompanied by considerable mortality (Joly, 1964). Scheffel & Wilke (1974) administered ecdysterone, which is closely related to ecdysone, to starved third instar larvae of *L. forficatus* which are normally unable to moult. They found that it caused apolysis (the separation of the old cuticle from the epidermis) although it only occasionally led to moulting. The ability of the epidermis to respond to ecdysterone is influenced by actinomycin D (AMD) to varying extents depending on the time of application. AMD treatment 15 min before ecdysterone application increases the rate of moulting to 70 per cent but AMD treatment 12–18 h after the application of ecdysterone leads to a remarkable decrease in the number of larvae showing signs of apolysis and no moulting occurs. The results suggest that the failing response of the epidermal cells of starving *Lithobius* larvae to ecdysterone is caused by specific inhibitor molecules whose biosynthesis may be blocked by AMD at the transcription level. Exogenous ecdysterone increases the rate of mitosis in the limb buds and the pre-anal segment forming zone in third instar larvae (Pollak, 1976).

Amputation of legs (Joly, 1966*a*) and removal of the antennae (Joly & Lehouelleur, 1972) in *L. forficatus* brings about premature moulting. Suppression of the visual function by painting over the eyes has the same effect. There is an antagonism between somatic growth brought about by moulting and oogenesis in *L. forficatus*. In nature moulting takes place in the spring, the growth of oocytes in the autumn (Herbaut, 1977*b*).

Joly, Descamps, Herbaut & Jamault-Navarro (1976) summarised the functions of the protocerebral neurosecretory cells. The *pars intercerebralis* exerts a weak stimulatory effect on the moulting cycle and gametogenesis. It controls the metabolism of the germinal cells and manifests itself, especially at the time of previtellogenesis and during the growth of the spermatocytes. The lateral protocerebral neurosecretory cells and the cerebral glands exert a moderating (inhibitory) action on the moult and in this respect, perhaps, resemble the X-organ of Crustacea. They have a stimulatory role on the growth of oocytes but inhibit spermatogenesis.

The protocerebrum of geophilomorphs contains both lateral and posterio-median neurosecretory cells (Fig. 106). The cerebral gland is situated beneath the protocerebral lobes (Figs. 74*a*, 106) in

Necrophloeophagus longicornis. The cerebral gland contains four types of axons originating in the protocerebrum which stain with paraldehyde fuchsin, and parenchymatous cells, the neurosecretory cells of the gland, that stain with Orange-G (Ernst, 1971).

There are two types of neurosecretory cells in the cerebral gland of *Scutigera coleoptrata* in addition to the axons of the protocerebral cells and interstitial cells. Immediately before ecdysis only neurosecretory cells with dense cytoplasm, well-developed endoplasmic reticulum and numerous mitochondria and Golgi bodies are present. In the middle of the moulting cycle both cell types are present. The protocerebral nerve consists of about 200 axons of which five types can be distinguished (Rosenberg, 1976).

Neurosecretory cells in the ventral nerve cord

Scheffel (1961) reported the presence of paired groups of anterior neurosecretory cells in the ganglia of the ventral nerve cord of *Lithobius forficatus* (Fig. 107). Their presence was confirmed, and further details of their nature were given by Prunescu (1970*c*). The anterior cells (Type A) are small, 8 μm in diameter, and in two groups of 2–3 neurons. The posterior cells (Type C) are large, 15–20 μm in diameter and in two groups of 5–6 cells. In addition there is a pair of cells (Type B) situated just in front of the Type C cells. Their axons run towards the centre of the ganglion

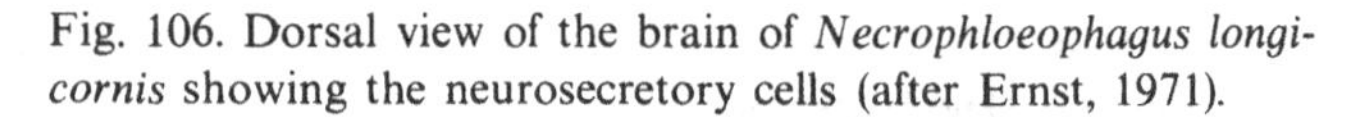
Fig. 106. Dorsal view of the brain of *Necrophloeophagus longicornis* showing the neurosecretory cells (after Ernst, 1971).

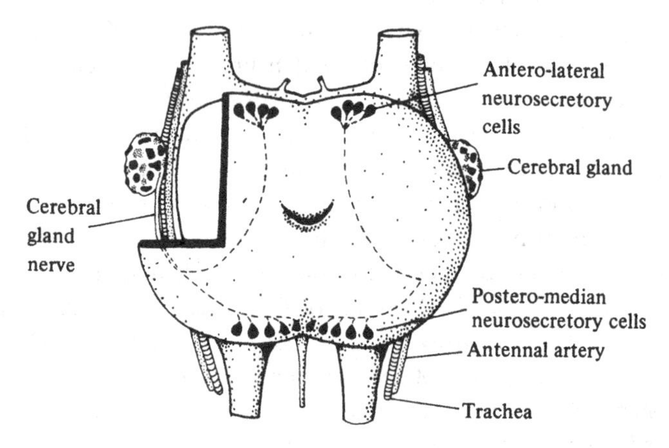

(Fig. 108*a*). Only Type A and B cells are present in *Scutigera coleoptrata* (Fig. 108*b*). In *Scolopendra cingulata* (Prunesco, 1970*d*) there are two groups of 7–10 Type A cells, 15–20 μm in diameter, a pair of Type B cells, 38–42 μm in diameter and two groups of 8–12 Type C cells, 20 μm in diameter (Fig. 108*c*). The scolopendromorph *Plutonium zweierleinii* Cavanna shows a similar arrangement of cells to *S. cingulata.* The geophilomorph *Dicellophilus carniolensis* lacks Type A cells but there are two groups of two or three Type C

Fig. 107. Diagram of the brain and ventral nerve cord of *Lithobius forficatus* seen from the ventral side, showing the position of the neurosecretory cells (after Scheffel, 1961).

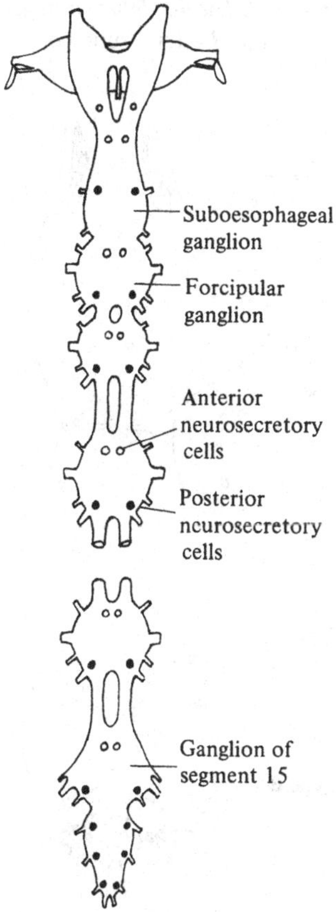

cells (Fig. 108*d*) and these send axons to the dorsal heart nerve via a lateral nerve.

In *Necrophloeophagus longicornis* anterior neurosecretory cells (Prunesco's Type A) are sporadically present and as in *Dicellophilus* posterior (Type C) cells are present near the point of origin of the segmental heart nerve (Prunesco's lateral nerve) (Ernst, 1971). The electron microscope reveals that the cells are of three types but Ernst does not appear to have seen Prunesco's Type B. The axons of the posterior neurosecretory cells pass into the segmental heart nerve, as may those of the anterior secretory cells. The segmental nerves lead to segmentally arranged pericardial organs which

Fig. 108. The distribution of neurosecretory cells in the ventral ganglia of centipedes. *a*, *Lithobius forficatus*; *b*, *Scutigera coleoptrata*; *c*, *Scolopendra cingulata*; *d*, *Dicellophilus carniolensis* (*a* and *b* after Prunescu, 1970*c*, *c* and *d* after Prunesco, 1970*d*).

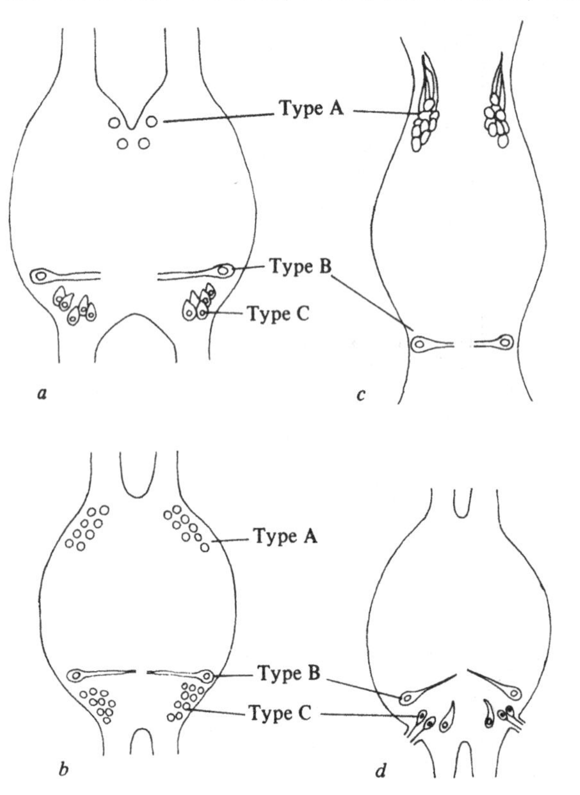

consist of numerous axons embedded in a matrix but which contain no neurosecretory cells. Some axons pass from the segmental heart nerves to the dorsal heart nerve: neurosecretory axons are found also in the adventitia of the heart and the myocardium.

The function of the neurosecretory cells of the ventral nerve cord is unknown.

Moulting glands

In the region of the first leg-bearing segment of *Lithobius forficatus* there is a pair of rather diffuse tissue strands. The cells are most numerous around the maxillary nephridia where they form strings of brownish coloured cells with highly refractory inclusions. They are also wrapped around the two large salivary glands (Scheffel, 1969). They were earlier termed *Lympstränge* or *cellules à carminate* as they were thought to be lymphocytes because of their tendency to accumulate carmine particles. Their distribution in other centipedes is described in Chapter 14.

Decapitation and ligation experiments with second instar larvae of *Lithobius forficatus* led Scheffel (1963, 1965*b*) to conclude that a moulting centre is situated between the head and the second leg-bearing segment. Implantation of cerebral glands from adult animals delays moulting (Scheffel, 1965*a*). Destruction of the salivary region by local ultra-violet irradiation causes a delay in or complete inhibition of the moult (Scheffel, 1969) but selective destruction of the small second maxillary gland does not lead to any delay in moulting and for this reason Scheffel suggested that the *Lympstränge* constituted the post-cephalic moulting centre. Examination of the mandibular and second maxillary glands of *Lithobius forficatus* showed that they are exocrine structures, thus the *Lymphstränge* must be the only moulting gland (*Glandula ecdysalis*) (Rosenberg & Seifert, 1975).

The moulting gland consists of a single layer of cells which are podocytes. Their ultrastructure changes as the moulting cycle progresses. They contain lipid droplets and numerous accumulations of glycogen before the moult and Golgi complexes, mitochondria and cytosomes appear to increase in number. After the moult neither lipid droplets nor glycogen are found and many of the cells disintegrate (Seifert & Rosenberg, 1974).

Other endocrine organs

In *Scutigera coleoptrata* a small paired endocrine organ of unknown function is situated behind the protocerebrum, lateral and somewhat dorsal to the fore-gut. It is spindle-shaped measuring 150 × 75 μm and is U-shaped in section. The cells are typical podocytes (Rosenberg, 1973). The ultrastructure of the gland is almost identical to that of the moulting gland of *Lithobius* (Seifert & Rosenberg, 1974) and the prothoracic gland of insects. It is probably the moulting gland of *Scutigera* (Rosenberg, 1974).

On the frontal region of the head of the late embryonic stage of *Scutigera coleoptrata* there is a pair of pigment spots, *Stirnocellen* (Fig. 109). These are absent in the adult (Knoll, 1974). In post-embryonic larvae the lumen of this organ increases and it assumes the characteristics of a neurosecretory organ. The ocellus nerve joins the brain between the protocerebrum and deutocerebrum.

Fig. 109. Late embryonic stage of *Scutigera coleoptrata* (after Knoll, 1974).

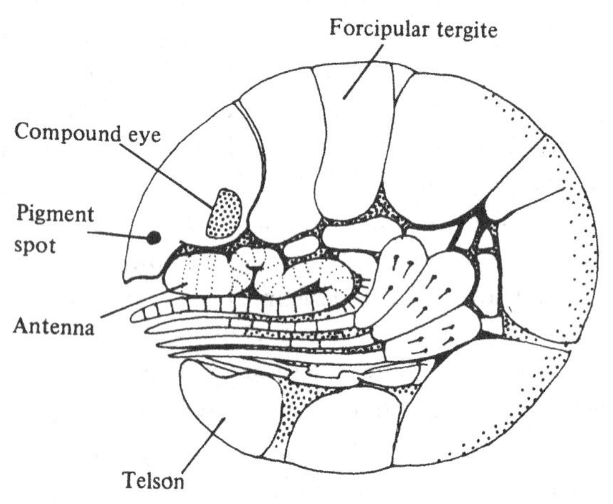

8

The alimentary canal

The alimentary canal of centipedes is a straight tube which is clearly divided into fore-gut, mid-gut and hind-gut. The fore-gut originates as an ectodermal invagination, the stomodaeum, and hence is lined by cuticle. It comprises the pharynx and oesophagus, the latter sometimes differentiated into crop and gizzard or proventriculus. The mid-gut is of mesodermal origin (mesenteron) and the hind-gut is formed from a posterior ectodermal invagination, the proctodaeum. A pair of Malpighian tubules originate at the junction of the mid- and hind-gut and run forwards towards the head (Figs. 110, 111). They are described in Chapter 16. The so-called 'salivary glands' are described in Chapter 15.

The pre-oral cavity is bounded anteriorly by the labrum, antero-dorsally by the membranous epipharyngeal surface, laterally by the mandibles and ventrally by the anterior portion of the hypopharynx, a delicate lobed structure which usually bears dense fields of hairs and the pores of unicellular glands. The mouth opens into the pharynx.

Geophilomorpha

The epipharynx is very poorly developed in geophilomorphs and the hypopharynx is reduced to a small bilobed structure covered with fine hairs. The pharynx is very short, its wall showing lateral longitudinal thickenings to which are attached the pharyngeal dilator muscles (Fig. 112) (Verhoeff, 1902–25).

The alimentary canal posterior to the pharynx has been described for *Haplophilus subterraneus gervaisi* (Plateau) and *Necrophloeophagus longicornis* (Fig. 110*a*) by Plateau (1878). The oesophagus is long and narrow, its folded cuticular lining lacking spines. The small diameter of the oesophagus of geophilomorphs ensures that no large food masses hinder anterior mobility and leaves more room for the locomotory muscles than would be the case if it were sac-like. It is centrally placed in the body thus suffering minimum deformity during burrowing activities (Manton, 1965).

Viewed by reflected light, the surface of the mid-gut shows a pattern of polygonal areas which are due to humps of cells carried

Fig. 110. The gut of geophilomorphs. *a*, *Necrophloeophagus longicornis*; *b*, hind-gut of *Haplophilus subterraneus gervaisi* (after Plateau, 1878).

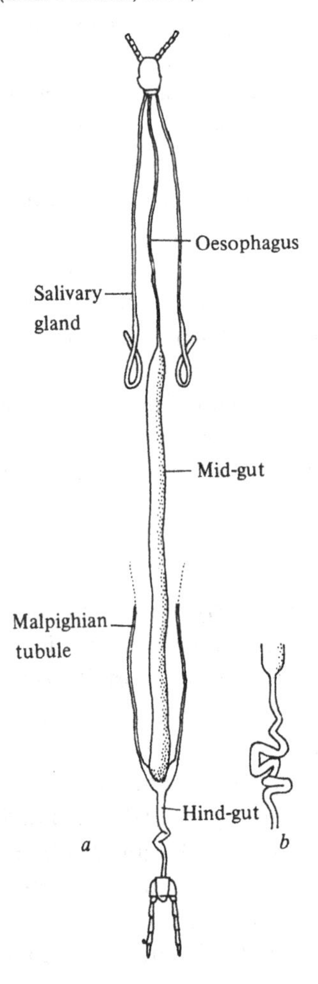

on the folded tunica propria (Plateau, 1878). Large groups of epithelial cells are budded off from the posterior half of the mid-gut in *Strigamia maritima* and these form a large proportion of the faeces (Lewis, 1960). Plateau showed the hind-gut of *Haplophilus* and *Necrophloeophagus* as convoluted (Fig. 110*a*, *b*). This may not normally be the case.

Scolopendromorpha

In *Scolopendra cingulata* there is a transverse row of gland pores along the anterior margin of the epipharynx which is

Fig. 111. The gut of *Cryptops savignyi* (after Balbiani, 1890).

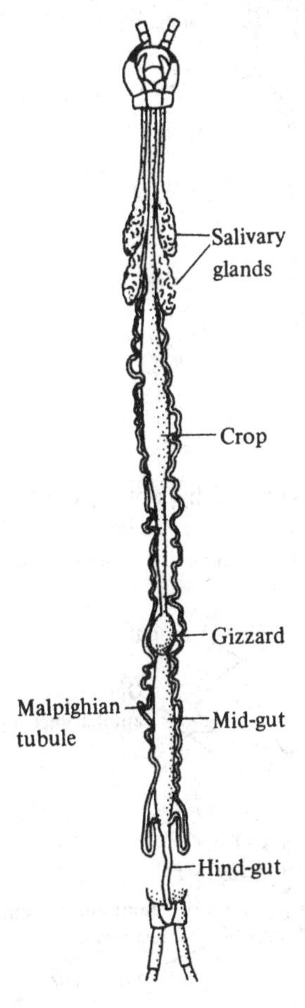

obscured from below by the labrum. More posteriorly there are a large central and paired lateral pore fields. The hypopharynx is bilobed and bears numerous fine hairs and pores (Verhoeff, 1902–25). The arrangement of these organs in *S. morsitans* is shown in Fig. 113. Behind the hypopharynx a semicircular *Pharynxdeckel* bears numerous small perforated pegs each supplied with a nerve

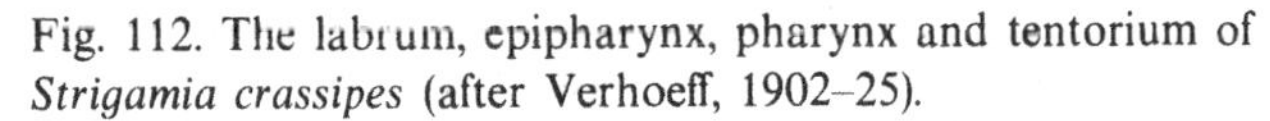
Fig. 112. The labrum, epipharynx, pharynx and tentorium of *Strigamia crassipes* (after Verhoeff, 1902–25).

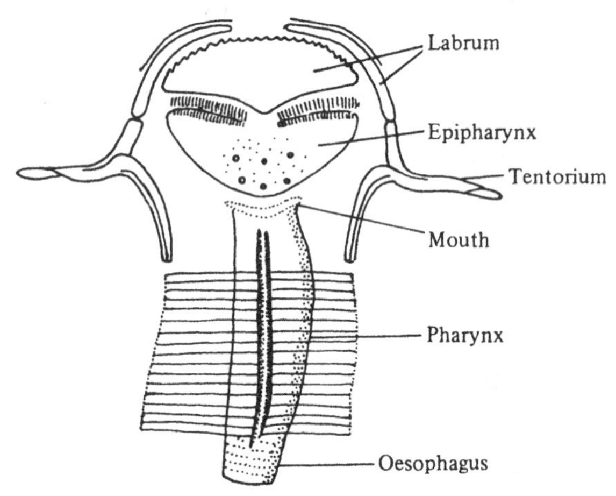

Fig. 113. The labrum, epipharynx and hypopharynx of *Scolopendra morsitans* seen in ventral view. The roof of the pre-oral cavity is flattened forwards to show the dorsal wall of the labrum (after Jangi, 1966).

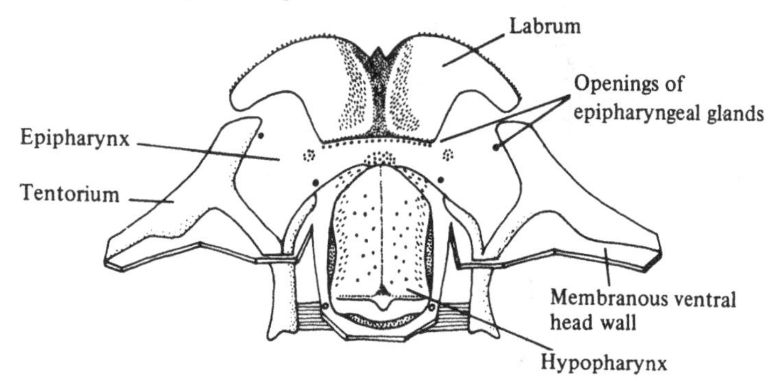

fibre that Verhoeff suggested were taste organs. Laterally this structure is clothed in a dense layer of simple and fringed hairs.

In *Scolopendra morsitans* the fore-gut extends as far as the thirteenth leg-bearing segment (Jangi, 1966). The pharynx is wide and has well-developed musculature consisting mainly of circular fibres, the longitudinal muscles being limited to the dorsal and ventral walls. The cuticular intima has broad dorsal and ventral folds so that the lumen appears H-shaped in transverse section. The anterior limit of the pharynx is marked ventrally by a spindle-shaped transverse oral muscle. The extrinsic musculature (pharyngeal dilators) consists of three ventral pairs originating on the tentorial apodemes and the transverse mandibular tendon and four pairs inserted on the dorsal wall of the head capsule.

The pharynx narrows posteriorly into a short oesophagus devoid of any extrinsic musculature. Its intrinsic musculature is composed almost entirely of circular fibres and its cuticular lining is thrown into folds. The largest part of the fore-gut is the crop which extends from the head region to the twelfth leg-bearing segment: its epithelium is irregularly folded and its musculature inconspicuous, the fibres appear to form an irregular network, some taking a longitudinal course, others running obliquely (Jangi, 1966).

In the fourth leg-bearing segment of *Ethmostigmus platycephalus spinosus* (Newport) the oesophagus bears a pair of small caecal outgrowths which are embedded in the salivary glands. The distal region of the caeca consists of small cells, some dividing. There follow columnar 'middle cells' with clear cytoplasm and, proximally, three types of columnar cells, two of which show intense alkaline phosphatase activity and may therefore be excretory (Rajulu, 1970*c*).

In the twelfth and thirteenth segments of *S. morsitans* the fore-gut forms a gizzard which Jangi termed the proventriculus (*gésier* of French authors). This is divided into anterior and posterior portions and the wall is thrown into strong longitudinal folds (Fig. 114*a*) which project into its lumen. The anterior proventriculus is characterised by armature in the form of cuticular plates and digitate lobes. The cuticular plates occur mainly on the ridges and bear backwardly directed spines (Fig. 114*b*). A circlet of forwardly directed digitate lobes marks the posterior limit of the anterior proventriculus. These carry minute spines. The posterior proventriculus is a funnel-shaped organ with longitudinal folds but without spines; it is invaginated into the anterior end of the midgut to form a circular cardiac or stomodaeal valve. The musculature of the proventriculus consists of inner longitudinal muscles and outer 'highly exaggerated' circular fibres. A similar arrangement is seen in *Scolopendra heros*, *S. cingulata* and *S. subspinipes*. In *S. heros* the cuticular plates bear a single spine, in *S. cingulata* the region is folded into an S-shape and bears cuticular plates with up to seven spines. Behind these there is a circlet of a dozen flaps with smaller secondary protruberances at their bases (Willem, 1889).

Jangi regarded the anterior proventriculus as a chewing apparatus and the posterior proventriculus as a valve. The longitu-

Fig. 114. *a*, the gizzard of *Scolopendra morsitans* opened longitudinally; *b*, a cuticular plate from gizzard (after Jangi, 1966).

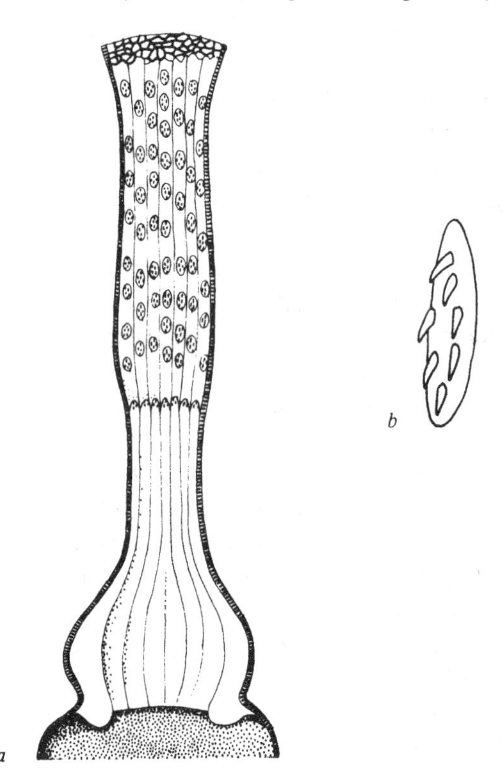

dinal muscles of the mid-gut are continued forwards over the proventriculus and are inserted on the crop. They are said to be the flexor muscles of the proventriculus, which promote chewing.

The mid-gut extends from the thirteenth to the nineteenth segment. Its surface is often marked by numerous transverse constrictions. There is an inner circular and an outer longitudinal muscle layer. The epithelium consists of columnar cells the distal ends of which are swollen. Between these cells there are small groups of regenerative cells. The epithelium shows two types of secretory activity: small cells charged, according to Jangi, with digestive enzymes are liberated into the lumen (holocrine secretion) and the distal ends of some of the epithelial cells are cut off (apocrine secrction). There is a continuous layer of granular material over the epithelium and the food is surrounded by a peritrophic membrane produced by the mid-gut epithelium. Jangi asserted that the peritrophic membrane was permeable in one direction to the digestive secretions and in the other to digested food.

The epithelium of the hind-gut consists of tall columnar cells and is thrown into folds. The wall consists of a thick coat of circular muscle fibres surrounded by a thin coat of longitudinal ones. There is an extrinsic musculature consisting of a dorsal and ventral pair of fan-shaped muscles inserted on the walls of the sclerites of the anal segment.

In *Ethmostigmus platycephalus spinosus* the peritrophic membrane consists of longitudinal and transverse strands forming a network with rectangular holes measuring 23×38 μm in the anterior third of the mid-gut and 46×58 μm in the posterior third. The strands of the membrane are in places frayed, suggesting that they are made of smaller fibrils. Analysis has shown 16 amino acids, suggesting the presence of protein: chitin is also present. The membrane may also contain a sulphated acid mucopolysaccharide which may protect it from digestive enzymes. The holes in the peritrophic membranes of the Dipolopoda and Onychophora are likewise rectangular, those of the insects circular and in arachnids and Crustacea they are hexagonal (Rajulu, 1971*a*).

The function of the peritrophic membrane is to protect the cells from damage by the gut contents. It acts as a barrier to the microflora so that infection is prevented (Chapman, 1971).

The alimentary canal of *Cryptops* has been described by Plateau (1878), Willem (1889) and Balbiani (1890). In general it resembles that of *Scolopendra*. The fore-gut is exceptionally long and the mid-gut exceptionally short. Balbiani gave a ratio of 38:12:8 for the three regions of the gut. The fore-gut of *C. savignyi* commences as a narrow tube, swells out into a mid-region where food is accumulated and terminates in a round or oval gizzard or *gésier* (Fig. 111). The gizzard in *C. savignyi* bears three types of forwardly directed appendages (Fig. 115*a*). Anteriorly there is a circle of brown conical projections bearing small spines. Posterior to these there is a circlet of lamellae which bear long bristles distally. There are 38–40 of these in adults, 20–24 in younger specimens. Finally there is a second circlet of 10–12 spine-bearing cones (Fig. 115*b*). In *Cryptops hortensis* there is only one type of appendage (Fig. 115*c*) of which there are about eight or nine rows. In *Scolopocryptops* the inner wall of the gizzard bears long tapering, forwardly directed, processes. The most anterior ones have their bases covered with small spines, these are followed by sharply angled processes: the most posterior processes are curved (Willem, 1889). Balbiani showed that the appendages, regarded as forming a straining apparatus, turn backwards as food passes into the mid-gut.

The fore-gut of *C. savignyi* has an outer layer of circular and an inner layer of longitudinal muscles. In the anterior part of the fore-gut the circular muscles are thin and well separated; in the posterior part they form a continuous layer and anastomose. This is also true of the longitudinal muscles. At the posterior end of the gizzard the longitudinal muscles pass through the circular muscles to form the striated outer longitudinal layer of the mid-gut (Fig. 115*d*). Posterior to the gizzard the circular muscles form a sphincter. The epithelium in the anterior part of the fore-gut is thrown into about 40 folds: posterior to the gizzard there are six large folds which form a 'cardiac valve' (Fig. 115*e*).

In the mid-gut the circular muscles are internal, the longitudinal muscles external. The latter anastomose at the end of the mid-gut and pass on to the hind-gut and also provide the musculature of the Malpighian tubules. There are two types of mid-gut epithelial cells: ordinary fusiform epithelial cells with a striated border and larger, round or oval 'mucus cells' which appear to move from the

Fig. 115. The structure of the gizzard in *Cryptops* spp. *a*, potash preparation of the gizzard in *C. savignyi*; *b*, detail of the appendages of the gizzard of *C. savignyi*; *c*, detail of the appendages of the gizzard of *C. hortensis*; *d*, the musculature of the hind part of the fore-gut of *C. savignyi*; *e*, a longitudinal section of the gizzard of *C. savignyi* (after Balbiani, 1890).

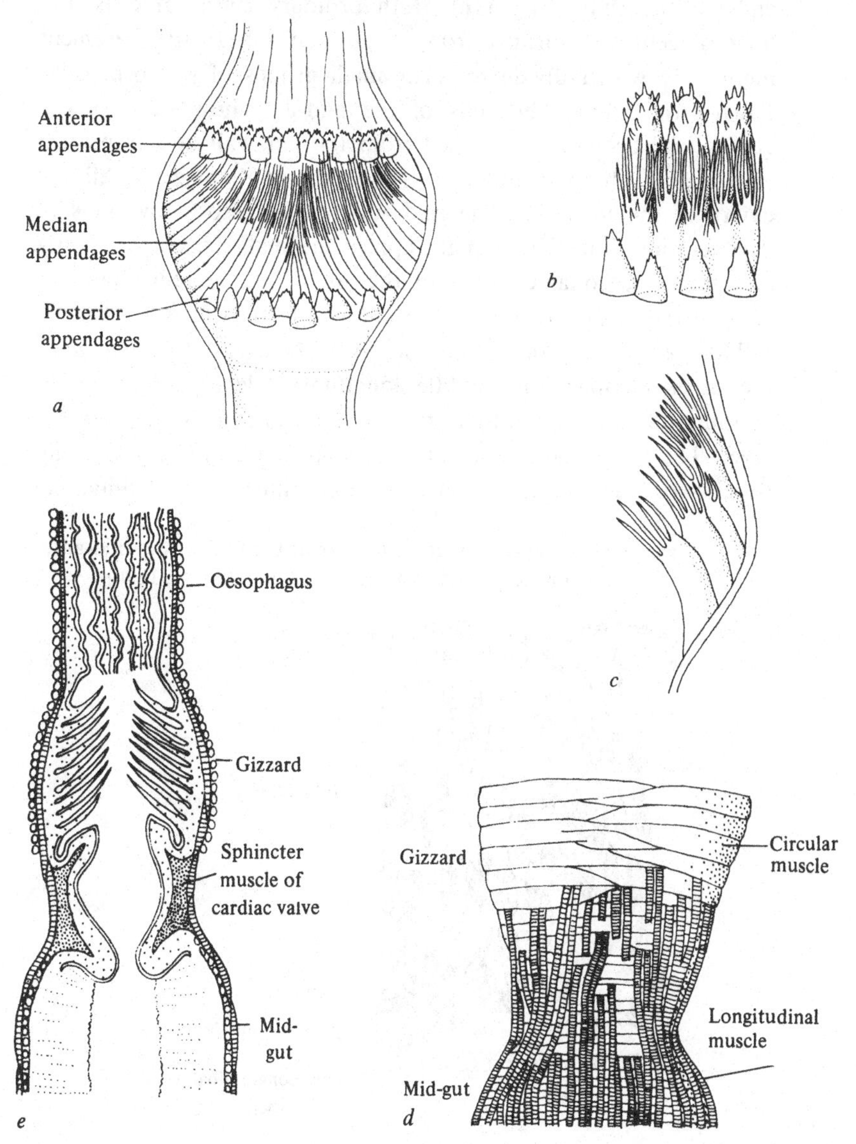

basement membrane to the outer region of the epithelium. These have a clear cytoplasm but in the autumn show three distinct zones: an upper homogeneous refractile region, a middle reticulate region and a basal clear region. At this time some sections show numerous vacuoles which form a continuous layer at the distal ends of the cells (Fig. 116). Both ordinary epithelial cells and 'mucus' cells are formed from small nuclei near the basement membrane which divide to form small groups of pyriform cells. During the winter the cells of the mid-gut show signs of degeneration, becoming detached from the gut wall and transform into an amorphous granular mass of refringent globules that fills the cavity of the intestine. The mid-gut muscles also show signs of degeneration but the animals appear normal. This is the only account of seasonal changes in the gut of a centipede. Balbiani showed that a peritrophic membrane was formed in the mid-gut.

There are three layers of muscles in the hind-gut of *C. punctatus*: the inner circular and middle longitudinal layer, which are a continuation of the mid-gut musculature, and an outer circular layer. The longitudinal and outer circular layers are very strongly developed. The circular muscles form a sphincter just behind the

Fig. 116. Transverse section of part of the wall of the mid-gut of *C. savignyi* sectioned in October (after Balbiani, 1890).

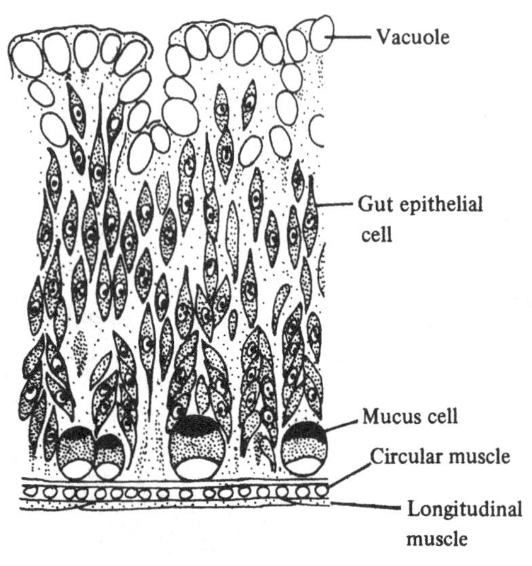

Malpighian tubules. The epithelium, which is sharply differentiated from that of the mid-gut is thrown into about 10 to 12 folds.

Lithobiomorpha

In *Eupolybothrus fasciatus* (Newport) the epipharynx bears a transverse row of pores in front of a central field which is bordered anteriorly and laterally by a dense field of hairs and behind which are sensory spines and groups of pores (Fig. 117). The hypopharynx is bilobed and richly setose.

The post-pharyngeal gut of *Lithobius* has been described by Plateau (1878), Kaufman (1961) and Rilling (1968). The last author gave the most detailed description. In *L. forficatus* (Fig. 118*a*) the fore-gut commences with a short pharynx with an approximately U-shaped lumen and a rod-like thickening along its floor (Fig. 118*b*). It has well-developed radial and circular muscles. The pharynx leads into the short thin-walled oesophagus whose wall is thrown into a number of longitudinal folds which are more pronounced in its posterior half where the cuticle is armed with forwardly directed spines. There is an outer circular and a sparse inner longitudinal muscle layer (Fig. 118*c*). The circular muscles form a sphincter at the posterior end of the fore-gut. There is no gizzard.

Fig. 117. Epipharynx of *Eupolybothrus fasciatus* (after Verhoeff, 1902–25).

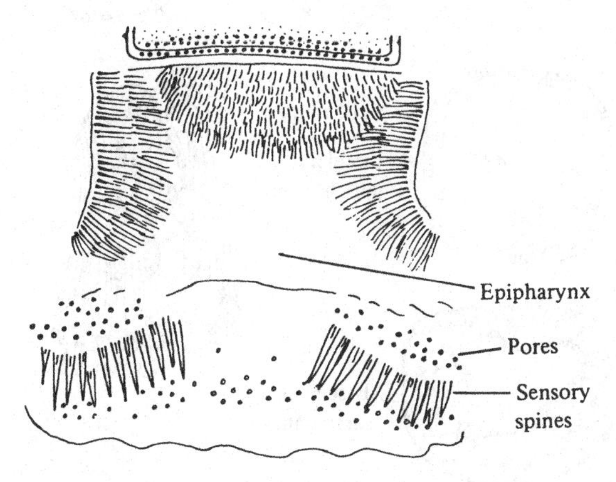

Fig. 118. Gut, Malpighian tubules and mandibular glands of *Lithobius forficatus*. *a*, complete alimentary canal; *b*, transverse section of the pharynx; *c*, transverse section of the oesophagus; *d*, transverse section of the mid-gut; *e*, transverse section of hind-gut. *f* and *g*, mid-gut and Malpighian tubule at greater magnification (after Rilling, 1968).

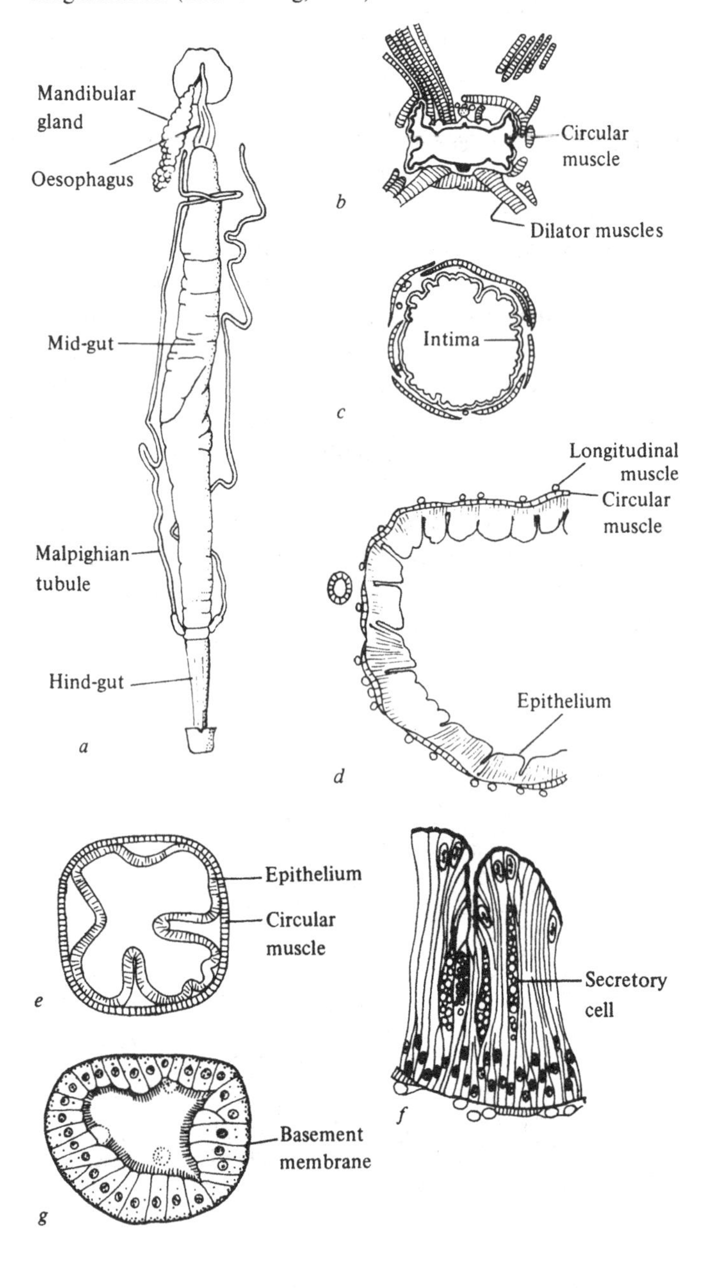

The cells of the mid-gut are tall and are in part secretory, in part absorptive (Fig. 118*d*, *F*). The mid-gut forms 70 per cent of the total length of the alimentary canal. Its weak musculature consists as in other centipedes of inner circular and outer longitudinal layers. Concretions appear in quantity at times in the mid-gut epithelium, cell lobes containing the concretions being nipped off and evacuated (Manton & Heatly, 1937). A peritrophic membrane is present (Plateau, 1878).

The hind-gut epithelium is thrown into longitudinal folds and surrounded by a thin layer of circular muscles (Fig. 118*e*).

Scutigeromorpha

The epipharynx, hypopharynx and pharynx of *Scutigera coleoptrata* have been described by Haase (1884*b*) and Verhoeff (1902–25). Manton (1965) gave a brief description of the pharynx. The epipharynx is supported by bars that are arranged in an approximately A shape, the apex of the A pointing posteriorly (Fig. 119). The anterior part of this structure formed of two side arms and a cross piece was termed *Labrumtrapez* by Verhoeff who regarded it as an important characteristic of the Scutigeromorpha. In the region of the side arms the epipharynx is densely setose.

The hypopharyngeal processes of the tentorium form lateral supports for the hypopharynx while a pair of dorsal sclerotisations converge to form a groove on the floor of the pharynx thus making a

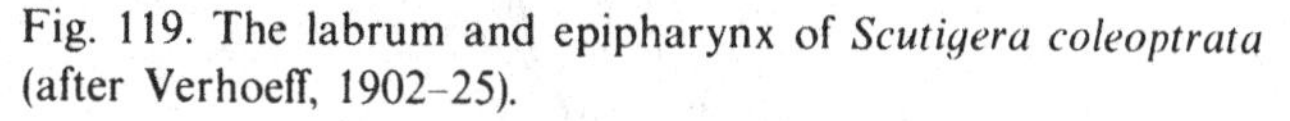

Fig. 119. The labrum and epipharynx of *Scutigera coleoptrata* (after Verhoeff, 1902–25).

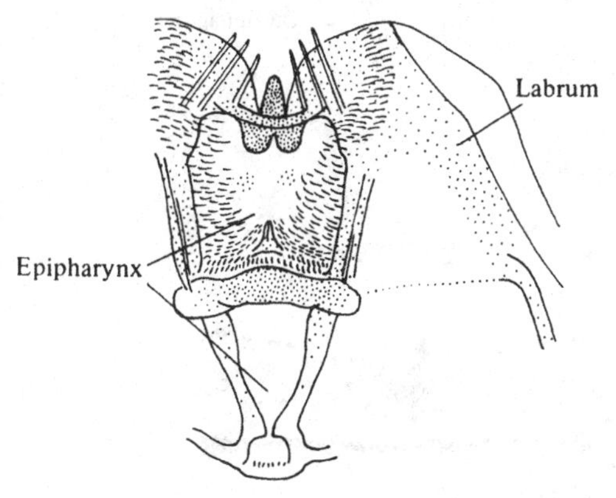

Y shape (*Pharynxgabel* of Verhoeff) (Fig. 120*a*). The central region of the pharynx is strongly sclerotised. Behind the tentorium the pharynx expands laterally into a pair of pouches with very thick cuticle on their dorsal lining. Verhoeff described all these structures

Fig. 120. The pharynx of *Scutigera coleoptrata*. *a*, viewed from below (after Verhoeff, 1902–25); *b*, transverse section (after Seifert, 1967a).

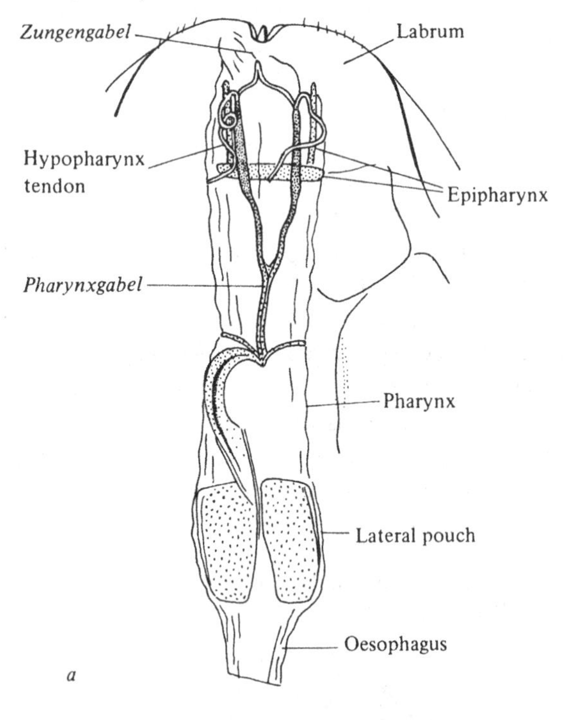

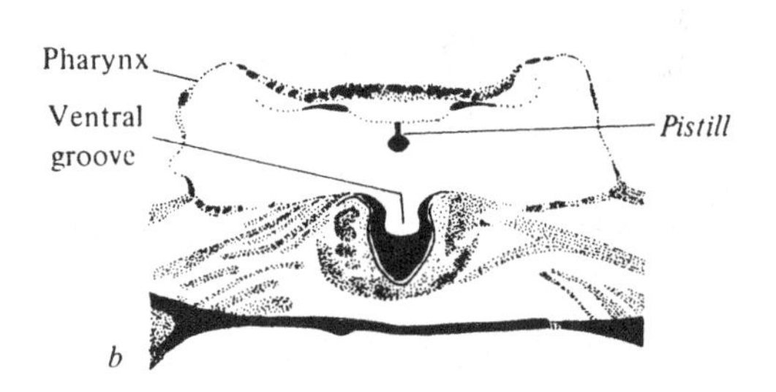

in considerable detail. He considered that the pharynx acts as a crop. The cuticular ridge on the dorsal wall of the pharynx bears a stalked knob (*Pistill*) in the region of the protocerebrum. This slides in a strongly sclerotised groove in the ventral wall of the pharynx (Fig. 120*b*) which may serve to sort large from small food particles and to grind the food (Seifert, 1967*a*).

Verhoeff (1902–25) stated that the structure of the gut of *Scutigera* was very similar to that of *Lithobius*. The oesophagus is a short opening by a funnel-shaped valve into the wide mid-gut at the level of the second leg-bearing segment.

Craterostigmus

In the legend to her Fig. 83 Manton (1965) wrote 'The short pharynx communicates with the long oesophagus behind the transverse mandibular tendon, the oesophagus is narrow but with folded walls permitting considerable expansion, the mid-gut is reached at the posterior end of leg-bearing segment 2 and is very wide.'

9

The poison glands

Newport (1844) was the first worker to recognise that the forcipules of centipedes, which he termed the mandibles, contained a poison gland. Despite this, many nineteenth-century workers confused the poison glands with other head glands (Duboscq, 1898): their true nature was recognised by MacLeod (1878). He examined *Scutigera coleoptrata*, *Lithobius forficatus*, *Cryptops savignyi* and various *Scolopendra* species and geophilomorphs noting the gland duct and pore. He demonstrated that whereas a bite from the poison claws of *L. forficatus* produces almost instantaneous death in flies, extracts of the 'salivary glands', when injected, did not.

Structure of the gland and discharge of poison in Scolopendra

The structure of the poison glands has been described for *Scolopendra subspinipes* by MacLeod (1878), for *S. cingulata* by Duboscq (1898), for *S. morsitans* by Pawlowsky (1913) and Dass & Jangi (1978), and for *S. viridicornis* by Barth (1967). Cornwall (1916) described the poison gland in *Ethmostigmus platycephalus spinosus* and Bücherl (1946) described the gland of a number of scolopendrids.

The gland is situated in the distal part of the trochantero-prefemur and extends into the poison claw on which its duct opens subterminally. It is innervated by a nerve from the suboesophageal ganglion and is well supplied with tracheae. In *S. cingulata* it is bluish white in colour. Transverse sections show a central duct surrounded for three-quarters of its circumference by elongated gland cells which open into the duct by pores (Fig. 121*a*). The remaining sector of the duct is imperforate and covered by a simple epidermis. The duct is a simple chitinous tube. The cytoplasm and nucleus is restricted to the distal region of each gland cell, the remainder being full of secretion (Duboscq, 1898).

With the doubtful exception of *S. subspinipes* the poison gland is

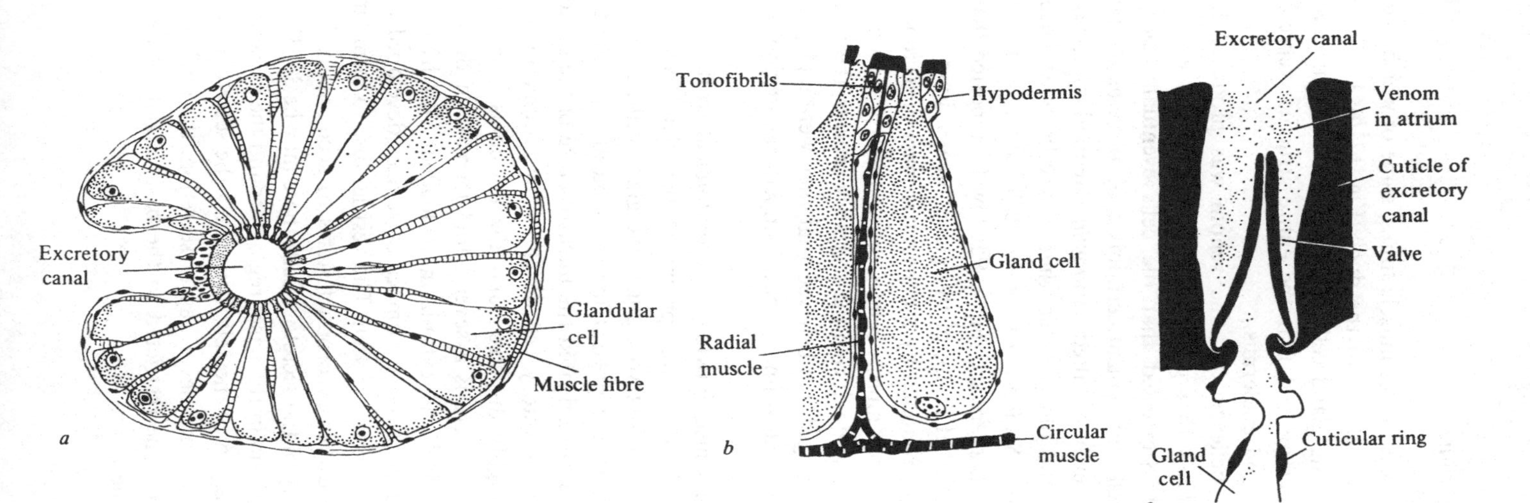

Fig. 121. *a*, transverse section of the poison gland of *Scolopendra cingulata*; *b*, diagrammatic longitudinal section of two gland cells of *Scolopendra viridicornis*; *c*, opening of a gland cell of *Scolopendra morsitans* (*a*, after Duboscq, 1898; *b*, after Barth, 1967; *c*, after an electron micrograph by Dass & Jangi, 1978).

surrounded by a meshwork of muscle fibres, some of which extend between the cells and attach to the duct (Fig. 121*b*). Pawlowsky (1913) showed that the gland cells of *S. morsitans* are surrounded by a delicate epithelium and that the entire structure represents invaginated cuticle and epidermis.

Duboscq (1898) believed that the venom was produced from vesicles in the nucleus that passed into the cytoplasm and were there modified. He suggested that the cells eventually die and pointed out that cell replacement certainly took place as the pores of the gland of a newly moulted *Strigamia* (Geophilomorpha) do not coincide with the old ones so that new cells must become glandular. In young *Scolopendra*, twelve cells are seen in cross sections but in adults there are thirty. Pawlowsky reported that in some cases the epithelial lining of the gland breaks down so that the duct becomes surrounded by a concentric mass of secretion in which are found isolated muscle cells. Cornwall (1916) showed that the gland cells disintegrate during the secretory process, their entire contents being discharged. He believed that they were replaced by young cells which moved round from the imperforate region of the duct.

Barth (1967) confirmed Pawlowsky's anatomical work and described the stages in the development of the gland cells in *S. viridicornis*. He showed that young inactive gland cells, active cells and degenerating cells are present in the gland at the same time. Dass & Jangi (1978) investigated the ultrastructure of the gland in *S. morsitans*: as the secretory cell develops the chromatin in the nucleus becomes highly dispersed and the nucleolus is active in ribosomal synthesis. There is a several fold increase in the endoplasmic reticulum and the cytoplasm is loaded with free ribosomes and polysomes. A large number of vesicles developed from the endoplasmic cisternae are seen. The vesicles fuse to form larger ones in which the excretory product starts to accumulate. At the height of the activity of the secretory cells there may be 30–35 dark bodies in the cytoplasm in the basal region around the nucleus and around the neck region of the cell. Ultimately the secretory bodies dissolve and the cell becomes a bag of secretion, the nucleus and other organelles become restricted to the basal region of the cell and undergo degeneration.

The opening of the secretory cell has been described by Pawlowsky (1913), Barth (1967) and Dass & Jangi (1978). It consists of a narrow passage around which there is a ring of cuticle in *S. morsitans*. This opens into a conical cuticular projection which may act as a non-return valve (Fig. 121*c*). The ring appears to be absent in *S. viridicornis*. The apex of the secretory cell is surrounded by a cluster of hypodermal cells (Fig. 121*b*). The muscles surrounding the gland cells are anchored to the collecting duct by tonofibrils. The typical banding pattern of skeletal muscles is not seen in *S. morsitans* by some striated fibres have been reported in other species.

Bücherl (1971) suggested that the secretion of the glands was of the merocrine type, the glands not being destroyed during the course of secretion, but the work of Barth confirmed by Dass & Jangi shows clearly that the secretion is holocrine. After releasing its secretion the cell degenerates and is replaced by another cell which is in the process of maturation. The new cells are produced by differentiation of the hypodermal cells.

Bücherl considered that the venom was expressed from the gland by the contraction of the forcipular adductor muscles but all other recent workers have assumed that it was by the contraction of the fibres surrounding the glandular cells. The venom is a clear, yellowish homogenous liquid and gives an acid reaction (Duboscq, 1898; confirmed by other workers).

Structure of the gland in other centipedes

In *Cryptops* the excretory canal is very short and perforated on all sides in the glandular region. The cells are inclined at an angle of 45° to the main duct. The posterior part of the gland is pear shaped and the gland is comparatively large (Duboscq, 1898). In geophilomorphs only the proximal end of the gland duct is perforate and since the distribution of pores is variable, Duboscq suggested that they should be used in taxonomic descriptions. The pores are often less numerous on the external face of the duct and the muscle fibres surrounding the gland are clearly striated and parallel (Fig. 122). The glands are normally situated in the trochantero-prefemur in geophilomorphs but in *Chaetechylyne vesuviana* they are sited in the trunk between the twelfth and eighteenth segments, the right gland being posterior to the left (Duboscq, 1898) and 'in the Aphilodontidae (-inae of Attems) the bodies of the glands are curiously displaced well back in the trunk; furthermore in such cases the bodies of the two glands are rarely side by side . . .' (Crabill, personal communication).

In *Lithobius* (Fig. 8) the gland lies in the trochantero-prefemur

Fig. 122. The poison gland of *Chaetechelyne vesuviana*. *a*, an isolated gland; *b*, transverse section through the inferior third of the gland (after Duboscq, 1898).

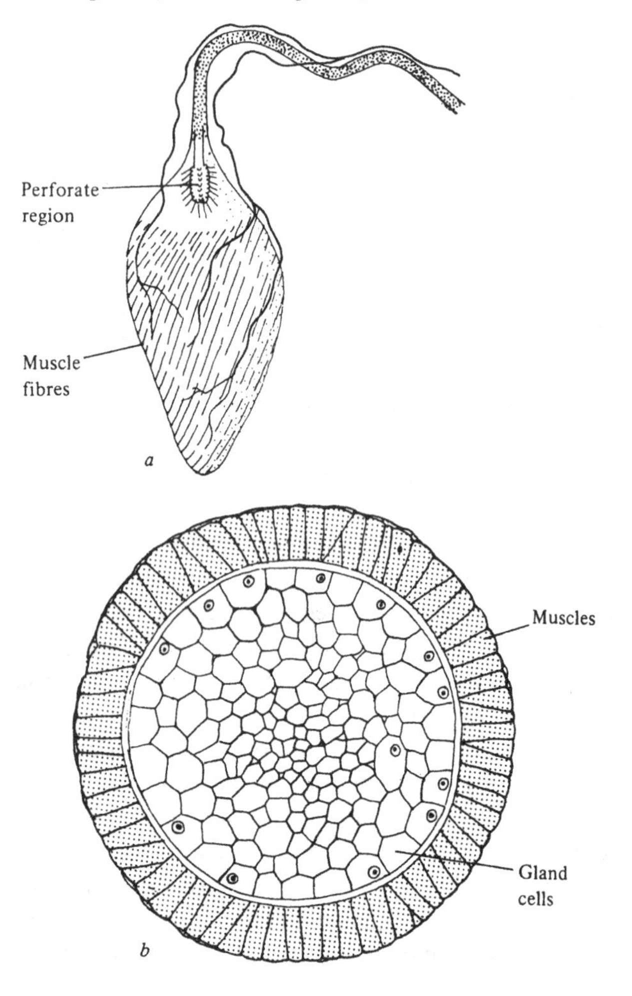

and runs into the claw. The muscular reticulum between the gland cells is poorly developed and the fibres are not striated. The arrangement of the gland, as seen in cross section, is similar to that of *Scolopendra* but the cells are relatively larger (MacLeod, 1878).

The gland of *Scutigera* is like that of *Lithobius* but more elongated, just reaching the coxosternum (Duboscq, 1898).

The effect of centipede bites on man

There is an extensive and scattered literature on the effect of centipede bites, recently reviewed by Bücherl (1971) and Minelli (1977). Unfortunately the former author does not give a complete bibliography of the work referred to in his text.

Instances of death from bites of scolopendrids were reported in the older literature (see Faust, 1928) but recent authors are inclined to deny the reports. Cornwall (1916) conducted a survey of the effect of centipede bites in Madras 'There were one or two descriptions of severe pain, swelling, eruptions, and desquamation in the affected limb leading to a week in bed, but the aggressor had not been actually seen, and so may have been a scorpion or a snake'. Baerg (1924) recounted the tale of a large centipede that crawled across the abdomen and chest of a Confederate army officer, leaving a number of deep red spots forming a broad red streak. Violent pains and convulsions soon set in, accompanied by excessive swelling of the bitten area; the man died in two days. The bite of the European *Scolopendra cingulata* causes pain, and at worst inflammation, oedema and superficial necrosis, but the pain goes quickly and the symptoms disappear in a few days (Pawlowsky, 1913). Klingel (1960*a*) was bitten some thirty times by this species. In 24 cases the only symptom was pain which disappeared in 20–30 min but in the other four cases the pain was severe, like a wasp sting, and the hand and arm became numb with some pain in the neck and chest. Local pain at the site of the bite and a slight numbness of the arm and hand disappeared in two to three days. Sometimes the animal bit painlessly, especially an hour after killing meal-worms.

In Arkansas a *Scolopendra heros* 135 mm long caused a sharp but strictly local pain which began to decrease after 15 min. After 2h there was only a slight swelling and all pain had disappeared within a day. The bite of *Scolopendra polymorpha* from New Mexico caused a sharp pain but no swelling. In 3 h the pain had practically disappeared (Baerg, 1924). In Southern California the bite of *S. heros* is said to cause intense pain and the animal also produces a red streak where it has crawled over the body (Herms, 1915). *Scolopendra subspinipes* produces intense pain, blistering and swelling, local inflammation, buboes and subcutaneous haemorrhage (de Fonesca, quoted by Turk, 1951). Bartmeyer & Schmalfuss (1933) recorded the effects of an *S. subspinipes* biting the arm of a sailor on a ship bound for Malakka. A severe appearance developed at once, similar to that produced by the bite of a poisonous snake. The wound was blue coloured and surrounded by a blue area which swelled and became extremely painful. The victim became anxious and depressed, and broke into a cold sweat. His heartbeat was fast and irregular. The arm became inflamed. He recovered in two days.

Remmington (1950) found the bite of *S. subspinipes* in the Phillipines unbearable at first, but it diminished in 20 min. Next morning the axillary region of the arm was severely swollen. A mildly painful condition persisted for some three weeks.

In three cases of bites by *S. subspinipes* in Sarawak, there was more or less intense pain but no other symptoms (Lewis, unpublished observations). Templeton (1846) reported that the bite of '*S. pallipes*' and '*S. crassa*' in Ceylon caused little pain unless the bite was in a non-fleshy region but swelling took place. Subsequently he reported pain and considerable swelling after a bite by *S. pallipes*. Unfortunately he failed to describe these species so they cannot be identified.

S. viridicornis in South America causes pain for eight hours and small superficial necroses healing after 12 days (Bücherl 1946).

Sebastiany (1870), quoted by Bücherl (1971), reported the effect of the bite of *S. morsitans* on an eight year old child and a forty-nine year old man: there was intense local pain, vomiting, headache, swelling of a large area around the bite which had a blackish centre. A case in the south of France led to considerable pain and a swelling of the hand. The swelling had gone by the third morning (Turk, 1951). The effect of the bite of *S. morsitans* in Nigeria is simply one of sharp pain which soon disappears (Lewis, unpublished data).

Of other scolopendrid genera, the Indian *Ethmostigmus platy-*

cephalus spinosus is not naturally vicious and it is difficult to get it to bite. When tried on a man's arm the subject winced but showed no local after-effects (Cornwall, 1916). A collector bitten by a large *E. trigonopodus* in Northern Nigeria experienced no pain although the poison claws had cut the skin (Lewis, unpublished data).

Rhysida sp. and *Otostigmus* sp. in India have small poison claws and can only penetrate very delicate skin (Cornwall, 1916). *Trachycormocephalus* from Eritrea causes pain, some swelling and subcutaneous bleeding (Lewis, unpublished observations); the region of the bite is still tender on the following day.

DeCastro (1921) quoted by Keegan (1963) found that recovery from centipede bites by patients in India, Burma and Ceylon was slow, sometimes as long as three months. Every case seen developed acute lymphangitis (inflammation of the lymph vessels) with oedema, as well as inflammation of the skin and subcutaneous tissues. In most cases a local necrotic process developed at the site of the bite and in some this progressed to a condition not unlike phagedenic ulceration (a rapidly spreading and sloughing ulcer). The most serious symptoms followed the bites of the Andaman species which can reach a length of 33 cm.

The only well authenticated fatality appears to be that of a child of seven years bitten on the head by a *Scolopendra* in the Philippines (Venzmer, 1932, quoted by Bücherl, 1971).

From the above data it seems possible that the effectiveness of the venom varies with species: *S. subspinipes* would appear to be one of the more unpleasant ones. There can be no doubt that the centipedes can vary the amount of poison they inject. Klingel (1960*a*) found that animals that had not fed for several days could bite harmlessly and therefore the lack of poison could not be ascribed to lack of poison in the glands.

Duboscq (1898) noted that the effects of the bite of *Scolopendra* vary with the time of year. In winter the bite causes, at most, a small pimple that disappears in an hour. In spring, when the centipedes are more active, a bite causes inflammation which can last for up to three days, a bite on the finger causing the hand and lower half of the forearm to become swollen.

The occasional serious symptoms caused when adults are bitten suggest that fatalities could occur amongst children but such cases must be extremely rare. The animals invariably escape after biting and therefore many reports must be, as Cornwall pointed out, unreliable.

Amongst the Cryptopidae, *Theatops spinicauda* (Wood), body length 48 mm, causes a sharp pain, all trace of which has gone in 30 min (Baerg, 1924). *Cryptops anomalans*, body length 35 mm, cannot inflict a very serious bite (Minelli, 1977).

There appear to be no records of humans having been bitten by geophilomorphs.

The bites of the lithobiomorphs *Eupolybothrus fasciatus* (Newp), body length 40 mm (Attems, 1926), are not serious. The American species *Lithobius mordax*, length 20–25 mm, is unable to puncture the skin of the little finger (Baerg, 1924).

The bite of *Scutigera coleoptrata* has been described as causing inflammation for 36 h (Kunckel d'Herkulais, 1911), severe pain with some swelling (Herms, 1939), and only minor swelling and nuisance pain (Johnson, 1952). *S. coleoptrata* generally fails to penetrate the thickened skin of the finger tips (Johnson, 1952). Ewing (1928) reported a case of a bite which caused only a slight burning sensation and a slight redness of the skin but in which a red weal developed two days after the bite. He concluded that the serious consequences of bites are probably due to subsequent infection.

Effect of centipede venom on animals

Duboscq (1898) summarised the results of his earlier work on the poison of *Scolopendra*. He found that arthropods and vertebrates were probably the only groups sensitive to the venom. Spiders and scutigeromorphs are very sensitive, as are carabid beetles, but scorpions and tenebrionid beetles are less so.

Norman (1897) investigated the effect of the bite of '*Scolopendra morsitans*' from Texas. Half to three-quarter grown mice bitten by specimens 150–190 mm long died almost immediately. Full grown mice died in about a day but two ground snakes 23 cm long showed no apparent ill effects. Jourdain quoted by Remmington (1950) showed that *S. morsitans* quickly killed small mammals and birds.

Cornwall (1916) believed that the poison of *Ethmostigmus platycephalus spinosus*, *Rhysida* sp. and *Otostigmus* sp. was largely digestive in function. After several experiments with rabbits he concluded that the bite of *Ethmostigmus* was unlikely to be followed by any more serious consequences than a little local irritation. He suggested that when severe local inflammation occurs it is chiefly due to septic infection. Baerg (1924) found that the bites of *Lithobius mordax* and *Theatops spinicauda* have little or no effect on rats and although the bite of *Scolopendra heros* and *S. polymorpha* obviously cause pain, bitten rats are normal in 3–5 h. Likewise Mathur (1926) found that the bite of *Scolopendra morsitans* had little effect on a lizard, a toad, a frog and a rabbit. He, like Cornwall, concluded that the secretion of the poison gland is largely digestive in function.

The effect of centipede venoms has also been investigated by injecting it into various animals. Briot (1904) injected the venom of French scolopendras into rabbits. Paralysis in the hind leg, oedema, then an abscess and death after 17 days was caused by 2 cm^3 of the venom. An injection of 3 cm^3 into a second rabbit caused death in 1 min. A white rat injected with 1 cm^3 showed paralysis in the hind leg but eventually recovered completely. Briot stated that the effect of the venom was like that of a viper, causing almost immediate paralysis and necrosis.

Bücherl (1946) made a detailed study of the effect of centipede venom on small mammals, expressing dosage as a fraction of the extract of one poison gland. Thus half (0.5) of the extract of the poison gland of *Scolopendra viridicornis* kills 20 g mice in 20–32 sec, 0.04 of the gland kills them in 3–7 h. The median lethal doses for mice are 0.03 gland given intravenously and 0.250 given intramuscularly. The extract of 2.5 glands of *S. subspinipes* killed an adult pigeon and an adult guinea pig within 8–16 h.

The venom of both these species acts powerfully on the central nervous system, with activation of all glands with smooth musculature, respiratory acceleration, abundant sweating, chiefly in the neck, loss of stability, vomiting, respiratory failures and progressive paralysis of the respiratory centres, convulsions and death (Bücherl, 1946). The author concluded that the venom was neurotoxic.

In *Otostigmus scabricauda* (Humbert & Saussure) the median lethal dose for mice is 0.012 gland intravenously and 0.070 intramuscularly. The figures for *Cryptops inheringi* Bröl. are 0.150 and 0.340 and for *Scolopocryptops ferrugineus ferrugineus* (Linn.) 0.160 and 0.390 respectively.

Lévy (1927*a*) demonstrated that an extract containing the venom of *Lithobius forficatus* caused reactions similar to those of intoxication when injected into crayfish (*Astacus fluviatilis*). The crayfish show a short period of agitation followed by violent contraction and cessation of respiratory movements followed by death. The blood remains liquid whereas it coagulates *in vitro*. The venom of *Cryptops anomalans* has the same effect. The lethal dose is a quarter of the venom of a large animal in both cases.

The venom of the scolopendromorphs *Scolopendra cingulata* and *Cryptops anomalans* has a haemolytic effect but that of *L. forficatus*, *Scutigera coleoptrata* and *Himantarium gabrielis* does not (Lévy, 1923*b*). The haemolytic action of *S. cingulata* is comparable to that of cobra venom (Lévy, 1923*a*).

Lévy (1927*b*) showed that *Lithobius* blood had an antitoxic effect: whereas venom alone kills *Astacus*, the animal recovers after an injection of *Lithobius* venom and blood. The blood of *Lithobius forficatus* also neutralises the effect of the venom of *Cryptops anomalans*.

Chemical nature of centipede venom

The venom of *S. viridicornis* contains 5-hydroxytryptamine (serotonin). This is an important pain producer and thus defensive though it causes also an increase in the rate of absorption of truly toxic components, usually proteins of low molecular weight in arthropods giving rapid immobilisation (Welsh & Batty, 1963).

10

Feeding and digestion

Geophilomorpha

Food

It has long been believed that earthworms are an important item in the diet of geophilomorphs. Newport (1844) and Wood (1865) stated that earthworms form the diet of the Geophilomorpha and Brehm (1877) figured a *Geophilus* coiled round a large earthworm. Brade-Birks (1929) recovered setae, probably of a very young lumbricid worm, from the gut of a specimen of *Haplophilus subterraneus.* Auerbach (1951) was unable to substantiate that *Geophilus rubens* Say or *Strigamia fulva* Sager fed on worms: both species refused a selection of small arthropods although the *Strigamia* accepted a small beetle larva. Weil (1958) found a *Necrophloeophagus longicornis* under a stone attacking a lumbricid twice its length. He found that large earthworms struggle free from geophilomorphs with ease and do not appear to be harmed by their bite and concluded that centipedes are seldom successful in overcoming earthworms larger than themselves.

Johnson (1952) showed that in the laboratory, *Geophilus rubens* accepted *Drosophila* larvae and adults, mycetophilid larvae, snails' eggs and small enchytraeid worms but not mites, elaterid or buprestid beetle larvae or earthworms. Weil (1958) reported that *N. longicornis* takes very small earthworms, weakly sclerotised insect larvae, young lithobiomorphs and occasionally enchytraeids, and entomobryomorph Collembola but not *Tomocerus.* It occasionally attacks newly moulted insect larvae: a teneral *Lacon murinus* was stabbed with one poison claw and opened between the scutellum and the elytra.

In the field, Crabill (personal communication) has found a specimen of *Strigamia bothriopa* Wood feeding on an ant, Weil (1958) saw three *Strigamia acuminata* feeding on a iuline millipede and *Necrophloeophagus lorgicornis* feeding on a lumbricid worm and a woodlouse. Gabbutt (1959) observed a juvenile *Geophilus carpophagus* feeding on a third instar larva of the orthopteran *Nemobius sylvestris.*

Field observations on littoral centipedes are numerous. Crozier, reported by Chamerlin (1920), observed *Hydroschendyla submarina* in Bermuda. It bit into the sides of leodicid worms 'licking up the juices and creeping off out of sight with one of the fragments into which the worms autotomise'. The best documented species is the European species *Strigamia maritima.* Pocock (1900) found this species feeding on a crustacean but it is not clear whether this was a woodlouse or a 'hopping sand-shrimp'. Blower (1957) observed *S. maritima* on the Isle of Man feeding at night. Both immature and adult specimens were found feeding on the barnacle *Balanus balanoides* and less frequently on the periwinkle *Littorina saxatilis.* As many a six individuals were found feeding on one barnacle, passing through a small gap between the opercular plates. The soft parts of the barnacles had been reduced to a slimy consistency which suggested extra-intestinal digestion. J. B. Hawthorne (personal communication) found *Strigamia* feeding on a lumbricid worm in the Isle of Wight. This, too, had been reduced to a slimy mass. Turk & Turk (1958) observed the species feeding on barnacles, winkles and top shells in Cornwall.

Lewis (1961) observed *Strigamia maritima* feeding in the field on an enchytraeid, on *Orchestia gamarella* (on one occasion 20 specimens were feeding on one specimen of this crustacean) and on the isopod *Sphaeroma* sp. All post-larval stadia of the centipede were found feeding on the isopod, either singly, or in groups of up to 20 individuals. In most cases the crustacean was attacked through its ventral surface, presumably because here the cuticle is much thinner. Specimens attacked through the dorsal surface had cracked or partially crushed terga. The author ascribed this damage to the pressure exerted by people walking over the shingle under which the *Sphaeroma* were found.

There can be little doubt that geophilomorphs are predominantly carnivorous but there have been a number of records suggesting that they may at times feed on plant material. Samouelle (1819) and Boisduval (1867) have reported *Geophilus carpophagus* from garden fruits and Verhoeff (1902–25) quotes early editions of Brehm's *Tierleben* as stating that *Necrophloeophagus longicornis* was found in the roots of bulbs, potatoes, parsnips and carrots. Brade-Birks (1929) obtained starch grains from the guts of four specimens of *Haplophilus subterraneus* that had been kept in a tin with slices of potato. Weil (1958) reported finding immature specimens of *Clinopodes linearis* in which the gut, viewed through the cuticle, was green and assumed that this was due to the presence of plant material but stated that this was exceptional: in captivity they only fed on plant material after they had been deprived of animal food for a considerable period. Eason (1964) stated that *H. subterraneus* sometimes turns vegetarian and causes damage to root crops. According to Dobroruka (1961) the digestive tract of Geophilomorpha often contains plant residues, and soil and plant remains are frequently found in *Schendyla nemorensis* (C. L. Koch).

Plateau (1878) examined the gut contents of at least 100 *Haplophilus subterraneus gervaisi* collected from the field at different times of the year and found them to consist of viscous material and, very rarely, small arthropod remains. Lewis (1961) examined 750 mounts of the guts of *Strigamia maritima* together with the faeces produced before they were dissected and found no plant or animal remains. Lewis (1960) found fragments of the alga *Enteromorpha* in the oesophagi of specimens of *Strigamia maritima* that had been kept in a container with the plant but it seems likely that the centipede took the *Enteromorpha* because it had been starved.

Geophilomorphs have, therefore, been observed to feed on a variety of annelids, molluscs and arthropods. It seems probable that they will attack any animals of suitable size provided that they can penetrate the cuticle. It would appear that plant material is unimportant as a food item to many geophilomorphs but is an aspect of their feeding biology that requires further investigation.

The feeding mechanism

Lewis (1961) described two feeding methods in *Strigamia maritima*. Small *Orchestia* (2–4 mm long) and *Drosophila* were seized with the poison claws and held with these and the first seven or eight pairs of legs during feeding, the anterior end of the centipede being arched over the prey (Fig. 123): the prey was reduced to very small pieces. Large *Orchestia*, 1 cm or more in length, were only attacked if damaged or dying. When attacking a large *Orchestia* the *Strigamia* levered up a tergite with its poison claws and, turning its head on one side, pierced the intersegmental membrane with one of these then slid the poison claw round the tergite making a transverse slit in the amphipod's body. The centipede next pushed its head and anterior segments into the body cavity and could be watched through the integument of the *Orchestia*. The poison claws were constantly in motion macerating the amphipod's tissues: the movement of fat droplets showed that there were irregular currents towards and away from the head of the centipede suggesting external digestion and suctorial feeding as did Blower's (1957) observations on the species. Unlike small *Orchestia* large specimens were left with their exoskeleton intact.

Other authors have reported geophilomorphs entering prey through intersegmental membranes: Crabill (personal communication) found a specimen of *Strigamia bothriopa* with the anterior part of its body thrust into the abdomen of an ant through the intersegmental membrane and Weil (1958) recorded the method by which *Strigamia acuminata* attacked a *Julus*. The centipede entered the millipedes from the ventral side just behind the head. It first ate the contents of the head capsule which fell off giving access to the rest of the body of the prey. *Geophilus carpophagus* is able to seize a blow-fly maggot with difficulty: the prey is torn open by the poison claws, and the centipede's head and poison claws inserted into the maggot, the tips of the antennae remaining outside (Manton, 1965).

Fig. 123. *Strigamia maritima* feeding on a *Drosophila* (after Lewis, 1960).

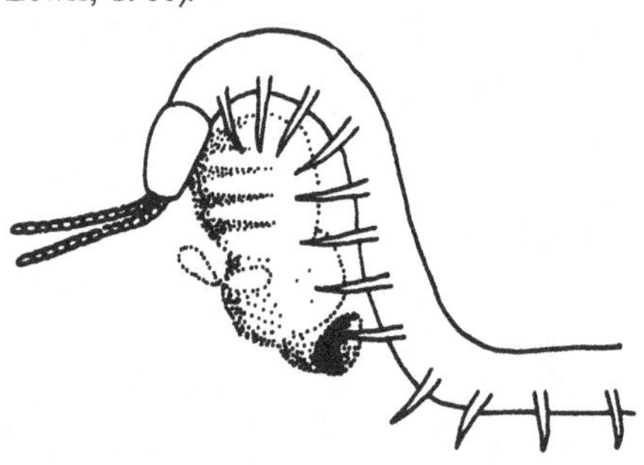

Group feeding has been observed only in the genus *Strigamia*; Blower's (1957) observations and those of Lewis (1961) on *Strigamia maritima* have already been referred to, as have those of Weil (1958) on *Strigamia acuminata*. It has been suggested (Lewis, 1961) that such behaviour is advantageous in that it makes prey, which would otherwise be invulnerable, available to small specimens. Alternatively, small specimens may be able to enter between the opercular plates of small barnacles otherwise inaccessible to larger specimens and bring about their opening. It is possible that pheromones may be involved in group feeding.

Manton (1965) has discussed the structure of the head and feeding mechanisms in geophilomorphs in some detail. She pointed out that the whole geophilomorph head shows specialisations correlated with burrowing, forming a suitable apex to the wedge-shaped anterior end of the body. She concluded that they are suctorial feeders with mandibles capable of weak rasping but which are probably mainly used to sweep semi-fluid matter into the mouth, aided by pharyngeal suction. The same type of inter-relationship may exist between the movements of the mandibles as that described for the scolopendromorph *Cormocephalus* (Manton, 1964). The poison claws may be used within the prey to pull tissues towards the mouthparts.

The considerable variation in the shape of the head capsule, the size and shape of the poison claws and the differences in mandibular structure must represent differences in feeding mechanisms that need to be investigated. In the South African *Philacroterium pauperum* Attems the forcipules are exceptionally small and the telopodite consists of only three segments (Fig. 124). In the genus *Mecistocephalus* the first pair of walking legs are greatly reduced in size.

When in the vicinity of adult *Tenebrio molitor* that had ben cut in half, individuals of *Strigamia maritima* show an increased rate of locomotion and violent swaying movements of the anterior part of the body. This behaviour is continued until contact is made with the beetle when feeding commenced immediately. This suggests that prey is detected at a distance. When *S. maritima* comes into contact with *Drosophila* it rapidly withdraws its body unless it touches the fly with it antennae. In this case the centipede begins swaying movements with the anterior part of the body and if it relocates the prey with its antennae, seizes it with its forcipules (Lewis, 1960).

Scolopendromorpha

Food

Laboratory observations suggest a very wide variation in the food of scolopendromorphs. *Scolopendra gigas* (possibly *gigantea*) from Trinidad at the London Zoo was fed principally on small mice (Cloudsley-Thompson, 1958*b*) and it has been reported that *S. morsitans* kills frogs and small toads in addition to cockroaches and termites (Jangi, 1955). *S. heros* takes a variety of nocturnal insects (Campbell, 1932) and *S. cingulata* will feed on nymphal cockroaches, spiders, flies, bees and wasps. Bees and wasps are taken in mid-air, the centipede rearing up the fore part of the body to snatch them with its poison claws as they fly past (Cloudsley-Thompson, 1955). This species has also been reported to feed on

Fig. 124. Ventral view of the anterior end of *Philacroterium pauperum* (after Attems, 1930*b*).

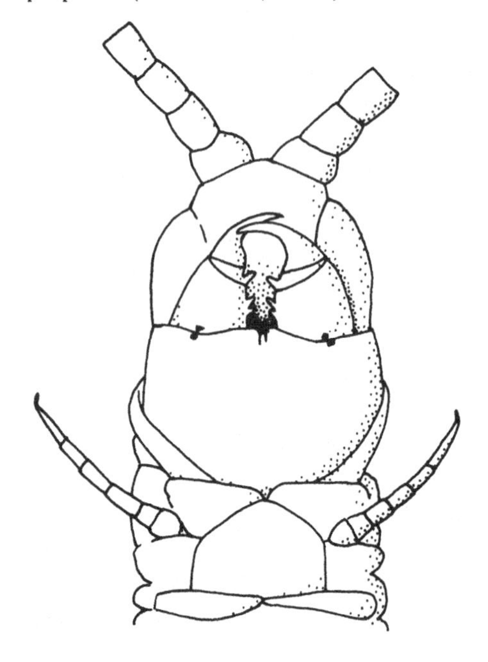

small worms and fleshy fruits (Heymons, 1901), and to 'lick' butter and eat pieces of bread and cake (Manton, 1964). Mathur (1926) fed *S. morsitans* on earthworms and coconut, Misra (1942) found the same species took apples, carrots and milk, Khanna (1977) found that it ignored ants, bugs and small cockroaches but accepted scarabaeid beetle larvae, termites and isopods, while Sharma & Vazirani (quoted by Khanna, 1977) found that it fed on the yellow-bellied house gecko (*Hemidactylus flaviviridis*).

Field observations confirm that scolopendrids attack vertebrates: an Indian species was observed eating into the side of a toad (Wells-Cole, 1898), a Burmese species eating a snake (Okeden, 1903), an Australian *Ethmostigmus* sp. attacking a lizard (*Diplodactylus*) (Whittel, 1883) and in the Persian Gulf, an unidentified species attacking a sparrow in its nest (Cumming, 1903). Shugg (1961) reported a case of a large Australian scolopendromorph, 16 cm long, attacking and feeding on a medium-sized mouse. The body of the centipede was wrapped round the rodent, its jaws clamped into the mouse's spine near the back of its neck: at this stage the mouse was still struggling. Later the centipede shifted its hold so that it lay along the mouse's spine, with some of its legs around the tail and hind quarters. The centipede had torn a hole in the animal and was devouring the flesh. A *Scolopendra heros*, 118 mm long, was seen in Texas carrying a freshly killed juvenile long-nosed snake (*Rhinocheilus lecontei*), 247 mm long. The snake was being carried head first under the full length of the centipede's body. A portion of the dorsal side of the neck and the side of the snake's head had been eaten (Goldberg, 1975).

There appear to be only two field observations of scolopendrids eating invertebrates: Lawrence (1934) observed a *S. subspinipes* with its mouth applied to a slug (*Veronicella leydigi*) and Remmington (1950) reported the same species feeding on various winged insects in the Philippines.

Lewis (1966) examined the gut contents of 37 *S. morsitans* from Khartoum, Sudan. Twenty-nine of these contained food material in the form of arthropods, including fragments of spiders, mites, centipedes, flies, staphylinid and carabid beetles, and ants. Later re-inspection of the material has shown that it also included termite mandibles.

Cryptops hortensis kills flies, young *Pieris* caterpillars, and small opilionids but does not devour these. When confined with *Anthomyia*, a *Cryptops* was observed to seize the fly with its terminal legs: the fly escaped with the loss of some appendages (Verhoeff, 1902–25).

Plateau (1878) observed *Cryptops savignyi* to kill and eat part of a *Haplophilus subterraneus*. Examination of the gut contents of *C. savignyi* showed that the species fed on young lumbricid worms in large quantity and also small mites, geophilomorphs and sometimes small spiders. The only recognisable remains in *C. hortensis* were of spiders. *Cryptops parisi* contained small insect larvae. Balbiani (1890) found grains of sand amongst the food of *C. savignyi*. He noted that they sometimes contained remains of their own species and sometimes numerous tyroglyphid mites of the type that is found on the body and legs.

The feeding mechanism

Manton (1965) considered that scolopendromorphs do not pounce upon their prey and do not apprehend it from a distance although Cloudsley-Thompson's (1955) observations on *Scolopendra cingulata* suggest that they can apprehend it from a distance but presumably not by sight. Remmington's (1950) observations on *S. subspinipes* led to the same conclusion: specimens fastened onto a tent near the ventilator hole with their anal legs and swung their bodies to one side or the other to seize insects that had alighted near by. Bücherl (1971) stated that 'specimens of *Scolopendra* and *Otostigmus*, *Cryptops* and others attack their prey with the last prehensorial anal legs; then the head is rapidly curved behind and the venom claws deeply and firmly buried in the body of the prey'.

The feeding mechanism of *Cormocephalus nitidus* Porat has been described in detail by Manton (1964). Typically, scolopendromorphs catch and eat living prey but Manton considered that they are also scavengers, though not hyaena-like. Their feeding mechanism is suited to soft fleshy contents of prey and small entire arthropods.

Prey is seized and bitten by the poison claws, whose movements are very strong, and held by the clawed tip of the telopodites of the second maxillae and the first few pairs of walking legs. The bite may be repeated several times, the prey being turned about by these limbs. When the prey has a firm exterior, a grip by the tips of the poison claws is followed by a sudden and strong flexure of these limbs and a forward movement horizontal to the body of the coxosternum of the poison-claw segment. The toothed coxal endities (coxopleural teeth) cut into the prey like a spade or a tin opener. The presence of lobes on the poison claw trochantero-prefemur (Fig. 23) suggests that these may be used to chew prey in conjunction with the coxopleural teeth. The oral field is then applied to the exposed flesh or the head is inserted into the lesion, or the poison claws may be thrust into the prey and the flesh drawn towards the oral field. Apart from such movements, the tips of the poison claws and the second maxillae are used only to hold the food in a convenient position while the mandibles and first maxillae carry out rhythmical feeding movements. Small *Scolopendra canidens oraniensis* Lucas can feed on blow-fly larvae using the 'tin-opening technique' but *Lithobius forficatus* of comparable size are unable to do this (Manton, 1965).

Manton considered the mechanism of ingestion speedy and efficient. Within $1\frac{1}{2}$ h intermittent feeding a young *Scolopendra subspinipes* 33 mm long and 2 mm wide consumed the contents of two blow-fly maggots each about 15×2.5 mm. The body of the

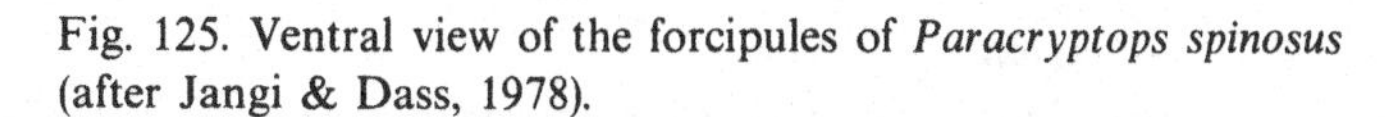

Fig. 125. Ventral view of the forcipules of *Paracryptops spinosus* (after Jangi & Dass, 1978).

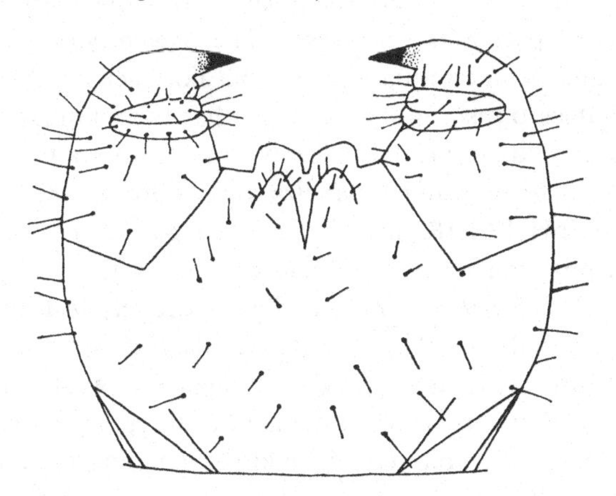

centipede was turgid after the meal and the cuticular body wall of the maggots was left clean and empty.

During feeding the rhythmical movements of the first maxillae and mandibles each last about 0.5 s. The first maxillae sweep a cone of flesh into which the mandibles bite. The head is often rhythmically raised at this point and so the mandibles as they bite drag up a cone of flesh which is again swept up and gripped by the first maxillae. The head is then lowered and the mandibles part. The mandibular movements are akin to those of a see-saw combined with the protractor-retractor movement.

Food enters the mouth of *Scolopendra* due to the tongue-like activities of the hypopharynx and the anterior lobe of each mandible as well as by suction generated through the musculature and armature of the pharynx. The extent to which external digestion may take place has not been investigated. Manton (1965) suggested that the secretion of the head glands could serve to lubricate the mandibles and cover the food with salivary juice.

The poison claws of *Paracryptops spinosus* Jangi & Dass (Fig. 125) are so small and set so widely apart that it is difficult to visualise how they could be used as prehensile organs. The last pair of legs of *Plutonium zweirleinii* Cavanna (and *Theatops* spp.) form stout forceps which may be used for catching food in crevices (Manton, 1965).

Digestion

The American species *Scolopendra heros* resembles insects in that it lacks pepsin-like proteases. The *Scolopendra* protease splits casein and bovine serum albumen but not haemoglobin. The enzyme resembles trypsin in its action but has two optima, one at pH 8.0, the second at pH 5.8. The protease also splits aminopolypeptidase substrate and resembles aminopolypeptidase in its pH optimum (pH 8.0) (Rajulu, 1967*a*). The pH of the fore-gut, anterior and posterior parts of the mid-gut, and of the hind-gut of *S. heros* are pH 6.5–6.9, 4.7, 7.8 and 7.6 respectively. An amylase resembling bacterial α-amylase is present in the 'salivary secretion'. The mid-gut also contains an amylase though it is weak and not similar to α-amylase. Sucrase and maltase are present in both salivary and mid-gut secretions. The mid-gut contains, in addition, a chitinase (Rajulu, 1970*d*).

Lithobiomorpha

Food

Newport (1844) concluded that the Lithobiidae were carnivorous, feeding on soft-bodied larvae, small insects and 'onisci' but stated that 'Mr. Westwood and some other naturalists believe that they feed partly on vegetable matter'. Plateau found that *Lithobius forficatus* refused young woodlice (*Porcellio scaber*), small millipedes (*Blaniulus guttulatus*), beetles (*Notiophilus* and small staphylinids) and small spiders but accepted house flies. He concluded that rather than feed on animals that were common in their cryptozoic habitats during the day, they actively hunted flies at night.

Sinclair (1895) reported that *L. forficatus* fed on worms and bluebottles, Jackson (1914) observed the species to carry off slugs at night and Britten (1920), whilst sugaring for Lepidoptera, saw it feed on small geometers and large noctuids. Brade-Birks (1929) dissected several *L. forficatus* and found arthropod remains and annelid setae; he also found a specimen carrying a small earthworm and later recovered setae from its gut. Bristowe (1941) said that 'in captivity centipedes (*Lithobius*) will greedily attack both eggs and small spiders. Small spiders are undoubtedly killed in large numbers'. Cloudsley-Thompson (1953, 1958*b*) saw *L. forficatus* feeding on woodlice in the field. Cole (1946) reported that it fed on Collembola, Auerbach (1951) concluded that lithobiomorphs probably ate any small soft-bodied form which they could catch. Simon (1960) found that adult *L. forficatus* accepted *Musca domestica*, *Sarcophaga* sp., lycosid and salticid spiders, *Lumbricus* and various Collembola in the laboratory. The Collembola were disregarded if larger prey was available. Collembola under 4 mm were not taken. Juvenile *L. forficatus* (stadia agenitalis and immaturus) did not take prey over 5 mm long. Simon regarded them as specialised Collembola feeders. *Tomocerus* and *Entomobrya* are preferred to *Orchesella* and *Sminthurus*. Mayer (1957) observed *Lithobius* feeding on Collembola in the field on four occasions. *L. forficatus* may be an important predator of apple maggot pupae (*Rhagoletis pomonella*) in Canadian orchards (Monteith, 1976*a*). *L. forficatus* fed on carabid beetle larvae and *Necrophloeophagus lorgicornis* in the laboratory (Weil, 1958).

Remains of Collembola were found in five of a sample of 12

Lamyctes fulvicornis Meinert (Sunderland, 1975). Murakami (1958*b*) fed *Bothropolys asperatus* on Collembola, *Drosophila*, small moths and butterflies.

Sutton (1970) investigated predation on woodlice by using a precipitin test. He found that three out of seven *Lithobius variegatus* taken in grassland gave positive results but none of the nine taken in woodland did. He pointed out that the precipitin test does not distinguish between living and dead bodies and for this reason the importance of lithobiids as predators cannot be evaluated since laboratory tests indicate that they are carrion feeders to some extent. He suggested, however, that they may well take an appreciable number of live woodlice.

Centipedes are the main predators of the symphylan *Scutigerella immaculata* (Waterhouse, 1969). *Lamyctes* sp. seizes the symphylan between the second and fifth segments, injecting poison. After the victim is motionless the centipede tears the head off and consumes it before feeding on the trunk. Adult centipedes consume 5–14 symphyla in 8 h. Individuals that had devoured nine or more symphylans became distended and died in 24–30 h. The number of centipedes in the field is too low to give adequate control of these pests of seedlings.

In an extensive ecological study Roberts (1956) examined the gut contents of five species of *Lithobius* from an English woodland (Tables 2, 3). Entomobryomorph Collembola (*Folsomia quadrioculata, Isotoma viridis, Podura minor, Orchesella cincta* and *Lepidocyrtus cyaneus*) are important in the diet of *L. variegatus*. The young stadia fed on small Collembola, immaturus and maturus stadia on larger species. During March and April approximately 70 per cent of the epimorphic and maturus stadia were filled with Collembola remains, probably *Podura minor*. Spiders, probably Linyphiidae, were important in epimorphic *L. variegatus*; other food items were parasitiform mites, including *Pergamasus crassipes*, Opiliones, Mollusca (slugs) and Oligochaeta. In the laboratory, *L. variegatus* accepted only nematodes, enchytraeids and Collembola, refusing onchiurid Collembola, adult beetles, larval carabids, diplopods and geophilid centipedes. In the field, *L. variegatus* was observed feeding on woodlice and opilionids.

The food of other species is shown in Table 3. The diet of *L.*

Table 2. *Percentage occurrence of the more important food items in* Lithobius variegatus (*from Roberts*, 1956)

Stadium	Foetus	Larva I	Larva II	Larva III	Larva IV	Agenetalis	Immaturus	Praematurus	Pseudomaturus	Maturus	Maturus senior
Sample size	146	244	205	119	98	24	156	142	89	598	76
Entomobryomorpha	—	61	54	33	38	65	67	51	66	54	47
Araneides	—	—	—	—	—	4	28	34	31	40	16
Opiliones	—	—	—	—	—	—	—	21	9	16	—
Acarina	—	4	5	13	6	12	9	17	27	20	18
Mollusca	—	—	—	—	—	—	4	11	3	7	—
Enchytraeidae	—	—	11	26	32	9	—	2	1	1	3
Lumbricidae	—	—	—	—	—	—	—	16	11	3	—
Nematoda	—	—	—	—	—	—	—	—	—	—	—
Other animal material	—	—	16	4	—	1	9	12	6	18	4

Table 3. *Percentage occurrence of the more important food items of four species of* Lithobius *from Hampshire woodlands* (*from Roberts*, 1956)

Species	*L. duboscqui*		*L. forficatus*	*L. muticus*	*L. lapidicola*
	Anamorphic stadia	Epimorphic stadia			
Sample size	356	293	157	217	132
Entomobryomorpha	61	37	43	—	11
Araneides	—	—	26	—	4
Opiliones	—	—	8	—	—
Acarina	9	22	33	—	52
Mollusca	—	—	—	—	—
Enchytraeidae	19	54	10	—	6
Lumbricidae	—	—	4	—	—
Nematoda	—	—	—	63	—
Other animal material	17	13	7	—	10

forficatus was similar to that of *L. variegatus*. Collembola were particularly important to the anamorphic *L. duboscqui* Brölemann and oribatid and parasitiform mites were important in the diet of *L. lapidicola* Meinert. *L. muticus* C. L. Koch contained large quantities of dead plant material but no obvious animal remains except shrivelled skins that seemed to be the exoskeletons of nematodes. When offered a wide variety of potential prey in the laboratory this species only accepted nematodes, Enchytraeidae and Isotomidae.

Roberts made the general comment that 'in the gut contents examined there were large quantities of plant material such as fragments of dead leaves, rootlets, fungal hyphae and spores, and bryophyte leaves and spores'. He presumed that this material was taken in with the prey. This he tested by offering *L. variegatus* and *L. muticus* dead leaves and fungi: the results of these tests were negative and he concluded that these centipedes were neither herbivorous nor saprophagous.

Lewis (1965) described the seasonal variation in the percentage composition of the gut contents of populations of *L. variegatus* and *L. forficatus* in a Yorkshire woodland with respect to arthropod remains and 'litter' which was similar in composition to Robert's 'plant remains' with the exception that it sometimes included mineral particles. In *L. variegatus* arthropod remains increased to a peak of about 70 per cent in midsummer, falling to about 10 per cent in the winter months. In December, January and February more animals contained litter than contained arthropod remains. Remains of Collembola, aphids, Diptera, spiders, mites, centipedes, molluscs and nematodes were found in the guts. No oligochaet setae were found and Lewis suggested that this reflected the paucity of worms in the study area. In *L. forficatus* a large percentage, 40 per cent or more, of individuals contained litter throughout the year and arthropod remains were rare in December and January. The animal food was similar in composition to that found in *L. variegatus.*

These results are at odds with Roberts' suggestion that the plant material in the guts of *Lithobius* spp. was ingested with the food.

L. variegatus is active on tree trunks at night. In summer and autumn they are present throughout the night but numbers are greatest before midnight. In winter they are present only during the early part of the night (Roberts, 1956). *L. forficatus* also ascends tree trunks at night (Britten, 1920; Cloudsley-Thompson, personal communication).

Nielsen (1962) showed that *L. forficatus* was able to utilise the sugar trehalose and suggested that this would make a diet of fungal spores rewarding since this sugar forms an important fraction of the spores: both Lewis and Roberts showed that fungal spores were present in lithobiid guts.

Several workers have commented on cannibalism in lithobiomorphs: Plateau (1878) found that *L. forficatus* collected from the field had eaten small individuals of the same species and Verhoeff (1902–25) found fragments of a lithobiid in sections of *Polybothrus fasciatus*. Cloudsley-Thompson (1945) reported that cannibalism took place in the laboratory and Simon (1960) observed a *Lithobius* attacking another in the field. Eason (1964) suggested that the control of density of centipede populations is largely controlled by physical factors and cannibalism.

Roberts (1956) carried out a series of experiments on the minimal food required for the development of *L. variegatus*. Larva III were fed *Folsomia quadrioculata* and the epimorph stadia *Podura minor*. Specimens given no food moulted once only but the majority of specimens that received one collembolan after each moult completed their development.

The feeding mechanism

L. forficatus captures prey with its poison claws, which hold it, whilst the first maxillae make rapid movements pushing the

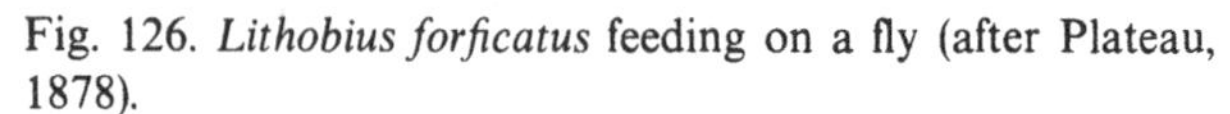
Fig. 126. *Lithobius forficatus* feeding on a fly (after Plateau, 1878).

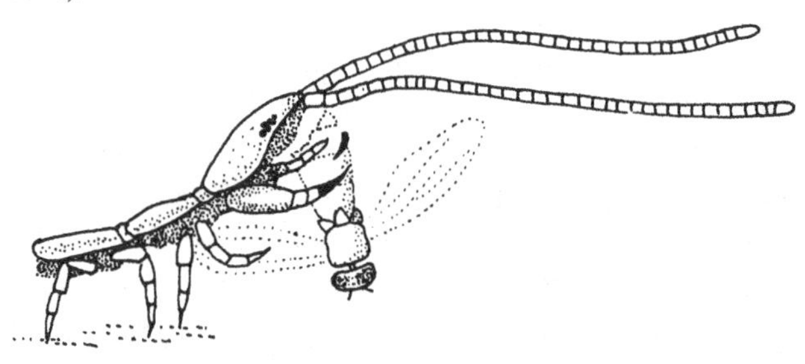

food towards the mandibles which cut it up. The food is pushed into the mouth by the mandibles and probably the first and second maxillae. When eating a fly, the head and first three segments of the body are raised off the ground at an angle of 45° (Fig. 126). The abdomen of the fly is eaten but the thorax merely 'hollowed out' (Plateau, 1878).

Rilling (1968) suggested that in *L. forficatus* the hypopharynx was drawn between the mouthparts in a tongue-like fashion. He pointed out that the numerous pinnate hairs of the hypopharynx, labrum, mandible, first maxilla and last segment of the telopodite of the second maxilla are especially suited to capillary assimilation of liquid food. The backwardly directed hairs of the epipharynx must prevent food particles from slipping forward, away from the mouth.

The stout poison claws are adapted for strong dealings with prey in the confined space below the mouthparts under crevice conditions. The coxopleural teeth (coxal endites) form a broad cusped tray on which the food is held when the claws are depressed, and against which the prey is gripped. This tray is much deeper than that of the scolopendromorph *Cormocephalus* (Manton, 1965).

Simon (1960) regarded *Lithobius* as a waiting animal, *Wartetier*, rather than as a hunter. The key stimulus is tactile or chemical. When potential prey runs into the centipede, catching behaviour is released and the prey is pursued and killed.

In *Paitobius zinus* (Chamberlin) from the mountains of SW Virginia the forcipules of males are enlarged and elongated (Fig. 127), somewhat resembling those of *Scutigera*, whereas those of females are normal (Crabill, 1960*b*). These differences may reflect differences in diet: it would be of interest to have more data on the biology of this species.

Scutigeromorpha

Food

A wide variety of prey items have been recorded for *Scutigera coleoptrata*. Plateau (1878) fed his specimens on flies, Marlatt (1914) reported house-flies, cockroaches, moths and any other insect inhabitants of dwellings as the food of the species in America. Cameron (1926) fed his specimens on flies, spiders, cock-

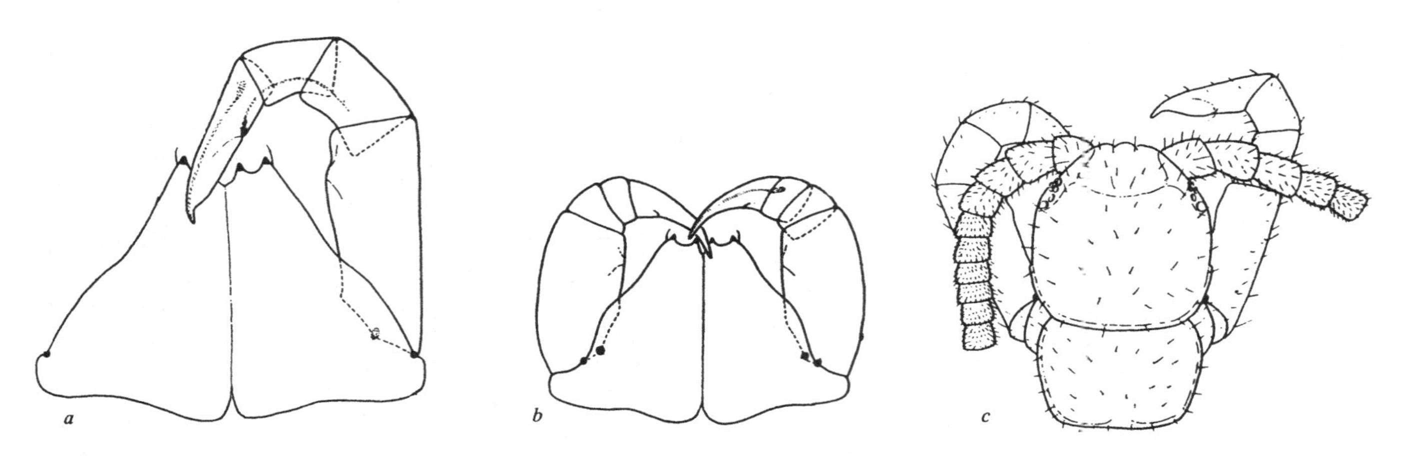

Fig. 127. *Paitobius zinus*. *a*, forcipules of male, ventral view with right telopodite omitted; *b*, forcipules of female, ventral view; *c*, dorsal view of head and first leg-bearing segment of male (from Crabill, 1960*b*).

roaches, bed bugs and banana pulp. Johnson (1952) fed *S. coleoptrata* on *Drosophila*, *Musca domestica*, *Calliphora erythrocephala*, *Blatella germanica*, *Gryllus assimilis*, adult spiders of the families Amaurobiidae, Lycosidae, Pisauridae, Salticidae and Argiopidae, a variety of unidentified noctuid moths and geometrid larvae. Pieces of freshly killed earthworms were also consumed. *Cylisticus convexus*, *Porcellio scaber*, *Melanoplus* spp. and *Phyllophaga* spp., 'formacids', elaterid larvae, staphylinids, Collembola and the larvae of the Isabella tiger-moth *Isia isabella* were rejected. Working on *Thereuonoma hilgendorfi* in Japan, Murakami (1958*a*) observed the centipede to feed on a locust nymph, a small moth, a small spider, a drone-fly, the lithobiomorph *Bothropolys* sp., the egg-sac of a spider and boiled rice. In the laboratory the species took a wide variety of insects including Collembola, silver-fish, earwigs, nymphal cockroaches, locusts and praying mantids, may-flies, damsel flies, nymphal termites and several moths and butterflies in addition to the centipedes *Bothropolys* and *Lithobius*, small spiders and sow-bugs (isopods).

Verhoeff (1938) conducted an extensive series of experiments on feeding in the laboratory of *Scutigera coleoptrata*. Only live animals were taken. House flies, *Calliphora erythrocephala*, *Bibio marci*, small *Tabanus* and spiders, *Salticus*, small lycosids, *Steatoda* and a small attidid were all killed and eaten. A woodlouse, *Armadillidium odhneri* was eaten after it had moulted. The heteropteran bug *Syromastes* was rejected probably as much for its glands as for its hard cuticle. Caterpillars of *Vanessa urticae* were likewise rejected as was the centipede *Lithobius erythrocephalus*. No attack was made on a large female house spider (*Tegenaria domestica*).

Bees and wasps were quickly bitten, left until all movement had ceased and then eaten. A small lycosid spider was prevented from escaping by blows from the centipede's anterior four to six pairs of legs; the spider was left for a short time after it had been 'overpowered' and then eaten. A *Dysdera* was attacked in the same manner but not eaten until it had discharged its silk.

Verhoeff noted that *Scutigera* appears to be able to recognise dangerous prey and modify its feeding behaviour accordingly. He recorded two cases of a male *Scutigera* being eaten by a female.

The feeding mechanism

Sinclair (1895) described how scutigerids in Malta 'come out from beneath great stones and run about rapidly on the ground or on the stones or rubbish with which the ground was covered, now and again making a dart at some small insect which tempted them, and seeming not to mind the blazing sun at all'. Most records, however, suggest that Scutigeromorphs are nocturnal. Forbes (1890) quoted Miss K. Rondeau Golcona of Illinois, 'the centipedes remained motionless for perhaps half an hour, when suddenly one, with a quick movement of one of its many legs, would catch an unwary house-fly that approached too near. Sometimes this would be eaten immediately, but sometimes held in a foot until two or three more flies were caught. I have seen a *Cermatia* eating one fly while holding two or three others'. Only the soft parts of the fly were eaten.

In France *S. coleoptrata* has been seen to catch flies (*Fannia scalaris*) attracted to light. The flies are caught with the anterior legs and bitten in the thorax to kill them. In this case too, three or four specimens were held with the front legs while another specimen was eaten. The soft parts of the body were eaten whilst the chitinous parts of the head, thorax, abdomen, legs and wings were rejected (Künckel d'Herculais, 1911). Verhoeff (1902–25) likened the legs with their much divided tarsi beset with cones and hairs to a 'snarelasso'.

Johnson (1952) reported that specimens of *S. coleoptrata* reared in the laboratory were generally found exposed on the surface of the litter during the day and night, and only when disturbed did they scurry with amazing speed underneath loose leaves and debris, leaving either the antennae or the last pair of legs exposed. They reappeared on the surface very shortly after being disturbed. *Drosophila* dropped into the jars in which the centipedes were kept were immediately 'spotted'. The centipedes caught them by half pouncing, half lassoing them: as many as five flies were captured at one time and held between the quivering, lashing appendages. Immobile while awaiting prey, the cursorial appendages maintain a constant fluttering tremor while the animal feeds. Manton (1965) noted that if a slab of rock, under which *Scutigera*, spiders and flies are hiding is moved, the *Scutigera* run and pounce on the prey immediately.

Johnson's and Manton's observations suggest that *Scutigera* detects its prey by sight but Klingel (1960*a*) maintained that it responds only to tactile stimuli and does not recognise flies unless it comes into contact with them: contact chemical stimuli are also important in eliciting feeding responses. The chemical and tactile sense organs are situated on the antennae, legs and mouthparts.

The head of scutigeromorphs is less flattened than that of any other centipedes. Manton (1965) maintained that its deep dome-shaped head capsule is not a primitive character but allows a large pre-oral cavity to accomodate the very large mandibles (entognathy) and that the feeding mechanism of *Scutigera* is the most specialised and advanced of all Chilopoda. The well-developed anterior glands are probably associated, not only with providing grooming fluids for the very long legs, but with supplying salivary juice used in lubrication and also digestion. A spider is held by the poison claws, the coxal spines and by the palps of the second maxillae. The 'paws' formed by the distal ends of the outer lobes of the first maxillae (Fig. 36) hold a leg or other part of the spider by apposition. The strong mandibles protrude momentarily far beyond the labrum as they abduct and then rapidly adduct and retract, biting off small pieces which are rapidly swallowed. The abdomen of the spider is eaten first, then the legs are detached and eaten one by one; the unpalatable tarsus is discarded. The whole body of the spider is consumed in this way, including the hard chelicerae: grooming follows.

This account appears to be at odds with those of other authors cited above who have found that only the soft parts of arthropods are eaten. The poison claws of *Scutigera* are not used for mastication as they are in Scolopendromorpha and *Craterostigmus* neither are they used for opening prey as they are in Geophilomorpha. Unlike *Lithobius*, *Scutigera* is unable to hunt or feed in shallow crevices but the leg-like mobility of the poison claws together with other specialisations allows it to utilise prey that are unavailable to other centipedes of comparable size.

A pair of blind vascular caeca open from the anterior aorta just behind the pharynx (Fig. 141*a*). It is possible that blood may be squeezed into these caeca on sudden expansion of the pharyngeal pouches. The caeca are backwardly directed out of the head and such a rearrangement of fluids might assist ingestion of fluid without disorganising the whole circulatory system: it is probably that the pharyngeal pouches act in a suctorial capacity (Manton, 1965).

Craterostigmus

Manton (1965) reported that termites are eaten with avidity by *Craterostigmus* and are expertly dug out of crevices in wood with the poison claws. In terraria fly maggots were accepted reluctantly but other foods on which most centipedes thrive were refused, such as adult flies immobilised by mutilation, spiders, small woodlice, small gastropods, beetle larvae, earthworm pieces and vertebrate liver. *Craterostigmus* possesses a long head capsule and large poison claws but the head–poison claw depth is not great and the animal, unlike the scolopendromorphs investigated by Manton, can pass through a shallow slot which will just allow the head to pass. This means that the animal can penetrate crevices in which there is no room for feeding. Termites and probably other insects, are pulled out and then consumed in more commodious spaces.

The poison claws can be projected well in front of the head, thus enabling this centipede to seize prey from in front of the head in a space so shallow that poison claws operating below the head only would be useless. The very flexible poison claws are relatively larger and longer than in all other centipedes and are used in prolonged chewing between the cusped endite at the base of the trochanter-prefemur and the coxal endite on either side. Copious secretions are poured out onto the prey. Strong licking movements are made by the hypopharynx and strong suction draws in the semi-digested contents of the prey. The mandibles are degenerate.

11

The respiratory system

The internal organs of all centipedes except the Scutigermorpha are supplied with oxygen by tracheae, spirally thickened chitinous tubules of ectodermal origin, which originate from laterally placed openings, the spiracles. In the Scutigeromorpha the spiracles are situated dorsally on the tergites and open into 'tracheal lungs'. Manton (1965) suggested that primitively each lateral spiracle may have had branching tracheae supplying its own segment, the head deriving its tracheal supply from the anterior pair of trunk spiracles. In addition a third, essentially pericardial respiratory system may also have been present, possessing a mid-dorsal spiracle from which tracheae extended into the pericardium dorsal and lateral to the heart and from which the scutigeromorph system evolved.

The terminology relating to the structure of the lateral spiracles has become very confused. The word spiracle (*Stigma* of German authors) will here be used for external openings of the tracheal system. The spiracle is often surrounded by a sclerotised rim or peritrema (*Stigmaring*) and leads into the spiracle cup or atrium whose wall is usually sculptured into trichomes, otherwise termed tubercles, pillars or cuticular lappets (Fig. 128). The tracheae, characterised by spiral thickenings (taenidia) may open directly into the atrium or, as in many geophilomorphs, into an inner atrial or substigmatic pocket by a slit (*Stigmamund*). In Scolopendridae the atrium is subdivided horizontally by flaps (valves) or a diaphragm.

Geophilomorpha

Spiracles

The spiracles of the Geophilomorpha are borne laterally on the stigmatopleurites of all leg-bearing segments except the first and last. The spiracular openings are round or oval in shape. In

Haplophilus subterraneus the anterior spiracles are about 30–50 μm in diameter, the posterior ones 20–30 μm (Curry, 1974). The spiracles of *Orya barbarica* are situated in grooves (Verhoeff, 1941). In *Himantanium gabrielis* (Linn.) and *Geophilus electricus* the atrium leads into a sub-atrial chamber. Both cavities are lined

Fig. 128. Transverse sections of the spiracles of geophilomorphs. *a*, *Strigamia acuminata*; *b*, *Necrophloeophagus longicornis* (after Füller, 1960).

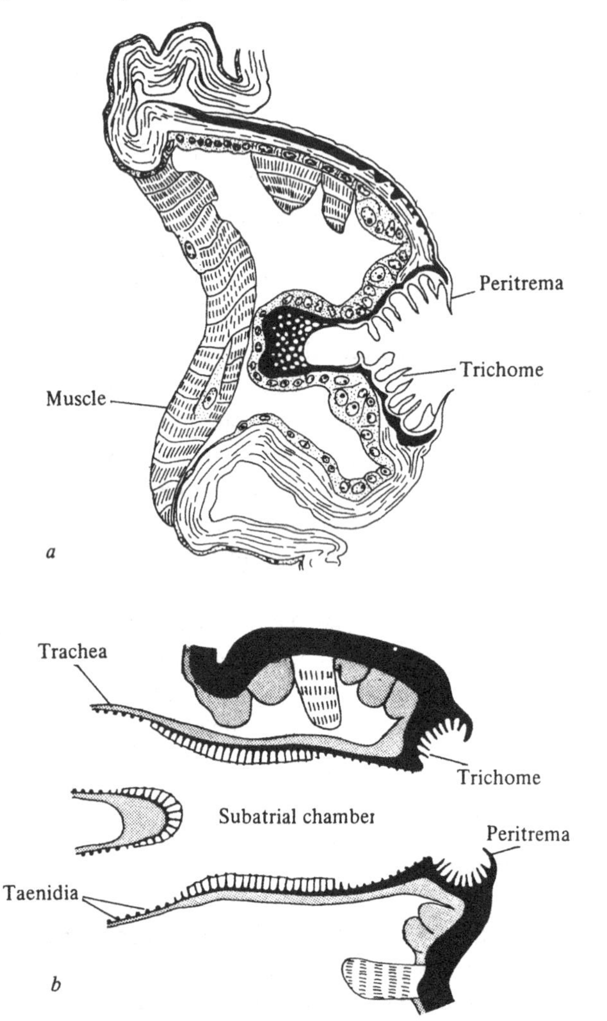

by trichomes (Haase, 1884*a*; Chalande, 1885). The tracheae of *Strigamia acuminata* open directly into the atrium (Fig. 128*a*) but in *Necrophloeophagus longicornis* a *Stigmamund* opens into a narrowed region with fine chitinous striations which itself leads to a sub-atrial chamber (Fig. 128*b*) whose narrow trichomes appear to be fused at their tips (Füller, 1960).

Manton (1965) in a discussion of the trichomes of Epimorpha based on *G. carpophagus* and the scolopendromorph *Cormocephalus calcaratus* stated that they are spaced at intervals of 0.0006 mm or less; solid and expanded distally they show a network of sclerotisations (Fig. 129*c*). She implied that the trichomes of all chilopods are similar although Füller (1960) had shown that *Strigamia* possessed flap-like trichomes (Fig. 128*a*) and that *N. longicornis* possessed two distinct types. In *Geophilus insculptus* Attems the trichomes are mainly cone-shaped and in *H. subterraneus* they are elongated plates (Curry, 1974).

Lewis (1963) drew attention to the differences in the structure of the spiracular atrium in four species of geophilomorph. In *H. subterraneus* there is a wide opening to the exterior and the trichomes are small and widely spaced. In *N. longicornis* the opening is smaller and the trichomes larger and more numerous and in *Strigamia maritima* the spiracular opening is even smaller and the trichomes larger and spatulate. In *Brachygeophilus truncorum* (Bergsö & Meinert) the spiracular opening is small but the trichomes are not as numerous as they are in *S. maritima*. The author suggested that a narrow spiracular opening and dense layer of trichomes might retard evaporation from the tracheae and showed that *H. subterraneus* had the lowest resistance to desiccation and *B. truncorum* and *S. maritima* the greatest but pointed out that there were also differences in the permeability of the cuticle and that water loss from the mouth might be important.

The walls of the atrium can be flattened thereby closing the spiracle: in *Geophilus electricus* the spiracles normally remain open but sometimes the subatrial cavity is obstructed by particles of solid material: the 'contractility' of the walls allows the animal to get rid of these. When a geophilid comes out of loose soil it can be seen to expel powdery material from the spiracles (Chalande, 1885). Manton (1965) observed that on longitudinal shortening of the

Fig. 129. Sections through Geophilomorpha and Scolopendromorpha to show features of the respiratory system. *a*, vertical longitudinal section through the dorsal intersegmental region of *Geophilus carpophagus* passing through the median dorsal atrium; *b*, parasagittal section through the deeper part of the lateral atrium of *G. carpophagus*; *c*, vertical section of the atrial cuticle of *Cormocephalus calcaratus* at the internal end of the atrium; *d*, the same, close to the spiracle; *e*, horizontal section of the atrial cuticle of *C. calcaratus* (from Manton, 1965).

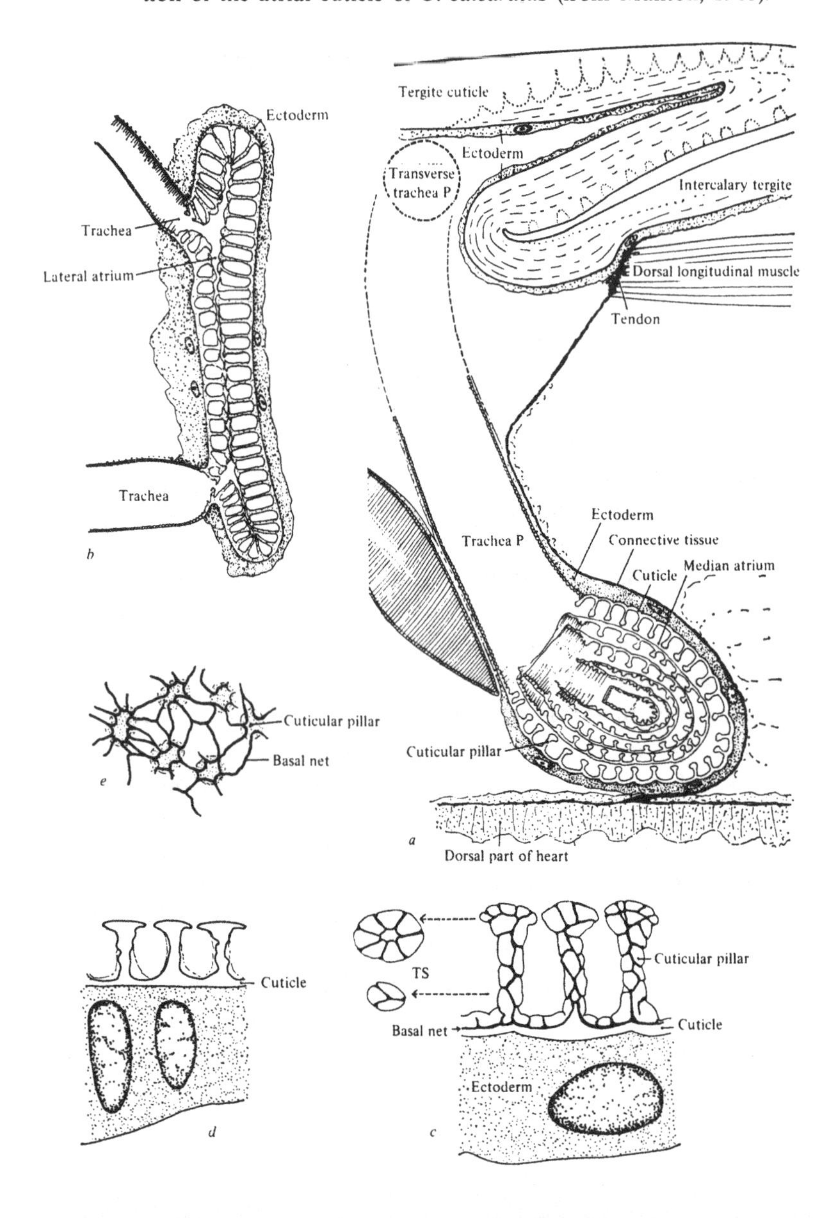

body the antero-posterior flattening of the atrium is virtual. The air space is then restricted to the air spaces between the trichomes. She suggested that these air spaces were admirably suited to serve respiratory needs under wet conditions, the cuticular pillars being practically unwettable. She was incorrect in her statement that 'the size and frequency of these columns is much the same in chilopods of different body sizes' which she suggested was related to physical properties such as surface tension of water and wettability of the cuticle. Nevertheless they are reminiscent of the more elaborate plastron respiratory structures in the eggs and pupae of many insects.

Curry (1974) concluded that the spiracular trichomes may reduce water loss but thought it more likely that they acted as a filter as Kaufman (1962) had suggested was the case in *Lithobius*. They also prevent the spiracle closing completely during muscular exertion and possibly act as a physical gill under wet conditions.

Tracheal system

Haase (1884*a*) described the arrangement of the tracheal system in *Geophilus proximus* C. L. Koch, *G. electricus, Pachymerium ferrugineum, Clinopodes linearis, Schendyla nemorensis, Strigamia maritima* and *Himantarium gabrielis.* Chalande (1885) gave a clear acount of the tracheal system in *G. electricus* and *H. gabrielis* and Demange (1942) gave simple diagrams showing the arrangement in several geophilomorph genera. The most recent discussion of the subject is Manton's (1965): she figured the tracheal system of *Orya* and *Haplophilus* (Fig. 130).

In the simplest arrangement seen in the geophilomorphs each spiracle gives rise to a number of branching tracheae (ventral plexus of Chalande) which supply the tissues and two pericardial tracheae, 'dorsal plexus', which are large, cylindrical and constant in diameter, having few side branches. The posterior pericardial trachea (Manton's trachea P) passes upwards and transversely or obliquely towards the mid-line. At the mid-line it descends deeply to join a median atrium (commissure of Haase and Demange). From the median atrium one anterior pericardial trachea (A) passes forwards on each side to join the posterior side of the spiracular atrium of the preceding segment (Fig. 130*b*). This arrangement obtains in many geophilomorphs and in a few posterior segments

Fig. 130. The tracheal system of geophilomorphs. *a*, segments 7–11 of *Orya barbarica* showing tracheae *A* and *P*, the finer tracheae are not shown; *b*, the tracheae of the mid-body region of *Haplophilus subterraneus* (from Manton, 1965).

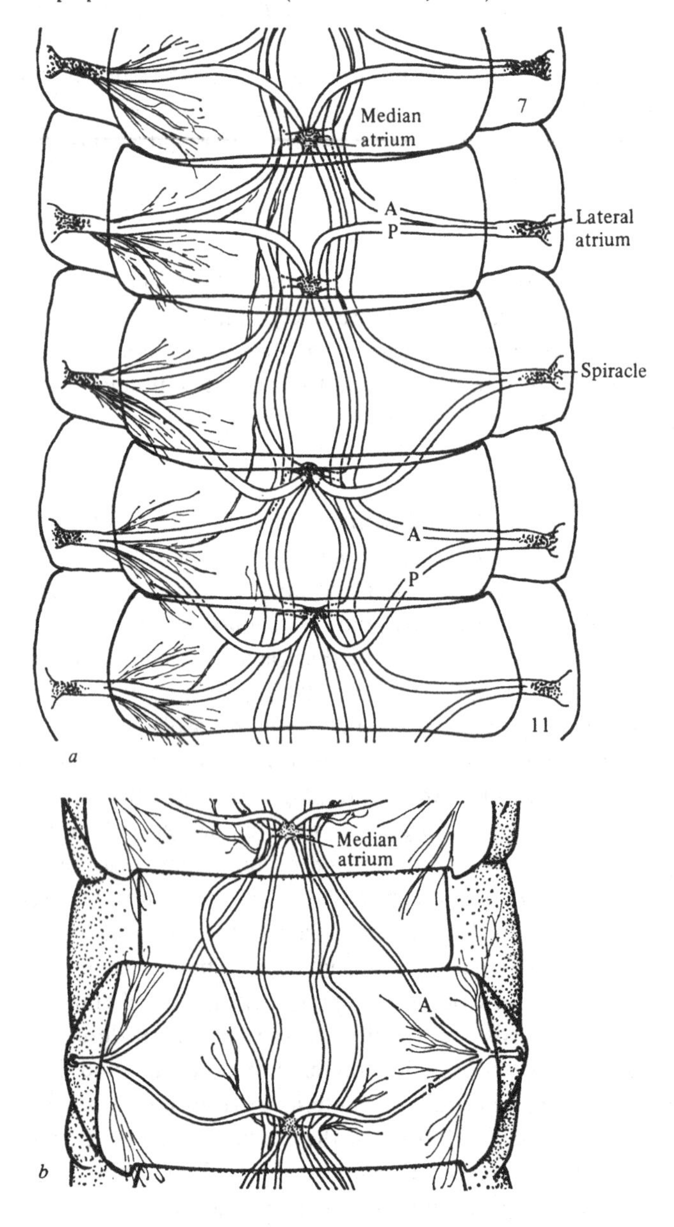

of the long bodied *Himantarium gabrielis, Orya barbarica* and *Haplophilus subterraneus*. In most segments of these latter species, however, the arrangement is more complicated, the anterior pericardial tracheae sending a branch to the spiracle of the preceding segment and then continuing forward to end in the median atrium of the second or third segment in front (Fig. 130*a*).

The median atrium (Fig. 129*a*) is unlike the tracheae in structure: the ectodermal layer is thick in contrast to the thin covering of the tracheae and the lining is raised into trichomes like those of the lateral atria but less expanded distally (Manton, 1965). Chalande (1885) observed that the wall of the median atrium of *G. electricus* resembled that of the spiracular subatrial cavity. At ecdysis the cuticular lining of the median atrium cannot be pulled out of the body and remains free in the lumen. Kaufmann (1960) suggested that the apparently complex internal organisation acted as a filter; it consists, in fact, of concentric layers of moulted cuticles. Chalande (1886) recognised the importance of the median atrium in the respiratory process (see below): his observations were repeated by Dubuisson (1928) on *Geophilus carpophagus*.

Manton saw the present arrangement of tracheae in geophilomorphs as having arisen to facilitate the burrowing habit from the supposed primitive arrangement with dorsal and lateral atria. Firstly the pericardial tracheae developed to link the atria and then the median spiracular opening was perhaps lost thus enabling the whole tergal surface to exert a strong thrust against the soil.

The Geophilomorpha is the only centipede order with long regular pipe-like tracheae and there is sufficient slack in these to allow for changes in the length of the animal during locomotion. The longitudinal and transverse linking of lateral atria allows gases to be drawn from a distance to supply any atrium momentarily blocked at the spiracle: it is from the lateral atria that gases are supplied to the tissues.

The muscles of the anterior third of the geophilomorph do more work in burrowing than do the more posterior ones. The oesophagus is narrow and the musculature more bulky in the anterior segments and, Manton suggests, gaseous exchange must go on here at a higher level than elsewhere. These needs are met by the larger size of the spiracles, the median atria and the pericardial tracheae.

The usually greater shape changes in this region are reflected in the anterior pericardial trachea extending through more segments in this region in the longer species.

Scolopendromorpha

Spiracles

As in the geophilomorphs, the spiracles of scolopendromorphs are borne laterally. In most genera with 21 leg-bearing segments they occur on segments 3, 5, 8, 10, 12, 14, 16, 18 and 20 but in several genera of the scolopendrid subfamily Otostigminae and in three genera of Cryptopidae spiracles are also present on segment 7: this is normally regarded as a primitive character. Crabill (1955) reported a case of a specimen of *Scolopocryptops sexspinosa* (Say) possessing poorly developed spiracles on segment 7. Normally, they are absent. As might be expected, in species with 23 leg-bearing segments spiracles are present on segment 22.

The spiracles are thus absent from the short tergites which have been developed to prevent undue body undulation during running (see Chapter 3). A notable exception is *Plutonium zwierleinii* (family Cryptopidae: subfamily Theatopsinae) which does not have long and short tergites and has spiracles on all leg-bearing segments except the first and the last.

The Scolopendromorpha exhibit a greater variation in spiracle structure than do other centipedes. In *Scolopendra* and *Cormocephalus* the atrium is divided into a vestibule (*äusseren Stigmenkelch*) and an inner cavity by a flap or diaphragm consisting of three valves (Fig. 131*a*, *b*). The spiracle opening is often triangular in shape with the apex of the triangle pointing forwards. In *Scolopendra cingulata* the vestibule and inner cavity bear small capped trichomes (Fig. 131*c*) and are separated by a ring of about 50 large, thick spines (Haase, 1884*a*). The trichomes are also present in *Cormocephalus* (Fig. 129*c*, *d*). In *Scolopendra* the openings of the tracheae are protected by long radiating spine-like hairs (Fig. 131*c*). These spine-like hairs are probably the same as the *panaches* or recumbent plumes described by Chalande (1885). When erect they block the entrance to the tracheae so preventing the entry of foreign bodies. When the spiracle is closed, these plumes project through to the outside. The cavity into which the tracheae

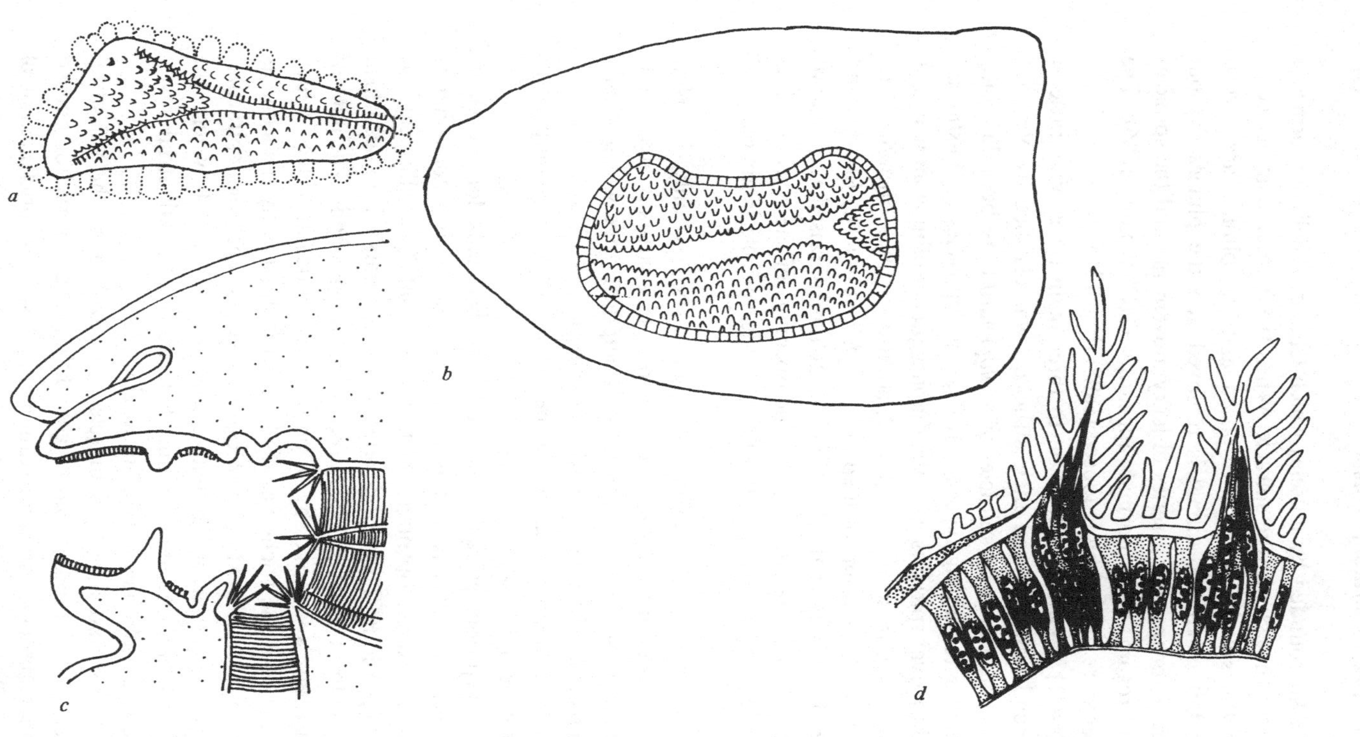

Fig. 131. Spiracle structure in Scolopendromorpha. *a*, the spiracle of *Cormocephalus hirtipes* (Ribaut); *b*, the spiracle of *Cormocephalus macrosestrus* Attems; *c*, longitudinal section of the spiracle of *Scolopendra cingulata*; *d*, plumes from the spiracle of *S. cingulata* (*a* and *b* after Attems, 1929; *c* after Haase, 1884*a*, *d* after Duboscq, 1898).

open is surrounded by muscles which Chalande believed formed a sphincter that closed the spiracle. When the animal is immersed in water the spiracular floor is raised and the plumes are turned down but when the animal is removed and the plumes dry out they rise abruptly by a series of jerky movements and the spiracles slowly regain their normal appearance (Chalande, 1886). The plumes were figured by Duboscq (1898) (Fig. 131*d*).

Haase (1884*a*) drew attention to three main types of spiracle in scolopendromorphs and their structure was clarified by Verhoeff (1941). The triangular spiracles of *Scolopendra* have been described above. The remaining types are the sac-like spiracles of *Otostigmus*: round or oval spiracles with the diaphragm containing an irregular opening and, in segments 1 and 2, additional small openings (Fig. 132*a*, *b*) and the cribriform spiracles of *Ethmostigmus*, *Rhysida* and *Alipes*. In this type the floor of the atrium is raised into humps between which are situated the fine openings of the tracheae (Fig. 132*c*, *d*). In *Ethmostigmus* the floor of the spiracle is almost level with the pleura.

In the genus *Campylostigmus* which occurs only in New Caledonia the outer vestibule of the spiracle is virtually absent and the stigmatopleurite is raised into an elliptical ridge which is only incomplete in the region above the spiracle (Fig. 133).

The functional significance of the different types of scolopendrid spiracles is unknown.

The cryptopid spiracle has been described only for the genus *Cryptops*. In *C. hortensis* the spiracle is borne on a cone-shaped hump. The outer opening which lies horizontally is elliptical and 70–120 μm long. The well-developed atrium bears trichomes which have the form of several sided rods with flattened ends. The floor of the atrium is covered by a network of cuticular ridges which continue into the trichomes (Füller, 1960) apparently the same as those described by Manton in *Cormocephalus*. The inner spiracular opening is a narrow cresent-shaped slit (Fig. 134) where the trichomes are very much reduced. It leads to a sub-atrial cavity where the trichomes are again developed.

A dorsal and a ventral muscle are attached to the sub-atrial pocket. Füller (1960, 1963*a*) suggested that their contraction opened the inner spiracular opening. Kaufmann (1964) made brief mention of two muscle bundles whose shortening opens the spiracular cavity.

Fig. 132. Spiracle structure in Scolopendromorpha. *a*, spiracle of *Otostigmus spinosus* Porat; *b*, diagrammatic cross section of the spiracle of *Cormocephalus*; *c*, spiracle of *Alipes multicostis*; *d*, diagrammatic cross section of the spiracle of *Ethmostigmus* (after Verhoeff, 1941).

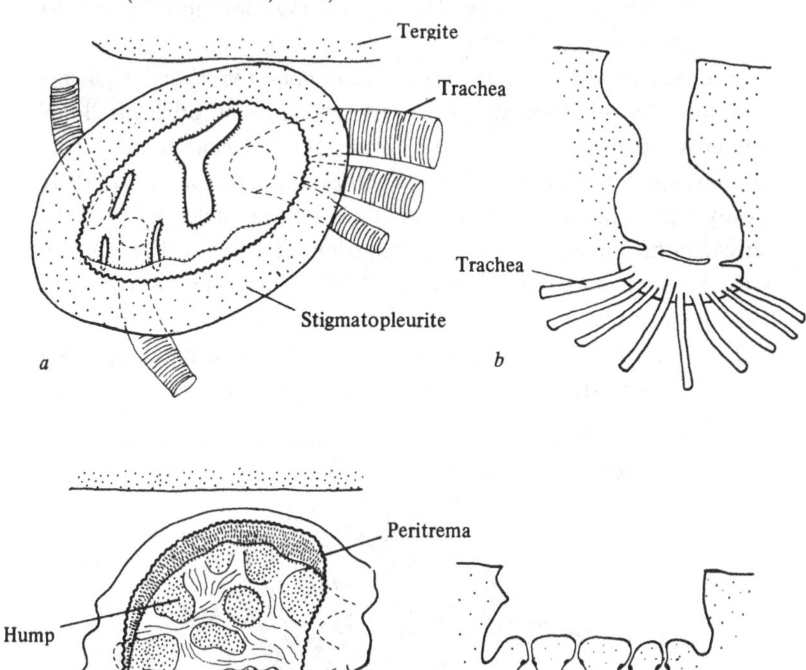

Fig. 133. Spiracle of *Campylostigmus biseriatus* Ribaut (after Attems, 1929).

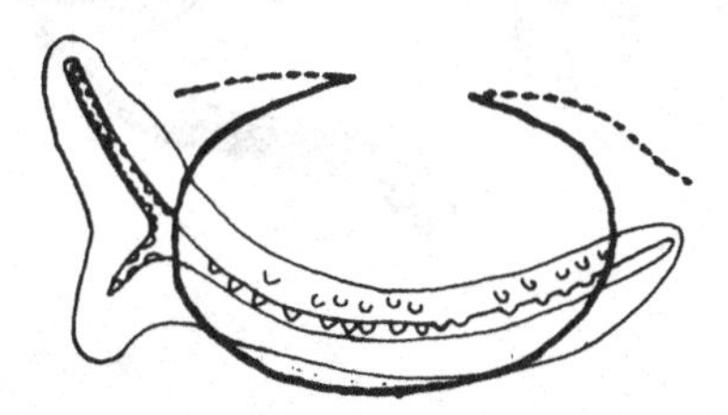

Tracheal system

Chalande (1886) described the tracheal systems of *Scolopendra* and *Cryptops* as forming a continuous plexus of anastomosing branches forming a web throughout the individual without order or symmetry. Haase's (1884*a*) descriptions suggest more regularity (Fig. 135).

In all scolopendromorphs except *Plutonium* the 'short' segments lack spiracles and are supplied with tracheae by adjacent 'long' segments. In many scolopendromorphs the tracheae arising from different spiracles are linked. In *Cryptops* and *Cormocephalus* there is only one transverse connective between the spiracles: it lies in the pericardium dorsal to the dorsal longitudinal muscles (Fig. 136*a*) it is in the same position as trachea P in the Geophilomorpha but there is no trace of a median atrium. The middle of the trachea is slightly narrower than elsewhere perhaps forming a breaking point at ecdysis. The transverse connective probably represents the pair

Fig. 134. The spiracle of *Cryptops hortensis*. *a*, surface view; *b*, transverse section (after Füller, 1960).

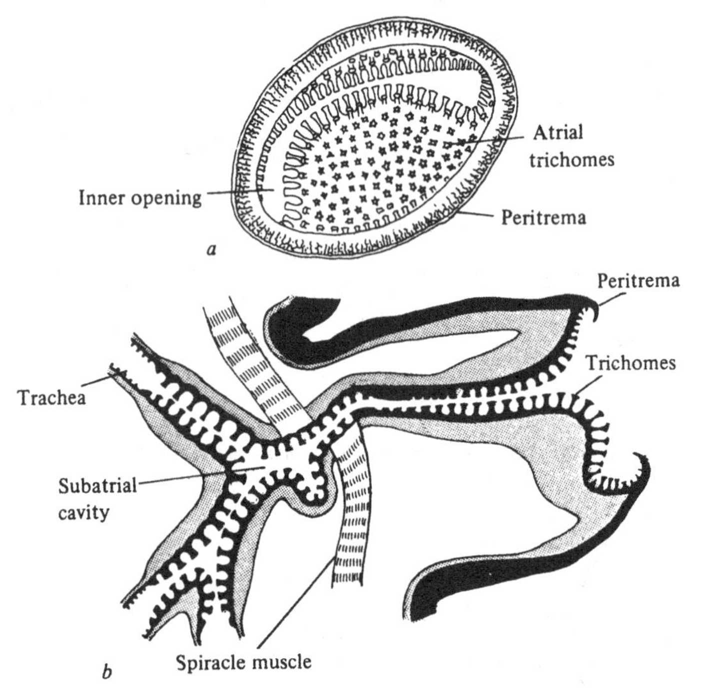

of tracheae P fused in the mid-line with disappearance of the median atrium.

With increase in size there is an increase in the number and relative dimensions of the tracheae. *Cryptops anomalans*, with a body width of 1.5 mm, possesses two tracheal trunks which pass forward to the head from each of the spiracles in segment 3. In *S. cingulata* with a body width of 4.5 mm there are ten such tracheae on each side. Larger species have many more branching tracheae arising from each spiracle than smaller species.

In many genera, but not *Cryptops* or *Cormocephalus*, longitudinal tracheae link successive spiracles. These tracheae are sinus-like, irregular in layout and diameter. The links between them are often very narrow (Fig. 136*b*, *c*).

Fig. 135. Tracheal system of segments 14–18 of *Scolopendra cingulata* (after Haase, 1884*a*).

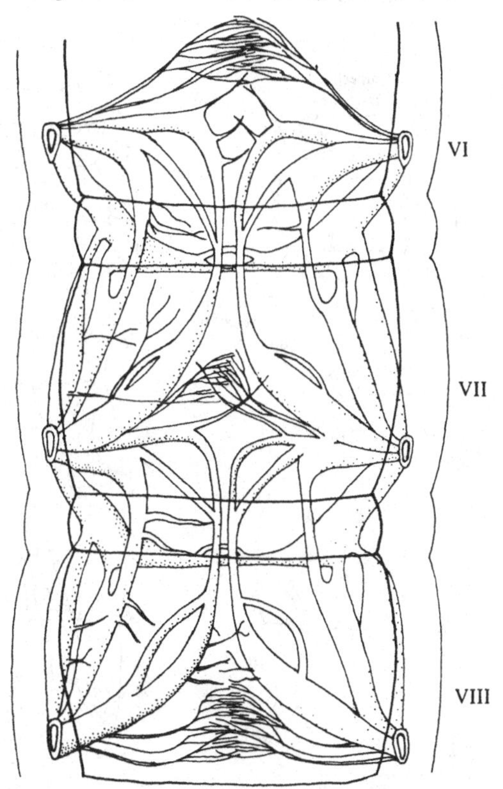

Each spiracle in *S. cingulata* usually gives rise to two sinus-like tracheae that pass backwards and one that passes forwards into the pericardium. In position they resemble trachea A of the Geophilomorpha and, Manton suggests, they appear to have grown forwards and linked up unilaterally with the spiracles. Branching tracheae corresponding in position with the geophilomorph tracheae A are present in *Cryptops* and *Cormocephalus*. In *Scolopendra* there is also a perivisceral tracheal system formed of sinus-like longitudinal connectives which give rise to a subneural

Fig. 136. The tracheal system of Scolopendromorpha. *a*, dorsal view of the anterior end of *Cryptops anomalans*; *b*, the more superficial (mainly pericardial tracheae) of *Scolopendra cingulata*; *c*, the deeper perivisceral tracheae of the same specimen (from Manton, 1965).

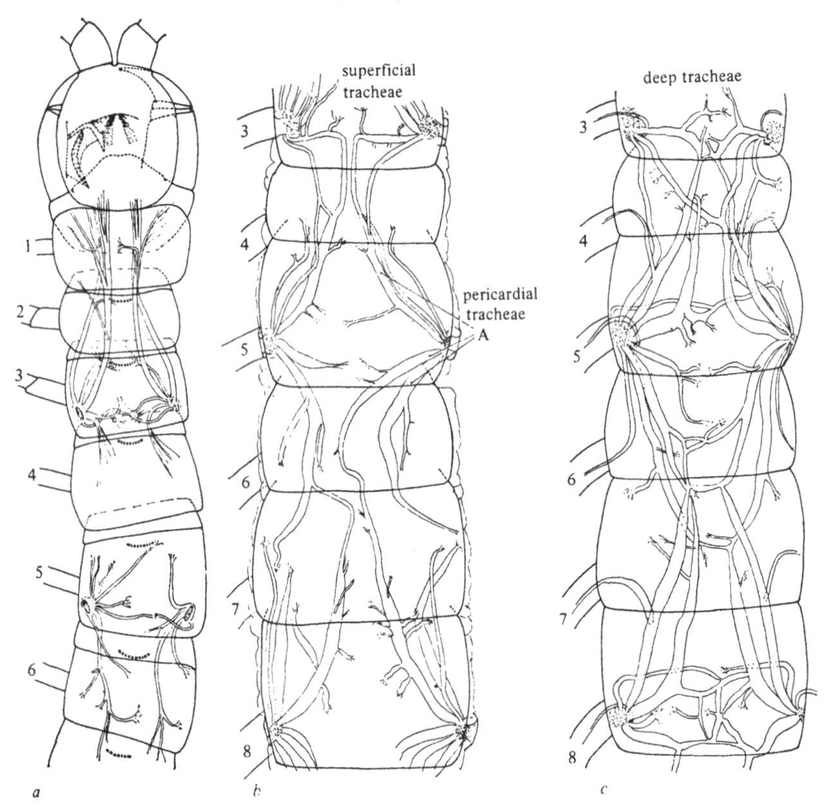

transverse network. Jangi (1966) described and figured the tracheal system of *S. morsitans*.

Manton suggested that the irregular sinus-like tracheae, elliptical in section and with narrow constrictions such as are found in *Scolopendra*, are likely to serve the needs of storage of gases under conditions of respiratory difficulty when the species has to go deep into the ground to avoid dry conditions in the surface of the soil. *Cormocephalus* which lives in well-watered forests and *Cryptops* have tube-like tracheae possibly because they do not have to store gases under adverse conditions, although it should be noted that Kaufmann (1964) stated that the tracheae of *Cryptops* have a tendency to form vesicular swellings, especially at branching points.

Haase's (1884*a*) figure of the tracheal system of *S. cingulata* which also appeared in Verhoeff (1902–25) and is reproduced here (Fig. 135) was assumed by Manton to be that of *Plutonium*. She thought that the numbers VI, VII and VIII on the figure referred to the segments which therefore each have a pair of spiracles. The numbers, in fact, refer to the spiracles and the segments shown are numbers 14–18. The figure is clearly diagrammatic and the short segments look like pretergites. The size of the tracheae may also have been exaggerated. Her conclusions concerning sinus-like tracheae in the probably subterranean *Plutonium* are thus unjustified.

Our knowledge of the structure and functioning of the scolopendromorph respiratory system is scant. This is particularly true of exotic forms.

Lithobiomorpha

Spiracles

The Lithobiomorpha bear lateral spiracles on segments 3, 5, 8, 10, 12 and 14. There are two exceptions to this pattern: *Lamyctes* has spiracles on the first leg-bearing segment and *Catanopsobius* bears spiracles on segments three and ten only (Crabill, 1955).

A clear description of the structure of the spiracle of *Lithobius forficatus* has been given by Füller (1960). Other descriptions have been given by Haase (1884*a*) Chalande (1885), Kaufmann (1960, 1961) and Curry (1974). The spiracles are borne on cone-shaped stigmatopleurites and are slit like, sloping diagonally upwards and backwards. In profile the lips of the spiracle are triangular (Fig.

137*a*). In *L. variegatus* they are curved (Curry, 1974). In a large *L. forficatus* the first spiracular opening can be 300 μm long. The first opening is twice as long as the last.

The spiracular atrium of *Lithobius* takes the form of a flattened funnel which tapers inwards and merges imperceptibly with the tracheae. The outer opening of the spiracle is bordered by large, incurved, chitinous lips (Fig. 137*b*). The regular parallel cuticular thickenings of the lips are continued into a reticulate pattern of thickening bearing trichomes in the atrium which in its turn merges into the spiral thickenings of the tracheae. Curry noted that in some sections of *Lithobius* 'muscles have been seen near to the base of the stigmatopleurite. It seems that these muscles are part of the body musculature, and have no function in closing the spiracle, but may facilitate ventilation and/or restrict the air passage.' Füller was unable to detect spiracle muscles. Kaufmann (1962) stated that there was a muscle at the base of the atrium which might by its contraction reduce its 'clear space'. The spiracle of *Lamyctes* is similar in structure to that of *Lithobius* (Haase, 1884*a*).

Tracheal system

The tracheal system of *Lithobius* is the simplest found in the chilopods for there are no transverse connections between the tracheae of each side and the tracheae do not anastomose. A group of tracheae originates at each spiracle and serves the adjacent region (Chalande, 1885). The arrangement of the tracheae is somewhat irregular. (Haase, 1884*a*; Kaufmann, 1961; Rilling, 1968). Rilling (1968) could find no spiracle-opening or sphincter muscles, but a pair of muscles inserted on the proximal wall of the stigmatopleurite could open the lips of the spiracle.

Scutigeromorpha

In scutigeromorphs the spiracles are median dorsal slits on segments 1, 3, 5, 8, 10, 12 and 14. Each slit leads to an atrium from which, in *Scutigera coleoptrata*, some 600 tracheae open. These branch and then end blindly.

In living *Scutigera* the tracheal system appears as a yellowish-

Fig. 137. The spiracle of *Lithobius forficatus*. *a*, seen in profile; *b*, longitudinal section; *c*, spiracle and tracheae of *Scutigera coleoptrata* (*a* and *b* after Füller, 1960; *c* after Haase, 1884*a*).

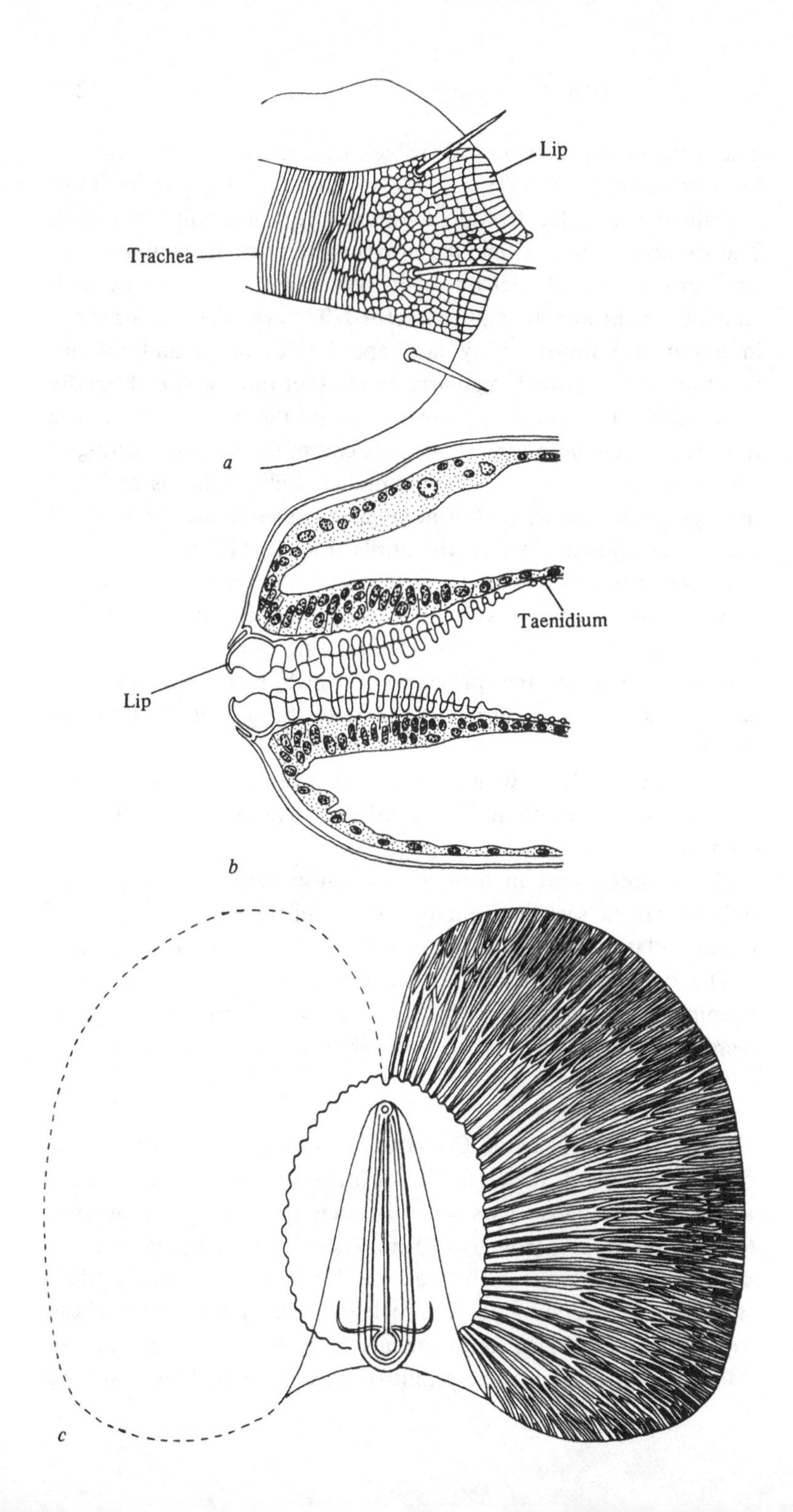
Lip
Trachea
a
Taenidium
Lip
b
c

gold kidney-shaped structure visible through the cuticle of each of the seven long tergites (Haase, 1884*a*). Each slit-like spiracle lies in a shallow triangular trough in the tergite, narrowing anteriorly. The spiracular slit is expanded into a small circular opening at its anterior end and a larger one at its posterior end (Fig. 137*c*). Both ends of the slit are strongly sclerotised. The tracheae are arranged in about five layers. They lack spiral thickenings and do not anastomose but branch regularly two to four times and end blindly in a mass of connective tissue close to the heart valves. The tracheae of *Scutigera longicornis* lack chitin, the walls consisting of a network of transverse and longitudinal collagen fibrils and acid mucopolysaccharides. A thin homogenous membrane seems to fill the square regions between the fibrils (Rajulu, 1971*c*).

Haase demonstrated that the tracheae were filled with air and blocking the spiracles with oil caused immobilisation and then death in one and a half hours. He found no evidence of a spiracle closing mechanism the presence of which had been suggested earlier. There are no muscles attached to the spiracle (Dubuisson, 1928).

In young *Scutigera* with only five spiracles the spiracular openings are oval rather than slit-shaped and have only 50 tracheae on each side.

Haase noted that in freshly autotomised legs of *Scutigera* air bubbles can be seen in the femur which inflate the 'conjunctiva' of the articulation. He was unable to ascertain the origin of the gas.

The coagulated material sometimes found in the respiratory openings of *Scutigera* and *Scolopendra* is probably a waxy secretion of adjacent glands (Haase, 1884*a*).

Craterostigmus

The tracheae of *Craterostigmus* differ from those of all other chilopods in their scarcity of branching. From each spiracle about 3000 tracheae pass inwards, their diameter predominantly 6–8 μm, but ranging from 6 to 30 μm. These sheafs of minute tracheae occupy far more space than do the large tracheal trunks which normally leave a chilopodan spiracle (Manton, 1965). The tracheae, arising from spiracles on segments 3, 5, 8, 10, 12 and 14 can, Manton argued, hardly be primitive since the reduction in spiracle number is presumably a secondary phenomenon. The only other arthropods with long almost unbranched tracheae (unless *Ethmostigmus* possesses them) are the Onychophora. A secondary evolution of numerous minute tracheae may be associated with the importance of hydrostatic movements to *Craterostigmus* in the feeding mechanism. Minute tracheae must be far more resistant to deformation by external hydrostatic changes than are larger tracheae. Prunesco's (1965*d*) account tallies with Manton's. The latter author figured a simple spiracular atrium lined with trichomes and noted that the trachea were lined with smooth as opposed to spirally thickened cuticle.

Spiracle closing mechanisms

Spiracular closing should reduce water loss and spiracular closing devices have been recorded from centipedes though they are less complicated than those of many insects.

Chalande (1886) concluded that the walls of the subatrial cavity of geophilomorphs were able to 'contract' and Manton (1965) that the atrium was flattened during locomotion. There appear to be no accounts of spiracular muscles in this order. Muscles attached to the subatrial cavity of *Scolopendra cingulata* cause the spiracular valves to open and close during violent movement (Dubuisson, 1928). Chalande (1885) stated that the spiracle of *S. cingulata* may be closed by the 'contractile' internal membrane, presumably the floor of the atrium and by the action of a sphincter muscle around the subatrial cavity.

In *Cryptops* a dorsal and ventral muscle are attached to the subatrial cavity and by their contraction open the inner spiracular opening (Füller, 1960; Kaufmann, 1964). Kaufmann (1962) mentioned the possibility of a similar arrangement in *Lithobius*. Rilling (1968) was unable to find spiracular muscles but mentioned a pair of muscles inserted on the stigmatopleurite which could open the lips of the spiracle, Lewis' (1963) statement that lithobiomorph spiracles possess a closing device should be discounted.

Ventilation of the tracheal system

In a large number of small insects diffusion is the only factor involved in oxygen transport through the tracheae, but in larger species there are ventilatory movements in which active compression of the abdomen, or more rarely the thorax, results in the expulsion of air from the air-sacs and tracheae whose walls are less resistant to collapse (Richards & Davies, 1977). Chalande (1886) observed no respiratory movements in the spiracles of centipedes. When *Geophilus carpophagus* is immersed in water the median part of the subatrial cavity 'contracts' regaining its original shape when the animal is dried off. This also happens in *Schendyla nemorensis* and *Himantarium gabrielis.* Movement of air in the tracheae is brought about by the movements of the heart which cause the median dorsal atrium to collapse. The pulse acts in a similar way on the subatrial pockets which are surrounded by blood lacunae. Both the median dorsal atrium and the subatrial cavity lack spiral thickenings. Chalande stated that the pulsation of the blood lacunae running alongside tracheae caused them to change shape. Tracheal pulsation has been observed in the littoral geophilomorph *Mixophilus indicus* (Rajulu, 1970*b*) (see Chapter 21).

Muscular contractions during locomotion and movements of the alimentary canal probably assist ventilation of the tracheal system.

In a struggling *Scolopendra* the spiracular valves open and close due to the movement of muscles attached to the substigmatic pocket and to the integument (Dubuisson, 1928). Rossi (1902) suggested that dorso-ventral muscles caused respiratory movements in *Scolopendra.* The irregular opening and closing of the stigmata in a struggling *S. cingulata* is synchronous suggesting that these movements play a part in ventilation. Dubuisson suggested that diffusion was adequate when the animal was at rest but respiratory movements were used when the animal moved rapidly. Jangi (1966) stated that 'A careful examination of an apparently inactive living *Scolopendra* shows that the breathing movements are confined to the spiracular region. The spiracular muscles control the entry or exit of air'. He did not describe these muscles.

When a spiracle of *Scutigera coleoptrata* is covered by a drop of water it is drawn in within 30–40 s suggesting that the ends of the tracheae are open. As the liquid level falls the meniscus oscillates and frequently the movements correspond exactly with that of the heart beat. At systole there is a considerable narrowing of the heart, from 675 to 250 μm, sending blood into the arteries. Blood is drawn through the pericardial cavity into the heart, passing en route over the tracheae. Rajulu (1969*a*) suggested that the blood of *Scutigera longicornis* contained haemocyanin.

12

The circulatory system

Insects, which have a very well-developed tracheal system, have a much reduced blood system but in the Chilopoda both systems are well developed (Fahlander, 1938). The dorsal tubular heart is connected by one or more pairs of commissures to a ventral supraneural vessel. The blood flows forward in the heart which is continued anteriorly as the anterior aorta. This vessel, the commissures, supraneural vessel and latero-dorsal arteries of the heart supply the organs with blood through open ended arteries. There is little evidence of a venous system. Blood returns to the heart through paired ostia which are typically arranged one pair to each segment.

The heart

In *Scolopendra cingulata* the heart is suspended from two connective tissue sheets which enclose the dorsal sinus (Fig. 138*a*). It is attached laterally on each side to a double sheet of connective tissue which encloses the lateral sinus containing the pericardial cells. The lateral sheets join and continue outwards as a single layer. Fan-shaped alary muscles which attach to the heart spread through these lateral sheets (Fig. 138*b*). The dorsal sinus communicates widely with the perivisceral coelom through large openings between successive alary muscles (Jangi, 1966). Herbst (1891), Duboscq (1898) and Fahlander (1938) regarded the cavity of the dorsal sinus and the lateral sinuses as representing the pericardium: Heymons (1901) considered that the pericardium consisted of two very large cavities enclosing the dorsal longitudinal muscles.

In *Lithobius mutabilis* L. Koch, Duboscq (1898) distinguished in addition to the sheets delimiting the lateral sinus, a sheet of connective tissue passing ventrally around the gut. In *L. forficatus* the ventral nerve cord is bridged over by a sheet of connective tissue that may represent the ventral diaphragm of insects (Rilling, 1968).

In *Geophilus* and *Lithobius* there is one pair of alary muscles per heart

Fig. 138. The heart in *Scolopendra cingulata*. *a*, transverse section through the heart region; *b*, dorsal view of a segment of the heart; *c*, diagram of a segment of the heart to show the ostia and valves; *d*, diagrammatic cross section of the heart (*a* and *b* after Herbst, 1891; *c* and *d* after Duboscq, 1898).

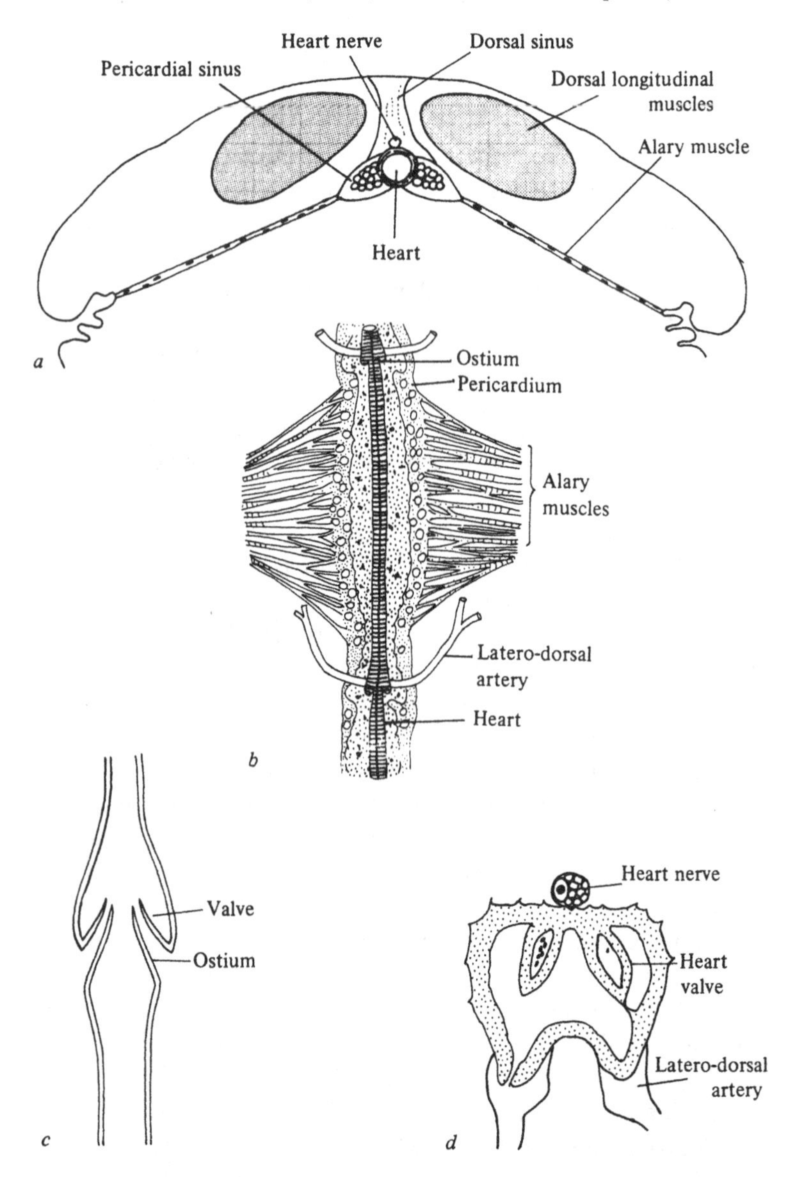

segment. The alary muscles of the forcipular segment are rudimentary in *L. forficatus* (Rilling, 1968). In *Scolopendra* the alary muscles appear to be subdivided: this is so in segments 4, 6, 9, 11, 13, 15, 17 and 19 in *Scolopendra morsitans* (Jangi, 1966) where there appear to be two pairs.

The pericardial cavity of *Scutigera* consists of a single cavity in which the heart is hung by connective tissue (Herbst, 1891; Fahlander, 1938). Its wall consists of a single layer of thick connective tissue which differs completely from the thin perforated membrane that surrounds other organs. The tracheal system lies within the pericardial cavity which has seven expansions to accommodate the seven tracheal lungs. From each expansion paired sinuses formed of thick connective tissue run obliquely forward to the body wall. These structures, termed lateral venous sinuses by Dubuisson (1928), merit the term vein (Fahlander, 1938). The seven pairs of alary muscles attach to the pericardial membrane.

In *Scolopendra cingulata* the heart is divided into 21 segmental chambers separated by valves which are the expanded lips of the 20 paired ostia (Fig. 138*c*). Newport (1843) reported 22 chambers, finding two in the last segment. The valves prevent the outflow of blood when the heart contracts (Herbst, 1891; Duboscq, 1898). Valves are also present at the point of origin of the paired latero-dorsal arteries and these prevent backflow into the heart when it relaxes (Fig. 138*d*). The first valve lies in the posterior part of the forcipular segment but there is no valve between this and the first leg-bearing segment (Fahlander, 1938). When one chamber is in systole, the preceding one is in diastole. At the moment of diastole, however, there is a slight backflow of blood because the auricular valves formed by the ostia do not completely separate successive chambers. Contraction of the alary muscles pulls the lateral pericardial cavities outwards, thus widening the dorsal sinus into which the blood flows (Herbst, 1891). In *L. forficatus* the first ostium is dorsal and the remainder latero-ventral.

Immediately above the heart there is an elongated ganglion sometimes called the heart nerve. This ganglion is responsible for the regular beating of the heart (Duboscq, 1898). Rates of heart beat per minute recorded for centipedes are *G. carpophagus*, 20–23, *Himantarium gabrielis* 19 (Chalande, 1886), *S. maritima*, 18 and *Cryptops hortensis*, 36 (Duboscq, 1898). Duboscq found the rate of heart beat to be very variable in *L. forficatus* but Auerbach (1951) recorded a rate of 83 per minute at 24.6 °C for the same species. The rate varies from 90 to 200 in *S. coleoptrata*, 90 to 100 being normal (Dubuisson, 1928).

Isolated *Scolopendra morsitans* heart beats at 48 to 54 per minute at 30–37 °C. The rate is accelerated by acetylcholine and adrenaline but depressed by histamine. The presence of striated muscle and the ganglion running the length of the heart suggest that the heart is neurogenic (Rajulu, 1966).

Brain and heart extracts of *Scolopendra morsitans* have a cardioexcitatory activity but the substance responsible is neither acetylcholine nor 5-hydroxytryptamine, the cardioexcitor neuro-hormones of vertebrates. The excitatory substance in *Scolopendra* is probably a protein. Bodies ranging from 9 to 25 μm in diameter, containing granules staining violet with chrome-haematoxylin-phloxin, and purple with Gomori's paraldehyde fuchsin, are present amongst the cardiac muscle fibres. After repeated stimulation of the heart these granules virtually disappear suggesting that the stainable material is neurosecretory (Rajulu, 1968*a*).

The arterial system

Geophilomorpha

Fahlander (1938) found it impossible to make a detailed study of the geophilomorph arterial system because of the small size of the specimens. The system resembles that of scolopendromorphs. A mandibular commissure is present in *Mecistocephalus*. *Strigamia* lacks a rectal commissure.

Scolopendromorpha

Newport (1843) described the blood system of *Scolopendra alternans* Leach and *S. hardwickei* Newport, and Duboscq (1898) that of *S. cingulata*. Fahlander (1938) worked on *S. cingulata*, *S. morsitans*, *Cormocephalus rubriceps* (Newport) and *Otocryptops rubiginosus* (L. Koch). In scolopendromorphs the heart and supraneural system are joined by forcipular commissures (*anneau aortique*, *Maxillipedbogen*, circumstomodaeal vessels). The latero-dorsal vessels supply the fat body and connective tissue of the body cavity and terminate in Kowalewsky bodies (see below).

The heart is continued forward as the *aorta cephalica* which runs ventral to the brain giving off six pairs of arteries including the cephalic, antennal and mandibular arteries. The latter send a pair of branches which end in the blood sinus of the hypopharynx (Fig. 139*a*). Newport described a pair of mandibular arches but these were not detected by later workers. Fahlander suggested that this might indicate considerable individual variation in *Scolopendra*. The supraneural vessel supplies each segment through a pair of arteries each of which branches to supply the leg and the fat body where it ends in Kowalewsky bodies. The artery to each of the last pair of legs gives off a small branch which subdivides to supply the coxal glands. The pair of arteries arising from the last heart valves run ventralwards and join the supraneural vessel forming the rectal commissure (Fig. 139*b*), Duboscq had made

Fig. 139. The arterial system of *Scolopendra cingulata*. *a*, anterior region; *b*, posterior region (after Fahlander, 1938).

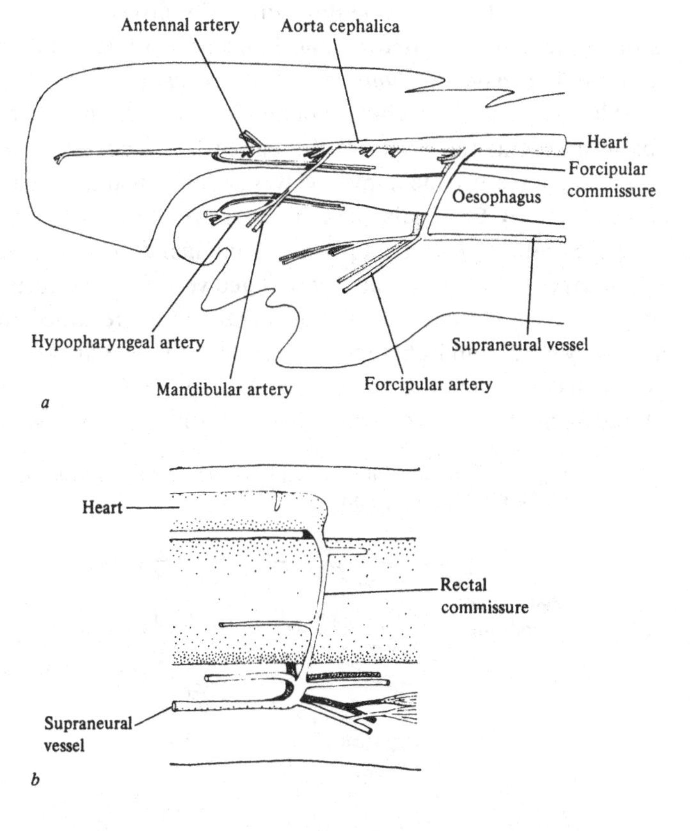

no mention of a rectal commissure and Fahlander considered that it was subject to variation. He was unable to find arteries from the subneural vessel supplying the gut or accessory reproductive glands.

Lithobiomorpha

The arterial system of *Lithobius* and *Lamyctes* resembles that of *Scolopendra* in general arrangement (Newport, 1843; Herbst 1891). Biegel (1922) found only the heart and supraneural vessel in *L. forficatus* but Fahlander (1938) showed that mandibular and forcipular commissures were present (Fig. 140) but could not establish the presence of fore- and hind-gut arteries and a rectal commissure with certainty. His results were confirmed by Rilling (1968). According to Rilling lateral dorsal arteries are only present from ostia 11 to 12.

Scutigeromorpha

Herbst (1891) and Dubuisson (1928) investigated the circulatory system of *S. coleoptrata* and Fahlander (1938) investigated in addition *Thereuopoda clunifera* and *Thereuonema tuberculata*. The heart has 13 pairs of ostia but is not divided into sections, valves being absent except at the origin of the paired latero-dorsal vessels and according to Herbst, incomplete valves are present at the origin of the *aorta cephalica*. Forcipular commissures are present. Blood vessels from the *aorta cephalica* supply the brain, mandibular segment and hypopharynx. Two blind sacs are situated ventral to the aorta which they join by a common duct (Fig. 141*a*). Herbst regarded them as accessory hearts but Fahlander thought that this was unlikely as they do not have a thick muscular layer. Manton (1965) suggested that blood might be squeezed into them on sudden expansion of the

Fig. 140. The anterior arterial system of *Lithobius forficatus* (after Fahlander, 1938).

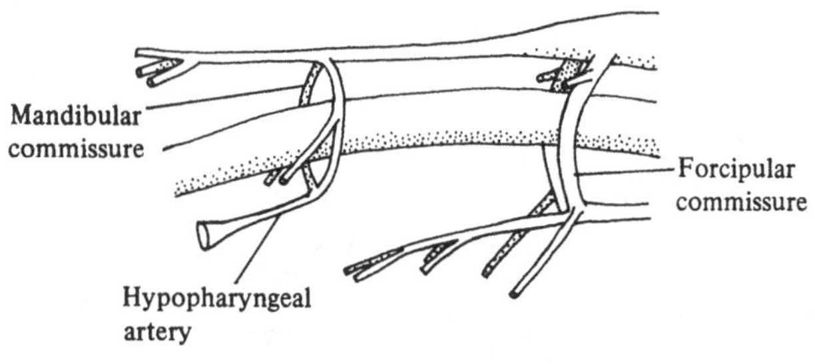

pharyngeal pouches during feeding. A pair of arteries from the supraneural vessel supply the first and second maxillae, the vessels to the former branch to form a blood sinus under each maxillary organ which causes their eversion.

The supraneural vessel sends branches to the legs in each segment and these branch to the fat body and the ventral ganglion. Herbst (1891) described a recurrent artery running forward along the rectum to the

Fig. 141. The arterial system of *Thereuopoda clunifera*. *a*, anterior region; *b*, posterior region (after Fahlander, 1938).

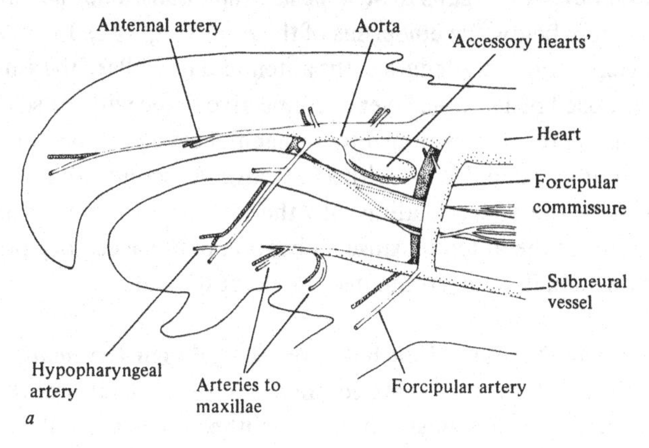

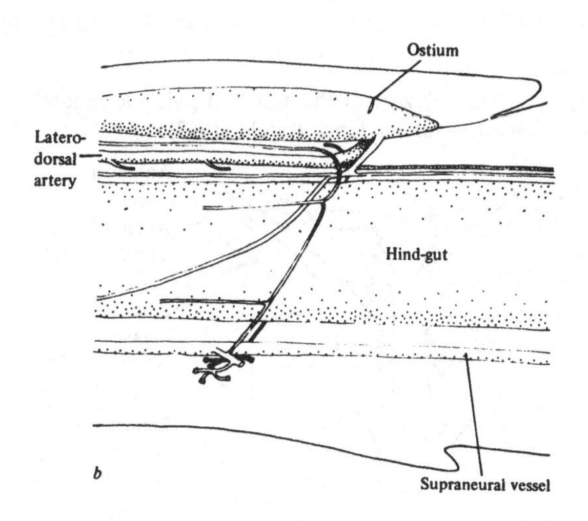

mid-gut. Fahlander showed that from the posterior end of the heart in segment 13 there proceed latero-dorsal arteries of very different sizes, the right being very large (Fig. 141*b*). There is no rectal commissure. The heart is absent in segments 14 and 15 but the supraneural vessel is present and sends paired vessels to the fourteenth and fifteenth pairs of legs and to the gonopods.

Accessory pulsatile organs

Rajulu (1967*b*) described a pair of small ampullae, the antennal pulsatile organs in *Scolopendra morsitans*. They are situated immediately below the epidermis of the *frons* (Fig. 142). From each a tube extends along the length of the antenna. The walls of the ampullae are composed of a double layer of connective tissue with muscle fibres between. There is an opening into the haemocoel on the median side of each ampulla through a slit with a flat valve. A second valve is sited at the base of the vessel leading into the antenna. Rajulu found that movement of the antenna caused a beating of the accessory pulsatile organs which, in turn, stimulated the heart to beat.

The structure of the blood vessels and their terminations

The blood vessels are composed of a simple layer of reticular cells which form a syncytium. No endothelium is present. In large vessels, aorta and supraneural vessel, the cells contain peripheral fibrils (Duboscq, 1898).

The arteries terminate in reticular connective tissue opening into

Fig. 142. Ventral view of the antennal pulsatile organs in *Scolopendra morsitans* (after Rajulu, 1967).

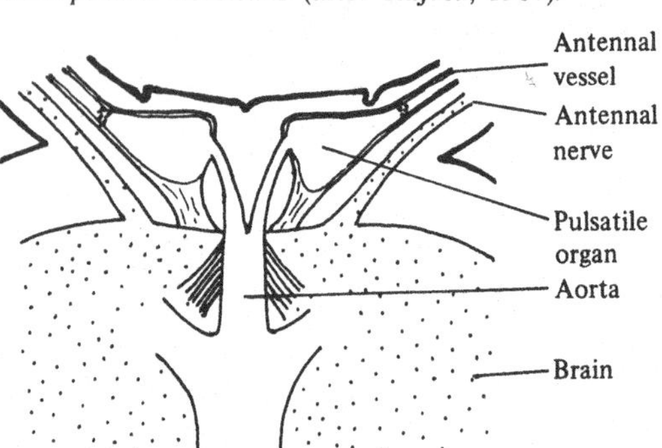

lacunae. The vessels do not anastomose except for those supplying the ovary in *Scolopendra.* Duboscq doubted Newport's (1843) representation of anastomoses in the branches of vessels supplying the Malpighian tulules of *Scolopendra.* The terminations of the vessels in the connective tissue around the intestine are rarely pointed, more frequently they comprise a funnel parting into three or six filaments. A third type of termination, the Kowalewsky corpuscle, is found only in scolopendromorphs. First described by Kowalewsky (1895), they are round or oval bodies embedded in the fat body at the end of branches from the supraneural vessel. Except for the first four segments there are six of these bodies per segment in *Scolopendra* and three in *Cryptops,* where they are only found in connection with the supraneural vessel. In *S. morsitans* they are bluish in colour and easily distinguished from the surrounding fat body (Jangi, 1966). Kowalewsky bodies measure between 0.2 and 0.3 mm in diameter and absorb injected vital dyes and bacteria. Duboscq considered that the tissue resembles embryonic connective tissue, being a mass of syncytial cells containing amoebocytes; many of the cells are dividing.

The blood

The blood of centipedes is usually colourless although it appears to have a violet tint in *Lithobius*; this may be due to the connective tissue pigment in solution (Duboscq, 1898). It does not change colour when exposed to air. Its reaction is distinctly alkaline and it contains albumins; in *Lithobius* and *Geophilus* oily granules are not infrequent.

Sodium chloride accounts for nearly all the blood osmolar concentration in *Lithobius* (Sutcliffe, 1963) and is found in high concentrations in *Cormocephalus rubriceps* (Newport). In the latter species amino acids represent 10 per cent of the osmotic concentration and a large contribution is made by unidentified organic substances, possibly organic acids (Bedford & Leader, 1975).

An electrophoretic study has shown that there are five blood protein fractions in *S. morsitans.* Numbers 1–3 are similar to human serum albumin and a_1 and a_2 globulins, protein 4 corresponds to γ-globulin but protein 5 has no equivalent in human serum proteins. There is no protein in the blood of *Scolopendra* comparable to fibrinogen (Rajulu, 1969*b*). The concentration of free amino acids in the haemolymph of

Scutigera longicornis and *Himantarium samuelraji* ranges from 277 to 337 mg/100 ml, forming 17–20 per cent of the total residues in the haemolymph as compared with 8–10 per cent in Crustacea. The haemolymph contains the sulphur-containing amino acids cystine and methionine which are absent in crustacean haemolymph, but lacks taurine which is present in Crustacea (Rajulu, 1973*a*).

In *Scolopendra* the living blood cells vary from 5 to 25 μm in diameter, the common size being 12 to 18 μm. Duboscq (1898) distinguished two main types of haemocytes: small cells with large nuclei and homogenous cytoplasm, probably formed in the Kowalewsky bodies and medium to large cells with acidophilic and occasionally (in geophilomorphs) basophilic granules in the cytoplasm. The cells were shown to be phagocytic. The larger blood cells are also able to accumulate soluble substances, for example ammonium carminate which is precipitated in the cytoplasm.

Rajulu (1970*a*) distinguished five types of haemocytes in *Ethmostigmus platycephalus spinosus* (Fig. 143):

Type 1 Round cells 5–7 μm in diameter with a large round nucleus resembling the prohaemocytes of insects.

Type 2 Cells measuring 15–20 μm in diameter commonly with central oval nucleus and cytoplasm devoid of inclusions. These occasionally send out thread-like pseudopodia. They ingest

Fig. 143. Haemocytes types 1–5 of *Ethmostigmus platycephalus spinosus* (after Rajulu, 1970*a*). For further explanation see text.

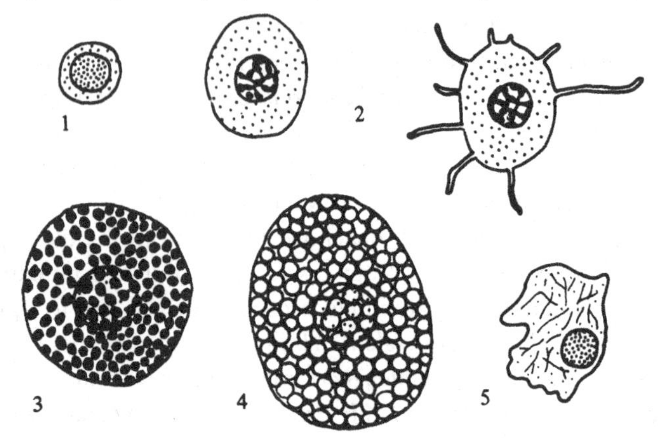

indigo carmine particles. These resemble the plasmatocytes of insects.

Type 3 Round cells 25–30 μm in diameter and filled with numerous round granules. They ingest injected carmine particles and resemble the granular haemocytes of insects.

Type 4 Oval cells 48–52 μm in diameter containing large spheroid inclusions which fill the whole cell. These cells are very unstable and resemble spherule cells of insects.

Type 5 Ovoid or irregular cells 20–22 μm in diameter with eccentric nucleus and cytoplasm and with a fine granular network, recalling the oenocytoids of insects.

In a later paper (Rajulu, 1971*d*), the haemocytes of *S. morsitans* were described. They are smaller than those of *Ethmostigmus* but the same types were recognised in addition to cells 8.5–9.5 μm in diameter containing fat droplets (adipohaemocytes). One striking feature was the presence of large numbers of non-cellular elements including nuclear-free cytoplasmic masses possibly derived from granular haemocytes or oenocytes and minute particles of 'glassy veils'. Prohaemocytes, plasmacytes, granular haemocytes and spherule cells may be interrelated. Rajulu pointed out that oenocytoids are common in insects and Onychophora but have not been reported from crustaceans or arachnids. Their presence in centipedes may be additional evidence for relation between the Onychophora, Myriapoda and Insecta.

13

Pigments

European centipedes are usually brown or yellowish in colour but tropical scolopendromorphs show a remarkable range of bright colours, *Ethmostigmus trigonopodus* (Leach) being almost black with orange legs, *Rhysida nuda togoenis* deep violet and *Asanda sokotrana* Pocock red.

The purple pigment of Lithobius

Plateau (1878) noted, that the mandibular glands of *Lithobius forficatus* were often violet and Duboscq (1898) noted that the connective tissue contained a violet pigment and the blood of this species had a violet tint. The pigment is water soluble and present in granular form in asteroid cells underlying the epidermis and following the tracheal system (Needham, 1960). When *Lithobius forficatus* is dissected the intensity of the violet colour increases implying that *in vivo* the pigment is in a colourless, reduced form (Needham, 1945). The reduced pigment is pale yellow. It is rapidly and reversibly reduced by chemical reducing agents, including ascorbic acid, but remains oxidised down to very low oxygen tensions (Needham, 1958).

The purified pigment has a molecular weight of about 4000 and contains no detectable copper so that the chromophore is not a copper-protein or biuret type compound as Needham (1960) had suggested. The absorption spectrum, the nature of the redox and pH colour changes and some other factors suggest that the pigment may be a hydroxyquinone (Bannister & Needham, 1971). For a quinone to be so soluble in water and so insoluble in all lipid solvents tested, as is this pigment, it would have to be very firmly bound to some very hydrophilic conjugant. Further work on the nature of the pigment is desirable.

In 0.1 M NaOH, purified pigment showed peaks at 545 nm and 585 nm with minimum absorption at 425 nm, but showed no resolved peaks in the UV range. Those previously recorded by Needham (1960) were probably due to an associated flavin.

The pigment, called *lithobioviolin* by Needham (1960), occurs in spherical granules or vesicles of 1 μm diameter, and fills elongated to stellate connective tissue cells throughout the body. It occurs very abundantly under the epidermis of certain regions. The density of the subepidermal layer of lithobioviolin is not very evident, however, because the violet colour is virtually complementary to the amber of the exoskeleton and therefore appears merely to darken the latter. There is little pigment under the ventral epidermis of the body and on the legs also it appears to assist countershading. A violet colour is usually visible through the pleural membranes and immediately after moulting an individual is a brilliant and deep violet. The pigment gives *L. variegatus* its dark patches.

The pigment cells appear to invest and permeate all the internal organs except, perhaps, the mid-gut. They are usually aggregated round tracheal tubes, particularly the finer branches.

The permeation of the fat body is particularly rich, especially in the segment behind the head and again towards the posterior end of the fat body. Here the cells are densely filled with granules as large as 3 μm or more and reddish in colour. The fat cells themselves contain a brownish pigment. Although red-purple within the granules, lithobioviolin becomes violet in aqueous solution. Around the skeletal muscles the pigment investment is rather sparse and the tracheal tubes running longitudinally between the individual muscle fibres appear to be virtually free of it. Larger greenish granules are found there, possibly in the sarcoplasm itself.

Contrary to Duboscq's (1898) and Attems' (1926) record of the blood of chilopods as sometimes violet, the blood of a normal *Lithobius* appears to be always colourless if obtained with minimal damage to the pigment cells, as by transecting the antennae or the posterior legs. The pigment shed at the actual site of a wound may play a part in forming a seal over it. This seal is dark brown and granular, and forms within a few hours.

Pigment cells are first evident in the young soon after hatching and increase in number with the darkening of the exoskeleton. The amount of pigment visible through the integument and the amount extracted varies between individuals. Males possibly contain more than females.

Occasionally pigment in solution is found in the gut lumen. Many large cells also are present so that the pigment may be of dietary origin.

Needham did not confirm unequivocally that the redox property of the pigment might be regularly exercised *in situ*. The ideal requirement is a visible fluctuation in the proportion of oxidised pigment in a living animal in correlation with changes in oxygen supply or in oxygen consumption. The deeper colour of the tissues of a dissected animal, and still better of a newly moulted living individual, would constitute such evidence, if the possibility could be ruled out that the deeper colour was due merely to release of pigment, already fully oxidised, from a compact granular form.

An animal immersed in liquid paraffin, in an air-free, sealed chamber, died in $\frac{1}{4}$–$\frac{1}{2}$ h, while one sealed in an air-chamber about forty times its own volume lived for 24–30 h. In both cases the subcutaneous pigment appeared to be scarcely changed until death when it began to dissolve out of the granules. The pigment released from the granules was fully reduced which implies that the animal had been able to survive until oxygen was completely exhausted. The average consumption of oxygen by such imprisoned animals was only 0.33 ml/g/h so they also resist anoxia by economical respiration.

Small individuals subjected to rapid anoxia by immersion in liquid paraffin did appear to become more translucent while alive possibly owing to the reduction of the deeper lying pigment. A decrease in the amount of oxidised pigment in such situations as the core of the antennae also seemed certain.

There is also ribitylflavin, a flavoprotein, in the exoskeleton of *L. forficatus*. It appears to be incorporated with the violet pigment so that both may continue to function in a redox capacity there (Needham, 1974).

Other Chilopoda

The purple pigment is not detectable in species of *Geophilus* but a mauve-pink pigment is distributed in the fat body of *Strigamia acuminata*. The exoskeleton is amber-coloured but thin enough for the tissue pigment to affect the gross colour which varies from orange to wine-red, possibly depending on the degree of oxidation of the pigment (Needham, 1960). Examination of young specimens of *Strigamia maritima* confirmed the findings on *S. acuminata*. Lewis (1961) observed that the eggs of *Strigamia maritima* appear pinkish when about to hatch owing to the pinkish-violet colour of the contained embryos. The pigment occurs in the gut cells of the larvae, the rest of the body being whitish and translucent. The gut darkens in the first adolescens stage losing its pinkish-violet coloration and the cuticle hardens. In six of 750 specimens of *Strigamia maritima* of both sexes and all post-larval stadia dissected by Lewis (1961) the mid-gut was violet instead of the normal orange-brown colour. All six specimens were maturus males. The pigment in *S. acuminata* would appear to be more independent of protein than that in *Lithobius* (Needham, 1960). Its absorption spectrum showed a peak of 560 nm and a trough at 500 nm.

In a specimen of *Scolopendra* sp. (collected in France) a prussian blue pigment was visible through the exoskeleton which is pale yellow. It was located mainly under the terga, near their anterior and posterior margins and in the tips of the legs. A more widespread distribution was evident after death and exposure of the dead tissues to the air gave an immediate and considerable increase in blue pigment. Extraction, as in *Strigamia*, yielded much more pigment than anticipated from the visible colour in life suggesting that *in situ* some is not evident because it is in a reduced state. As in *Lithobius*, the subcutaneous pigment cells formed a reticulate pattern.

The pigment initially extracted from the tissues was blue-green. It reduced to a deeper yellow than that of *Lithobius* but was contaminated by a second, yellow pigment, possibly corresponding to the contaminating pigment in *Strigamia* and lacking a redox colour change. It probably came from the fat body which is brown. The absorption spectrum of the oxidised pigment on the alkaline side of its indicator point showed a peak at 600 nm and a trough at 520 nm.

Pryor (in correspondence with Needham) reported a similar pigment in tropical scolopendromorphs and *Scutigera*. From the variation in colour of the chilopod examples studied, Needham concluded that a class of pigments rather than a unique compound has been developed in centipedes. Rajulu (1968*b*) described a bluish-green pigment which occurs in granular form in the hypodermis of *Scolopendra morsitans*. The absorption-spectrum of the oxidised form in neutral aqueous medium showed a peak around 580 nm and a trough around 495 nm. The properties of this pigment resemble those of lithobioviolin.

In addition to the bluish-green pigment of *Scolopendra morsitans* the species contains orange regions in the integument. This resolves into two bands when chromatographed. The first has an absorption spectrum in hexane indicating the presence of β-carotene, the second fraction showed a single absorption peak at 468 nm in hexane, suggesting the presence of an astaxanthin ester (Rajulu, 1968*b*). Extracts of yellow regions of the cuticle gave a UV absorption spectrum in good agreement with the absorption spectrum of pteridine.

Rajulu (1969*a*) demonstrated the presence of copper containing pigments in the blood of *Scutigera longicornis* which is normally faint indigo blue in colour. He suggested that it might be a haemocyamin. No mention of this work is made in Needham's (1974) monograph.

The function of the lithobioviolins

Needham regarded the precise function of lithobioviolin as presenting a challenging problem. It may act in continuous respiratory mediation but as it is extremely difficult to decrease substantially the visible amount of oxidised pigment in the body by reasonable physiological methods, Pryor (quoted by Needham) concluded that it could have no normal respiratory function. The case might be similar to that of integumental ommochromes in the isopod *Asellus* which also can be reversibly oxidised and reduced *in vitro* but show no detected variation *in vivo*. Parasitisation of *Asellus* by echinorhynchid larvae causes an intense deposition of integumental ommochrome in the gill-operculum, where there may be special redox conditions. Perhaps the concentration of pigment in the mid-gut of some maturus male *Strigamia maritima* is a similar phenomenon. An autoxidisable pigment like lithobioviolin could either act as an oxygen carrier or as a terminal oxidase within the cell. Its restricted distribution, only in the connective tissue, and the resistance of the redox property to boiling and other drastic treatments shows that it is not enzymic. If, as seems probable, the tracheal system supplies all the cells in the body, then there is a virtually complete screen of violet pigment between the air and the tissues and it could mediate virtually all gaseous exchange. Its enclosure in granules is not necessarily a bar to such activity, since mitochondria of about the same size-range function actively in respiratory metabolism (Needham, 1960). The epidermal cells might be expected to obtain enough oxygen by diffusion through the exoskeleton and this may explain why the subcutaneous component of the pigment is distributed only where required for the purpose of countershading.

Needham regarded it as unlikely that a special agent of this kind would be required except in the emergency of low oxygen supply. *Lithobius* must frequently experience low oxygen tensions in narrow crevices under bark, stones, coal and leaf mould in which it lives, partly due to competing oxidative processes. Since it appears to have no active ventilation, any aid from fortuitous muscular movements must be very seriously curtailed in restricted spaces. Through its high affinity for oxygen the pigment could concentrate oxygen from an otherwise ineffectual tracheal system. Pryor suggested that the pigment is released from its cell under hypoxic stress and that as a more labile mediator in solution it transfers oxygen via the blood to the tissue cells, or actually enters the latter. Muscle cells which have the greatest oxygen requirement take up the pigment from solution very readily. Release of the pigment from its mother-cell is strongly indicated by its appearance in the faeces of individuals under anoxia. If a critically low tension of oxygen in the tracheal system is the effective trigger for the release of pigment from the mother-cell, this could account for the observed distribution of the mother-cells. In further support of this alternative mechanism the vascular system of chilopods seems rather better developed than in insects which have a particularly efficient tracheal system.

The amount of pigment in *Lithobius* including the reserves in the fat-body investment, seems surprisingly large. Pryor observed that the amount released in an animal killed by asphyxia is enough to colour the whole animal brilliantly when re-exposed to oxygen. The pigment may function in the oxidative process of hardening and darkening in the exoskeleton, the brilliant coloration immediately after moulting being due to mass-dissolution of stored pigment. There is some evidence that the pigment is involved in the formation of the dark escher-material over a wound.

14

Connective tissue and fat body

Connective tissue

The body cavity of centipedes is lined by reticulate connective tissue consisting of an outer somatic or parietal layer and an inner splanchnic or visceral layer around the gut (Fig. 144). In *Lithobius* the connective tissue contains a violet pigment but in *Scolopendra* it is colourless. The tissue shows a varying degree of development of *conjonctive*, elastic and muscle fibrils and grades into adipose tissue (Duboscq, 1898).

Fat body

Vogt & Yung 1883 (quoted by Duboscq, 1898) first described the distribution of adipose tissue in *Lithobius* as irregular masses in the form of strips running in all directions, covering the muscles and separating the internal organs. A large number of fine tracheae run through the tissue which is often blue-violet in colour. It is most commonly found in the immediate region of the ventral nerve cord, under the lateral edges of the tergites, along the flanks

Fig. 144. Transverse section of *Lithobius* in the region of the oesophagus (after Duboscq, 1898).

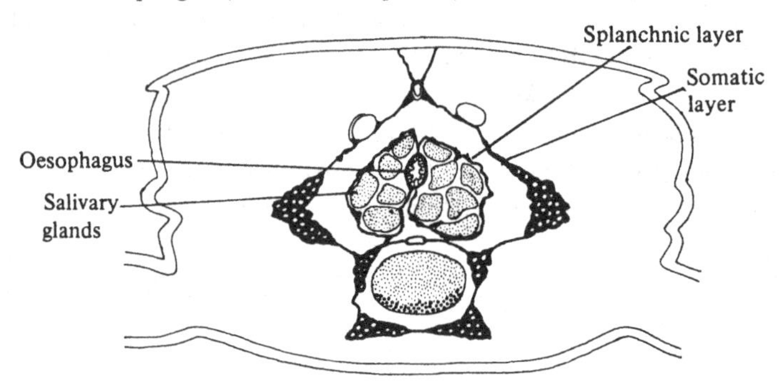

and often at the base of the legs. In adult *L. forficatus* the entire somatic layer of connective tissue is adipose in nature but in immaturus individuals there are three constant masses: pericardial, lateral and ventral. The same disposition of fat is seen in *Strigamia maritima* (Fig. 145) with the exception that the ventral masses are replaced by *cellules à carminate* (nephrocytes).

The adipose tissue has the same structure in all centipedes being syncytial, with scattered nuclei and formed into lobes or globules. In some species the globules fuse to form conglomerations (*Himantarium gabrielis*), in others (*Scolopendra*) the globules are joined to form chains. In *Scolopendra* there are two rows of rosettes of adipose tissue situated above the gonad which Duboscq considered might have a special role.

In addition to fat the cytoplasm of the adipose tissue contains pigmented bodies yellow in colour, probably proteinaceous and giving the fat body its characteristic colour. In *Chaetechelyne vesuviana* this pigment is black and increases with the age of the specimen, being absent in very young animals. Duboscq (1898) made brief mention of brown bodies in the fat. These are brown or yellow in colour, resistant to acids and he presumed were storage secretions.

The pericardial cells of *Scolopendra* are a modified adipose tissue that take up the dye indigo carmine to a greater degree than other adipose cells (Kowalewsky, 1895). The pericardial cells are probably not homologous with the pericardial cells of insects: they are

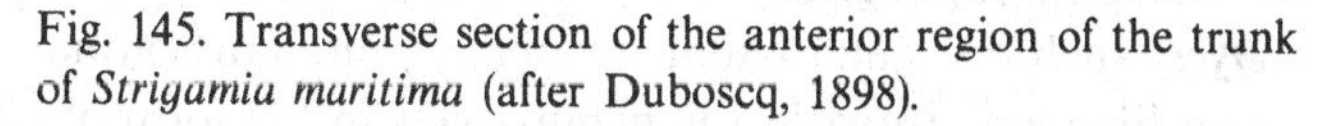

Fig. 145. Transverse section of the anterior region of the trunk of *Strigamia maritima* (after Duboscq, 1898).

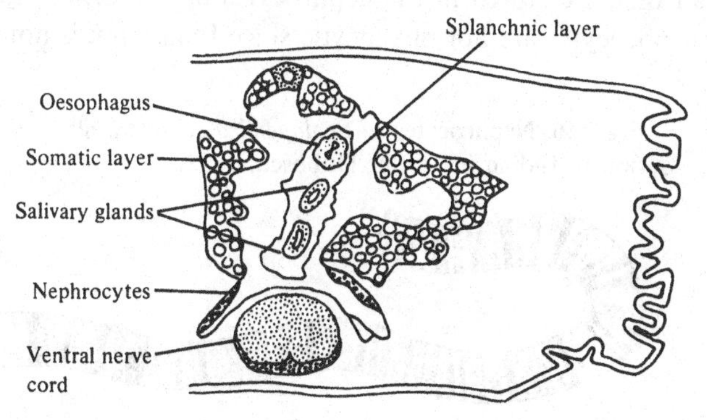

not very rich in fat droplets and absorb small amounts of vital dyes (Palm, 1953).

Duboscq (1898) considered that the fat body was involved in excretion. There are no concretions of uric acid in the fat body as there are in millipedes.

Nephrocytes

In all centipedes there are cords or groups of vesicular cells alongside the adipose tissue which take up injected ammonium carminate. They were first described as *cellules à pigment* by Plateau (1878). Further descriptions were provided by Herbst (1891) and Kowalewsky (1895): Duboscq (1898) gave them the name *cellules à carminate.* In *Scolopendra cingulata* they are arranged in filaments like piles of coins (Fig. 146) surrounding the Malpighian tubules. In *Lithobius* they form lobules around the salivary glands and oesophagus with two or three nuclei per lobule.

In *Geophilus* and *Strigamia maritima* they are found around the salivary glands and on both sides of the ventral vessel where the somatic layer of connective tissue is transformed into nephrocytic tissue (Fig. 145). In *Chaetechelyne vesuviana* the cells form a sinus over the ventral glands, in *Haplophilus subterraneus* they form filaments bordering the fat body and in *Himantarium gabrielis* they are scattered through the fat body.

Cellules à carminate are absent in *Scutigera* but one of the 'metameric glands' takes up ammonium carminate and the nuclei of the respiratory organs take up the stain.

Dyes such as ammonium carminate, lithium carminate and trypan blue are stored in the nephrocytes in the form of granules. The nephrocytes are not phagocytic since Indian ink is not usually

Fig. 146. Nephrocytes of *Scolopendra cingulata* after an injection of Indian ink (after Duboscq, 1898).

taken up. The reactions of the nephrocytes of centipedes are practically the same as those of the pericardial cells of insects. In both, the permanent storage of trypan blue and carminate occurs. Palm (1953) was in no doubt that the nephrocytes play a role in cleaning the body fluids, absorbing waste products and storing them in a harmless insoluble form. Most probably some, at least, of the pigment found naturally in these cells is derived from waste products, this is indicated by the gradual increase in the amount of pigment in the cells with the age of the animal. When trypan blue is injected into a *Lithobius* an estimated amount approaching 90 per cent is stored by the nephrocytes.

Rosenberg & Seifert (1975) showed that the *cellules à carminate* of *Lithobius* form the moulting gland.

Phagocytes

In addition to phagocytic haemocytes *Scolopendra* and *Cryptops* have special phagocyte containing structures the Kowalewsky bodies (see Chapter 12). These structures are not present in other orders. In *Lithobius forficatus*, *Pachymerium ferrugineum* and *Necrophloeophagus longicornis* phagocytic cells are found almost anywhere in connection with the connective tissue. It appears that free haemocytes in the blood can attach themselves to connective tissue and remain as fixed haemocytes (Palm, 1953) haemocytes that have this ability are undifferentiated and non-granular. Such fixed haemocytes are phagocytic and take up injected Indian ink, they also take up trypan blue and lithium carminate. The fixed haemocytes are far less common than the nephrocytes and are probably of little importance in the removal of vital dyes from the blood.

Free haemocytes are phagocytic but do not take up injected dyes to more than a negligible degree.

15

Head glands

There has been considerable confusion over the nature of the glandular structures of the anterior region of centipedes: Duboscq (1898) pointed out that no two description of the anterior glands of *Scolopendra* were in accord and from one to three pairs of salivary or venomous glands had been figured or described for the genus. The most detailed accounts of centipede head glands are those of Herbst (1891) and Fahlander (1938). Both single and multicellular glands occur, some of the latter forming a metamerically arranged series.

Scutigeromorpha

The Scutigeromorpha have the largest number of head glands. Fahlander (1938) investigated *Scutigera coleoptrata*, *Thereuopoda clunifera* and *Thereuonema tuberculata*, describing from these species seven pairs of multicellular anterior glands. He distinguished two pairs of buccal glands filling the greater part of the head antero-dorsal to the mouth, the medial pair opening into the anterior region of the pharynx, the lateral pair by short ducts into the oral cavity (Fig. 147*a*). The mandibular or hypopharyngeal glands lie largely in the hypopharynx, their ducts opening on the hind wall of the buccal cavity. In contrast to other head glands, the lobes of these glands are long and narrow and radiate in all directions.

The first maxillary glands open immediately anterior to the basal part of the first maxillae. The secretory part of the glands lies ventral to the nerve cord and stretches from the region of the first maxillae to the second leg-bearing segment (Fig. 147*a*). The duct is wide and in the secretory region of the gland without spiral thickening. The lobes of the gland are short and broad and lie more or less free from each other. The second maxillary glands lie between the forcipular segment and the first leg-bearing segment and consist of a large number of lobes. The glands open on the side of the head behind and under the coxa of the second maxilla next to the opening of the maxillary nephridium (Fig. 148).

In the anterior segments there are two pairs of the so-called vesicular glands. The first pair consists of small glandular sacs belonging to the forcipular segment on whose sides they open somewhat postero-dorsally to the opening of the maxillary kidney (Fig. 148). The second pair belong to the first leg-bearing segment.

In addition to the 'multicellular glands' Fahlander described masses of unicellular glands around the hypopharynx and in the head. He also described a complex of 'gland packets' (gl 5). This complex is grouped round a pair of ducts that Fahlander termed 'apodemes'. These are spirally thickened and it is difficult to see how they differ from the ducts of other head glands. The ducts have very wide lumina and many short branches and bend in a caudal

Fig. 147. *a*, diagrammatic sagittal section of the head and anterior segments of *Scutigera coleoptrata* to show the head glands; *b*, gland gl 5 of *Scutigera*. (after Manton, 1965).

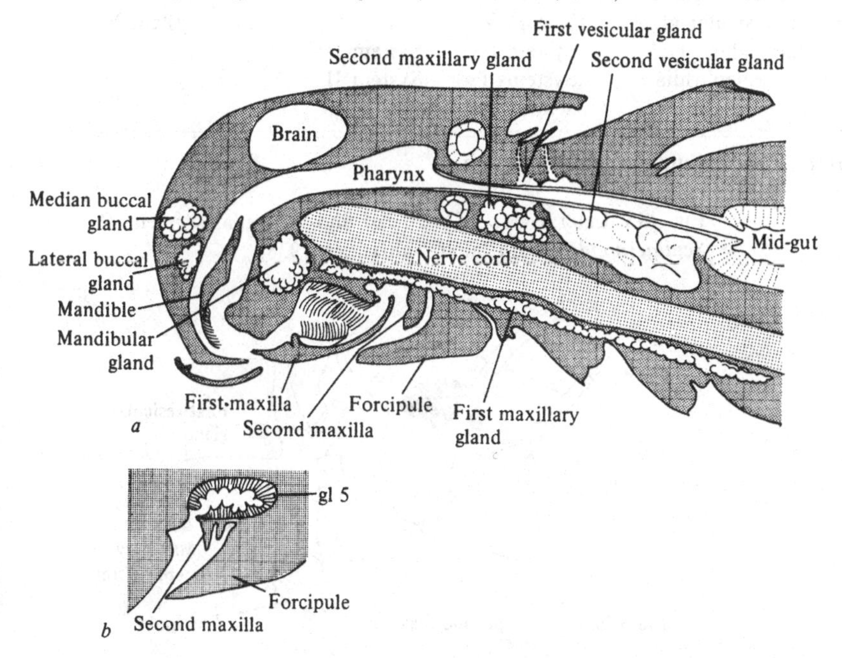

direction. The ultrastructure of mandibular gland and the second maxillary gland show that they are exocrine in nature (Rosenberg & Seifert, 1975).

Manton (1965) figured the head glands of *S. coleoptrata*, following Fahlander's terminology. Herbst (1891) gave a detailed account of the anterior glands of *Scutigera* but his terminology is somewhat confusing; it is compared with Fahlander's in Table 4.

Table 4. *The homologies of the head glands described by Herbst* (1891)

	Herbst's terminology		
	Scutigero-morpha	Lithobio-morpha	Scolopendro-morpha
Median buccal gland	GUF 1		System II
Lateral buccal gland	GUF 1		System I
Mandibular gland	GUF 2	System I	System III
First maxillary gland	GUF 3		
Second maxillary gland	GUF 4		System IV
First vesicular gland	System IV		
Second vesicular gland	System V		System V
Maxillary gland gl 5	System III	System III	
Maxillary nephridia	Systems I and II	System II	

GUF = Gewebe unbekannter Funktion.

Fig. 148. Lateral view of the head and forcipular segment of *Scutigera coleoptrata* (after Fahlander, 1938).

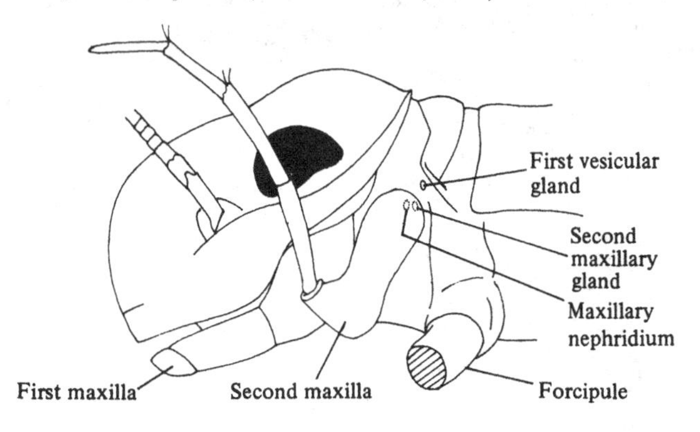

Lithobiomorpha

Lithobius forficatus possesses two pairs of buccal glands, the median lying mostly between the antennal lobes of the brain, the smaller lateral pair are located more ventrally. The paired mandibular glands lie in the posterior part of the head, the forcipular segment and the first leg-bearing segment. These were termed System I by Herbst but are not homologous with his System I in Scutigera (Table 4).

First maxillary glands are absent in *Lithobius* but there are second maxillary glands lying between the forcipular segment and the first leg-bearing segment. They open at the base of the second maxilla (Fig. 149). In adult *L. forficatus* the mandibular gland is 3.0 mm long, the maxillary gland 0.3 mm long. Vesicular glands are absent in lithobiomorphs (Fahlander, 1938).

Unicellular glands are abundant on the coxal processes of the first maxillae and on the inner side of the mandible. A pair of spirally thickened ducts, 'apodemes' serve a several layered group of unicellular glands opening on the ventral side of the hypopharynx. Immediately medial to the cranial openings of the maxillary nephridia there is another pair of ducts on which open another complex gland (Herbst's System III for *Lithobius*) which appears to be the same as the gland gl 5 of *Scutigera.*

Geophilomorpha

Only two or three pairs of head glands are present in the Geophilomorpha. Buccal glands are absent in *Strigamia hirtipes* (Attems) but median buccal glands are present in *Mecistocephalus smithi* Pocock, *Geophilus proximus* and *Pachymerium ferrugineum* (Fahlander, 1938). The glands are relatively small, lying under the brain. Their ducts run posteriorly.

The mandibular glands of geophilomorphs extend far back into the body. They open on the hypopharynx. Plateau (1878) and Chalande (1905) described simple sac-like 'salivary glands' in *Necrophloeophagus longicornis* (Fig. 110). These, Fahlander asserted, were either mandibular or second maxillary glands. He did not find this type of gland in the species he studied. In *Scolioplanes hirtipes* the excretory canal with spirally thickened cuticle expands, the cuticle becomes thinner and numerous gland cells lie between the undifferentiated (epithelial) cells. Posterior to this sac-like secretory region the duct again narrows, the gland cells disappear and the cuticle becomes thick. From this section branches pass to a secretory region of exactly the same structure as that of other orders. Unfortunately Fahlander did not give a figure to illustrate this unusually complicated arrangement.

There is considerable variation in the structure of mandibular glands. Binyon & Lewis (1963) described the gross morphology of the 'salivary glands' of six species of geophilomorphs. These were in all probability mandibular glands as they extend far back into the body. In the terrestrial *Schendyla nemorensis* (C. L. Koch) the glands are tubular, in the littoral species *Hydroschendyla submarina* the gland is vesicular. In *Strigamia acuminata* (Leach) and the littoral species *S. maritima* the glands are lamelliform. In *Haplophilus subterraneus* the glands are again long, thin and tubular, in the littoral species *Henia bicarinata* they are tubular but rather thicker and more contorted.

The second maxillary glands of geophilomorphs open onto the coxosternum of the second maxillae or caudomedially from then: the chitin ring round the opening (Fig. 17) is of taxonomic significance. Fahlander said of this gland only that its glandular part is extended backwards into the body to a greater or lesser extent.

The unicellular glands of the head of geophilomorphs are very large but their frequency is not so great as in the Anamorpha. The cells are most common on the hypopharynx.

Fig. 149. Second maxilla of *Lithobius forficatus* (after Fahlander, 1938).

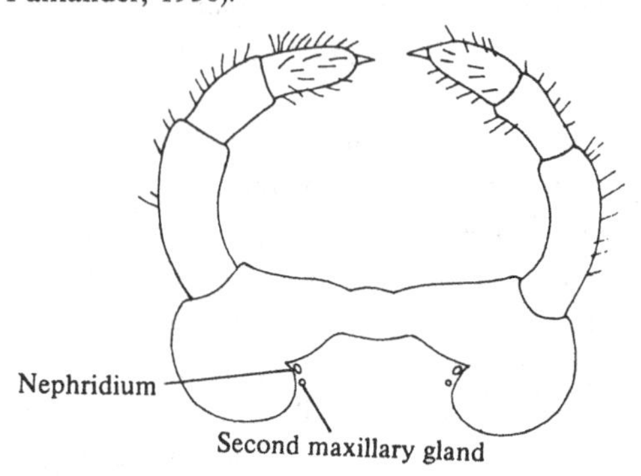

Scolopendromorpha

The lateral buccal glands of *Scolopendra cingulata* are smaller than and ventral to the median pair. Their ducts are very short, opening latero-cranially from the media pair. Jangi (1966) termed them epipharyngeal glands in *S. morsitans*.

The mandibular glands lie on either side of the gut in the anterior part of the body, their long ducts open on the hypopharynx. In *S. subspinipes de haani* Brandt the glands are situated in segments 3–7 and in one case the right gland was much larger than the left and surrounded the gut (MacLeod, 1878). The ducts have a spiral thickening and the glands form a mulberry-like structure. These are Duboscq's *glandes antérieures* (Fig. 150). Jangi termed them labial glands: in *S. moristans* they extend from the fifth to the seventh segments.

First maxillary glands are absent in *Scolopendra* but second maxillary glands (*glandes moyennes* of Duboscq) are present with their glandular portion situated in the third and fourth leg-bearing segments. They are similar in appearance to the mandibular glands and it is difficult to separate them.

Scolopendra possess only the second pair of vesicular glands (system V, Herbst, *glandes posterieures du premier segment*, Duboscq). They open on the first leg-bearing segment immediately under the tergite and in front of the leg insertion (Fig. 150). There is a sphincter muscle at the base of the duct. The glandular region lies mainly in the second leg-bearing segment. It is composed of 30 to 40 vesicles each with a canal opening into a common excretory duct (Duboscq, 1898). Jangi's (1966) term coxal glands is confusing: in *S. morsitans* the gland consists of about 15 vesicles. The epithelium of the main duct appears to be syncytial and is lined by a cuticular intima as are the vesicles.

In *Cryptops* the gland duct widens into a simple thin-walled sac which extends backwards to the fifth leg-bearing segment. The wall of the sac consists of a glandular epithelium similar in structure to that of the vesicles in *Scolopendra.* Balbiani (1890) figured two pairs of 'salivary glands' in *Cryptops savignyi* (Fig. 111).

In scolopendromorphs unicellular gland cells are common around the maxillae (Fahlander, 1938).

Cornwall (1916) described two pairs of 'salivary glands' in

Ethmostigmus platycephalus spinosus as conspicuous white multilobular masses on segments 2–6. These are clearly the maxillary and mandibular glands. A third pair of small translucent glands lying in the fat body laterally and somewhat dorsally to the two salivary glands extend from segment 2 to segments 7 or 8. These open in front of the first pair of legs and are the vesicular glands.

The distribution of head glands in Anamorpha and Epimorpha is summarised in Table 5.

Craterostigmus

The head glands of *Craterostigmus* resemble those of the

Fig. 150. Diagrammatic representation of the head glands of *Scolopendra cingulata* (after Herbst, 1891).

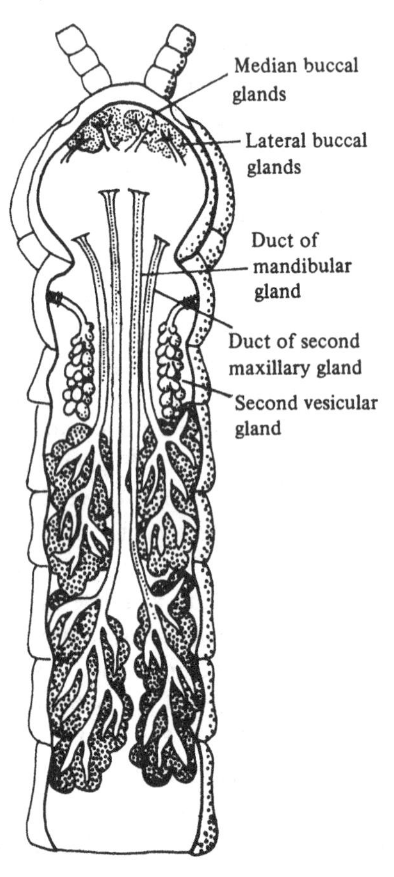

Scolopendromorpha (Manton, 1965). Two pairs of buccal glands open into the pre-oral cavity. The mandibular and second maxillary glands lie in the trunk: they have very wide ducts. The mandibular gland lies in leg-bearing segments 5–6 and the maxillary gland extends from segment 3 to segment 5. A very large second vesicular gland with a large lumen and folded walls extends from the middle of segment 1 to the middle of segment 2. Its duct opens under the edge of tergite 1.

The homology and function of the head glands

Heymons (1901) showed that the anterior glands are of ectodermal origin. The mandibular and maxillary glands I and II would appear to be a metameric series of which the buccal glands may or may not be members.

Fahlander was of the opinion that the vesicular glands might not be serial homologues of the head glands but might belong to another series. This opinion is supported by the fact that the vesicular glands are eosinophilic whereas the head glands are basophilic suggesting that they have different functions.

Duboscq (1898) believed that the head glands had a digestive function. Fahlander also inclined to this view but pointed out that if this was the case, the position of the opening of the second

Table 5. *The distribution of head glands in centipedes*

	Scutigero-morpha	Lithobio-morpha	Scolopendro-morpha	Geophilo-morpha
Median buccal gland	+	+	+	(+)
Lateral buccal gland	+	+	+	–
Mandibular gland	+	+	+	+
First maxillary gland	+	–	–	–
Second maxillary gland	+	+	+	+
First vesicular gland	+	–	–	–
Second vesicular gland	+	–	+	–

maxillary glands high on the lateral side of the head in *Scutigera* is surprising. He believed that the position of the vesicular glands ruled out the possibility that they were connected with digestion and suggested that they might be defensive as they had the same form as the repugnatorial glands of millipedes.

Cornwall (1916) showed that extracts of maxillary, mandibular and vesicular glands in *Ethmostigmus platycephalus spinosus* are slightly acid and when injected into a frog had no effect but caused complete lysis of the erythrocytes of man, dog and other mammals. The extract contains a substance that prevents the coagulation of human blood but extracts of *Rhysida* sp. and *Otostigmus* sp. did not possess this property. The extracts also contained diastase, invertase and proteolytic enzymes. Cornwall stated that experiments suggested that the third pair of glands were largely concerned with the production of lysin and anticoagulin.

Auerbach (1951) and Palmen & Rantala (1954) suggested that brooding centipedes might coat their eggs with a fungicidal secretion produced by the 'oral glands'.

Bennett & Manton (1963) pointed out that *Scutigera* groomed its legs at frequent intervals by passing them, one after the other, between the palps of the second maxillae. It is possible that the nephridia and the second maxillary glands and first vesicular glands (maxillipedal head glands of Bennett & Manton) provide the grooming fluids. *Lithobius* grooms its much shorter legs less frequently, and the flatter head contains fewer accessory glands. Leg grooming is rarely observed in epimorphic centipedes and these, Bennett & Manton assert, readily become infested with the resting stages of tyroglyphid mites, which form a carpet covering their scutes and legs. Such infestations were not found to occur in *Lithobius* or *Scutigera*. The fuller series of head glands in *Scutigera* in comparison with other centipedes, segmental in position but not homologous with segmental organs, may be correlated with the long legs, running habits and grooming needs, rather than representing a primitive feature as suggested by Fahlander.

Binyon & Lewis (1963) suggested that the 'salivary glands' of littoral geophilomorphs might secrete excess salt.

16

The Malpighian tubules and nephridia

Malpighian tubules

In centipedes the Malpighian tubules are a pair of long, forwardly running, blind tubules which originate at the junction of the mid- and hind-gut (Figs. 110, 111).

Lithobiomorpha

The Malpighian tubules of *Lithobius forficatus* have been described by Plateau (1878), Palm (1953) and Rilling (1968). The tubules open into the gut by way of a distinct urinary bladder or ampulla, the inner portion of which is a narrow tube piercing the intestinal wall (Fig. 151*a*). Palm failed to identify distinct sphincter muscles but thought it probable that the muscular coat of the intestine which surrounds the inner portion of the ampulla functions as a sphincter. Rilling (1968) stated that the ampullae show a typical mid-gut musculature but annexe the circular muscles of the hind-gut. In some specimens the tubules terminate in a small thin-walled transparent vesicle, Palm was unable to determine whether this was caused by physiological conditions or whether it was a morphological variation.

The epithelium of the ampulla has a strong basement membrane along which run very thin longitudinal muscle fibres; a few circular and oblique fibrils are also present. In many places connective tissue cells occur around the ampullae, either situated directly on the basement membrane or connected with it by fine cytoplasmic processes. The ampulla cells are tall and narrow with basal nuclei. The apical parts of the cells contain indistinct granules and more or less prominent vacuoles. There is no cell cuticle or brush border. At the junction of ampulla and intestine the ampulla cells are curved outwards with their apical ends nearer the intestinal lumen. At the distal end of the ampulla the cells curve over the epithelium of the tubule (Fig. 151*b*). The epithelium shows little sign of cellular activity. Palm saw no rhythmic contractions of the ampullae although weak and irregular peristaltic movements take place. Short muscle fibres pass from the connective tissue coat of the ampulla to the mid-gut and a short distance anterior to the

Fig. 151. Structure of the Malpighian tubules of *Lithobius forficatus*. *a*, section showing the opening of the tubules into the intestine; *b*, longitudinal section of a tubule showing the region of transition to the ampulla; *c*, transverse section of the distal portion of the tubule (after Palm, 1953).

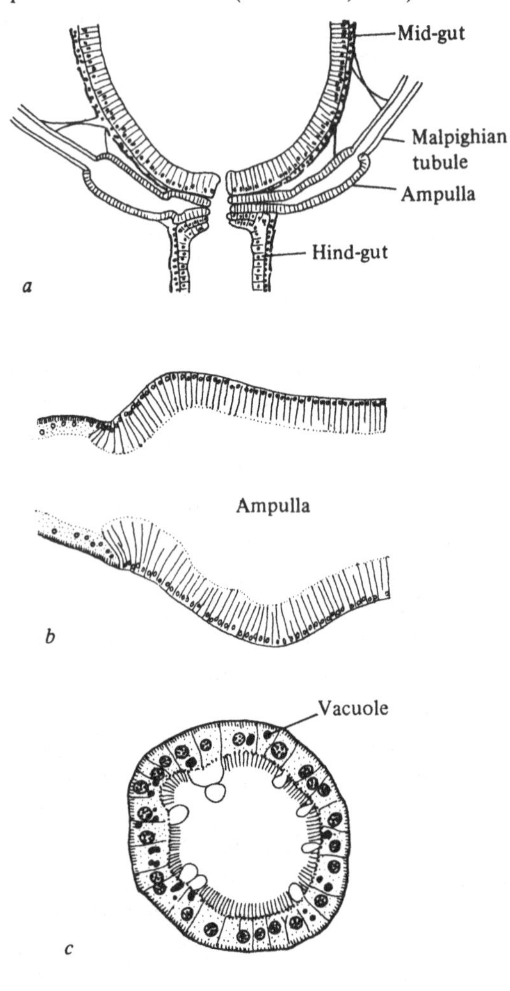

ampulla two long thin fibres stretch from the tubules to the mid-gut (Fig. 151*a*). The fibres consist of two cross striated muscle fibres connected by a thin structureless fibre. Similar fibres connect the ends of muscle fibres with the connective tissue of the Malpighian tube and the mid-gut. These structures show irregular contractions causing some movements of the lower portion of the tubes and the ampullae.

The main part of the Malpighian tubules in *Lithobius* have a histological structure similar to that of the tubules in insects with the exception that the number of cells found in a cross section of the tubule in *Lithobius* is much greater (Figs. 118*g*, 151*c*). The tubule consists of a single layer of cuboidal cells sometimes with distinct cell borders, more often without. The inner surface of each cell has a brush border with rather long filaments. In the part of the tubules near the ampullae the brush border is absent being replaced by a fine granular zone with indistinct perpendicular striae. The end portion of the tubules may also be without a brush border. The nuclei of the cells of the proximal part of the tubule are usually poor in chromatin and have a large and distinct nucleolus: in the distal part the nucleoli are more fragmentary and the nuclei are fairly rich in chromatin (Palm, 1953).

The epithelium of the tubules of *Lithobius tricuspis* Meinert consists of a ring of about 8–12 low cells with a brush border and numerous crystals in the lumen of the tubules (Bertheau, 1971).

A partial or total disintegration of the brush border often occurs in the central portion of the tubules. Immediately beneath the brush border there is a zone filled with fine but distinct granules which stain bright orange with Azan stain. Palm was in no doubt that these were the granules that Verhoeff (1902–25) claimed were excretory. Palm considered that they were connected with the formation of the brush border or a zone where regeneration could take place. Vacuoles are often present in the cells perhaps indicating resorption of water from the lumen of the tubule. Sections show small vesicles being discharged from the lumen with simultaneous disappearance of the brush border (Fig. 151*c*).

In living material the contents of the tubules are partly clear, partly granular. The solid substances are often aggregated to form cylindrical bodies: sometimes the sediment contains cellular debris.

When examined in saline the bodies of solid matter can be seen to be moved towards the proximal end of the tube a short distance at a time but contractions of the tubules are not directly visible under the microscope and no distinct rhythm can be seen.

Electron microscopical studies confirm, in general, the results of earlier work. The basement membrane probably consists of collagen fibres. The plasma membrane of the basal part of the cells of the tubule is folded in a complex manner. The middle region of the cell contains the nucleus and Golgi apparatus and the inner surface is produced into microvilli (brush border) rich in mitochondria. The lumen contains concentrically stratified bodies (urosphaerites) 1.5 μm in diameter, which probably consist of calcium urate and other calcium salts (Füller, 1966).

Other orders

Plateau (1878) figured the gut and Malpighian tubules of the geophilomorphs *Haplophilus subterraneus gervaisii* and *Necrophloeophagus longicornis* (Fig. 110). An ampulla is present in *Necrophloeophagus* but not in *Haplophilus*. In *Clinopodes linearis* and *Pachymerium ferrugineum* the tubules are circular in section, the distal region consisting of a ring of about 10 low cells, the proximal region of a ring of about 20 columnar cells (Bertheau, 1971).

The Malpighian tubules of *Cryptops savignyi* were figured by Plateau (1878) and Balbiani (1890). They lack ampullae (Fig. 111). The cells resemble those of the hind-gut (Balbiani, 1890). In cross section the tubules are seen to consist of a ring of from 4 to 9 low cells with a brush border (Bertheau, 1971). The tubules of *Scolopendra subspinipes* show undulations and loops and are about one and a half times as long as the body. In transverse section they are seen to consist of 40–75 columnar cells without a brush border. There is no trace of muscles (Wang & Wu, 1948). The cells of the Malpighian tubules of *Scolopendra cingulata* are similarly long and narrow but have brush borders. They resemble the mid-gut cells in their structure (Fig. 152). About 80–90 cells are seen in transverse section. There are a number of circular and longitudinal muscle fibres outside the basement membrane (Bertheau, 1971). In *Scolopendra morsitans* the tubules which originate in segment 19 run posteriorly for a short distance before turning forwards and pursuing a sinuous course as far as segment 2. The tubule consists of a ring of only 16 columnar cells which lack brush borders (Jangi, 1966).

A transverse section of the Malpighian tubule of *Scutigera coleoptrata* shows a ring of some 15 cuboidal cells (Knoll, 1974).

Excretory products and function of Malpighian tubules

The breakdown of proteins leads to the formation of carbon dioxide, water and ammonia. As ammonia is toxic it has to be excreted as a dilute solution. Alternatively, it may be converted into a less toxic substance which in insects is usually uric acid. Uric acid is also formed during the breakdown of nucleic acids. Both uric acid and ammonia have been reported as excretory products in centipedes.

Davy (1848) reported an abundance of ammonium urate in the faeces of *Scolopendra morsitans.* Plateau (1878) allowed the tubules of *Lithobius forficatus* to desiccate and found that small crystals of uric acid were produced. He concluded that the species produced very little uric acid. The desiccated Malpighian tubules of *Cryptops* sp. contained small indeterminable crystals that may have been uric acid. The tubules of *Haplophilus* contained no crystals or con-

Fig. 152. Transverse section of the mid-region of the Malpighian tubule of *Scolopendra cingulata* (after Bertheau, 1971).

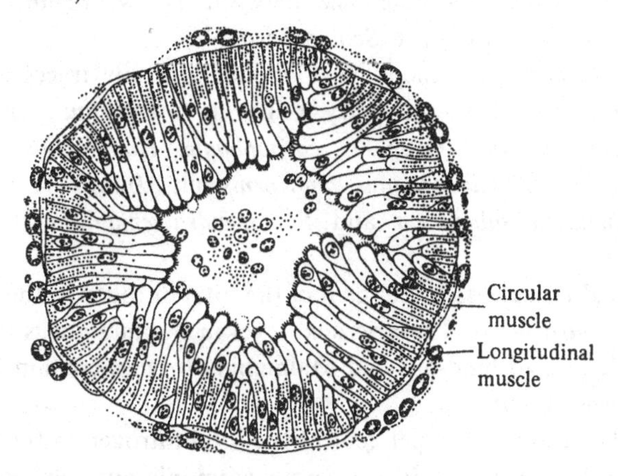

cretions when fresh but rarely some indeterminable crystals were produced on desiccation.

High concentrations of uric acid are found in the haemolymph of *Scolopendra subspinipes* but concentrations in the intestinal tissue are very low (Wang & Wu, 1948). Rather high values were obtained for the mid-gut contents, values for the hind-gut were double these. There is very little or no uric acid in the distal portion of the Malpighian tubules: the concentration in the central portion is lower than that of the haemolymph, the concentration in the basal portion almost twice as much.

Uric acid concentrations were higher in the tubules if the animals had been starved for some weeks. The authors suggested that this was due to the uric acid being chiefly derived from the 'degradation of protein in the tissue and not directly absorbed from the food'. Palm (1953) pointed out, however, that the rate of excretion drops when no new food is ingested so that the flow of urine is less and more water can be reabsorbed leading to the formation of uric acid crystals. Bennett & Manton (1963) investigated uric acid production in *Scolopendra cingulata*, *Lithobius forficatus* and *L. variegatus*. The bodies of the centipedes were divided into four parts, namely the head, fore- and mid-gut and the hinder part of the salivary glands, the hind-gut and Malpighian tubules, the remaining parts of the body and legs. The uric acid values of the extracts ranged from 0.14 to 144.00 mg uric acid/g dry weight. In all cases the hind-gut/ Malpighian tubule extracts gave significantly higher yields than those of the heads, the difference, between the two regions being greater in *Lithobius* than in *Scolopendra*.

Whole *Lithobius melanops* Newport contain only traces of uric acid. There are 3.0–4.0 mg/g dry weight in *L. forficatus*, *Cryptops hortensis*, *C. savignyi*, *Geophilus carpophagus*, *Haplophilus subterraneus* and *Necrophloeophagus longicornis* but 8.0 and 9.0 mg/g in *Hydroschendyla submarina* and *Scolopendra cingulata* (Hubert, 1968, 1977).

Uric acid represents 12 per cent of the total nitrogen in the faeces of *Scolopendra heros*. The uric acid level in the mid-gut is 0.9 per cent while pooled hind-gut and Malpighian tubules contain 17 per cent (Horne, 1969).

In *Lithobius* only 1–8 per cent of the total nitrogen in the faeces is uric acid nitrogen: 50–60 per cent is ammonia nitrogen (Bennett

& Manton, 1963). The authors concluded that 'although the quantities analysed were small, ammonia appears to form the greater part of the nitrogenous excretory material of *Lithobius*'.

Analysis showed that the head, where the nephidia are situated, is not a major site of ammonia excretion. The ratios of body ammonia to head ammonia in *Lithobius* vary from 0.9:1 to 5.5:1 (mean 2.2:1).

Hubert (1977) showed that the faeces of *L. forficatus* contain 150 mg ammonia/g dry weight, those of *L. melanops* 250 mg/g. The faeces contain 40–50 per cent water.

Uric acid may be converted via allantoin, allantoic acid, urea (plus glyoxalic acid) to ammonia. The enzymes controlling this metabolic pathway are uricase, allantoinase, allantoicase and urease respectively. Hubert & Razet (1965) investigated the activity of these enzymes in six species of centipedes and reported uricase activity in all of them but found no trace of allantoinase or allantoicase activity: urease activity was very weak. Uric acid was found in the haemolymph, tissues and faeces but glyoxalic acid, allantoin and allantoic acid were not found and there was no appreciable evidence for the presence of urea. In contrast, ammonia was found in all species. Despite results obtained *in vitro* the authors were unable to confirm that uric acid was degraded by uricase *in vivo*. It was concluded that centipedes are both uricotelic and ammoniotelic. The Malpighian tubules appear to concentrate uric acid and, perhaps, ammonia.

Palm (1953) examined the reaction of the Malpighian tubules of several species of lithobiomorphs and geophilomorphs to the injection of vital dyes such as neutral red, indigo carmine, methylene blue and Janus green B. In most cases no traces of dyes were found in the tubules. He concluded that the ability of chilopods to eliminate vital dyes was very low, especially when compared with the Malpighian tubules of insects.

Nephridia

Both lithobiomorphs and scutigeromorphs have glandular structures in the head quite unlike the other head glands and clearly homologous with the nephridia of millipedes and apterygote insects.

Lithobiomorpha

The nephridia of *Lithobius* were first described by Herbst (1891) as System II of the head glands. Fahlander (1938) gave a detailed account of the structures. Each has two openings: the first medial of the first maxilla, the second posterior and median to the coxae of the second maxillae near the opening of the second maxillary gland (Fig. 149). Neither Verhoeff (1902–25) nor Attems (1926) mentioned that the nephridium had two ducts: Fahlander suggested that they probably doubted Herbst's account.

The anterior opening leads by a short canal to a large folded sac, the utriculus or urinary bladder which fills most of the space between the ventral nerve cord and the ventral and lateral body wall in the region of the first maxilla. The utriculus is lined by cuticle which is raised into small tubercles (Fig. 153*a*), the cytoplasm of the epithelial cells contains numerous vertical fibrils. Gabe (1967) reported that the bases of these cells were much folded. A narrow canal lined by a continuation of the utricular epithelium joins the utriculus to the dorsally situated labyrinth (*Endsack* of Herbst). It has a low secretory epithelium and is not lined with

Fig. 153. The nephridium of *Lithobius forficatus*. *a*, optical section of the right nephridium (after Fahlander, 1938); *b*, dorsal view of the left nephridium showing the dilator muscles (Dil 1–8) (after Rilling, 1968).

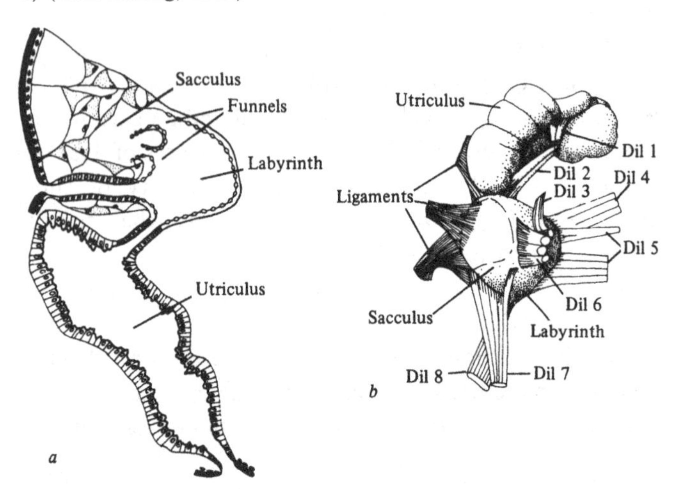

cuticle. The labyrinth connects with the exterior by a duct similar to that joining the labyrinth with the sacculus, its distal portion formed of invaginated body wall and opening near the opening of the second maxillary gland (Fig. 149). The third part of the nephridium, the sacculus, was taken for fat body by Herbst. It consists of a cavity traversed by a network of connective tissue. Fahlander was of the opinion that the sacculus was of the same structure of that of diplopods (he figured *Antichiropus variabilis*) in which the wall of the sacculus consists of fine canals which hang into the lumen and in which the blood circulates. In *Lithobius* it is difficult to distinguish the cavities of the lacunae from the cavity of the sacculus. Lymphocytes are often found in the lacunae in large clumps. Lymphocytes are also found in the cavity of the sacculus as well as in the labyrinth. Fahlander inclined to the view that this represented their natural distribution rather than being the result of disruption during death or fixation.

The labyrinth has a well-developed system of dilator muscles. Fahlander was unable to identify a muscle running to the apodeme between the maxilliped segment and the head as described by Herbst but found four sets of muscles.

There are two funnel-like openings from the labyrinth into the sacculus which are clearly made of labyrinth epithelium. The sacculus lies lateral to the labyrinth, surrounding the duct of the latter (Fig. 153*a*).

Rilling (1968) redescribed the nephridium of *Lithobius forficatus*. His description agrees substantially with Fahlander's. He figures three ligaments and eight dilator muscles attached to the sacculus and labyrinth (Fig. 153*b*) and regards the cavity of the sacculus as a network of lacunae. Dilator 8 attaches to the cuticle between the head capsule and the forcipular tergite and is probably the one described by Herbst.

Scutigeromorpha

The nephridia of scutigeromorphs were also first described by Herbst (1891) for *Scutigera coleoptrata*: he described two pairs of head glands – systems I and II. System I opens, according to him, between the first maxillae and consists of a pair of glandular tubes each having two branches which are attached to but not fused with system II. System II consists of a pair of sacs, each folded double and situated in the second maxillary segment. These sacs open by short ducts, on the lateral side of the head. Fahlander (1938) maintained that the two systems were connected in *Scutigera* as they are in other genera of Scutigeromorpha (*Thereuopoda* and *Thereuonema*).

According to Fahlander, a short cuticle-lined duct leads laterally from the opening medial to the first maxilla and from this a glandular tube leads postero-dorsally to about the level of the *aorta cephalica*. From this tube lead three lateral tubes (Fig. 154).

The most ventral branch originates in the region of the supraneural system and first runs postero-dorsally and then runs towards the body wall, joining a thick-walled sac, the utriculus (Herbst's system II). The utriculus is horseshoe-shaped and opens by a short duct near the opening of the second maxillary gland and

Fig. 154. Optical section of the right nephridium of *Thereuonema clunifera* (after Fahlander, 1938).

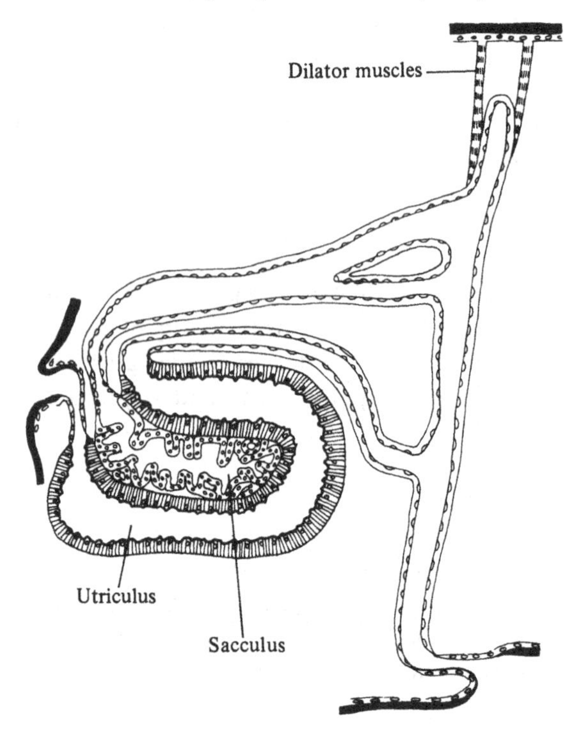

is lined by rather thick cuticle-bearing tubercles. The two dorsal branches join and run more or less parallel with the ventral branch, turning ventrally by the body wall to open into the sacculus which is situated between the two shanks of the utriculus.

The labyrinth has only two weak dilator muscles: these are attached to the head shield (Fig. 154).

The sacculus is limited by connective tissue beneath which is the saccular epithelium. The latter is thrown into folds and produced into ramifying outgrowths. The epithelium does not form canals as it does in the Diplopoda and the cavity does not contain lymphocytes. The epithelium is syncytial, it is not eosinophilic as is the epithelium of the labyrinth but basophilic.

Homologies and function of nephridia

Fahlander (1938) was in no doubt that the glands described above were true nephridia. They correspond in position, anatomy and histology with the maxillary glands of the Crustacea and the labial kidneys of Diplopoda and apterygote insects and he considered them to serve as excretory organs.

The fact that there are two pairs of ducts in the two orders and two openings between the labyrinth and sacculus in the Lithobiomorpha suggest that the glands represent two pairs of fused nephridia, those of the first and second maxillary segments, the sacculus being formed from the coelom of both segments. This situation is not unique, Fahlander demonstrated that the Symphyla possessed separate segmental organs, one pair in each maxillary segment. In the Lithobiomorpha the utriculus belongs to the first maxillary segment, in the Scutigeromorpha to the second.

Active secretion take place in the labyrinth. The distal part of each epithelial cell protrudes into the lumen of the labyrinth and becomes constricted to form a sphere. Similar activity is seen in the sacculus of Scutigeromorpha but not in the sacculus of Lithobiomorpha. The waste material may be carried away by lymphocytes. Around the labyrinth in both orders there are found the so-called nephrocytes which pick up waste material from the blood. They form balls or strings surrounded by a connective tissue membrane whose cells are not clearly distinguishable from fat cells.

Geophilomorphs and scolopendromorphs show no certain traces of nephridia. Fahlander (1938) however, suggested that lymphoidal tissue that appears and then disappears during the development of *Scolopendra* and which Heymons (1901) considered to be formed mainly of mesoderm of the intercalar segment might be a vestige of the nephridium.

Palm (1953) repeated and confirmed Fahlander's histological work on the nephridium of *L. forficatus* and also examined the excretion of Indian ink, carmine suspension and nine vital dyes by this organ. Only when a dye is found in the lumen of the labyrinth or utriculus can real excretion be said to have occurred: when the dye is found in the sacculus, the absorption must be regarded as vital staining.

Carmine suspensions were only found to be absorbed by the haemocytes of the sacculi and the same was true of Indian ink with the exception of two specimens in which a small amount of the ink appeared in the lumen of the utriculus. Mixtures of Indian ink, lithium carminate and indigo carmine were injected, the carminate was found in the nephridia but not the other dyes. Congo red caused some staining of the sacculus tissue and was found in the lumen of the utriculus in a few *Lithobius*. Neutral red also produced some vital staining. The basic aniline dyes (methylene blue, brilliant cresyl blue, Janus green B and toluylene blue) were not excreted by the nephridia and caused little or no vital staining. Palm concluded that the nephridia 'possessed some ability to excrete injected vital dyes although their activity cannot be regarded as very efficient; that the organs really are chiefly excretory remains to be demonstrated, although it seems fairly certain'.

Palm postulated that the arrangement of sacculi indicates that they form a filter apparatus belonging to the same type of structures as the glomeruli of vertebrate kidneys even though the flow of blood in the sacculi must be very low. Gabe (1967) considered that the histological characters of the sacculus of *Lithobius* and *Scutigera* are compatible with the formation of urine by filtration and that the labyrinth could be the site of reabsorption.

The function of chilopod nephridia as excretory organs has been questioned. Bennett & Manton (1963) drew attention to the fact that whereas nephridia (segmental organs) are major sites of nitrogenous excretion and osmoregulation in Crustacea, they provide copious salivary juice and grooming fluids in Onychophora, Diplopoda and Thysanura. These great functional differences are not associated with histological differences of the end sacs (sacculus). Thus statements about function based upon histology alone, such as Fahlander's assertion that they are excretory organs, were not justified. They further pointed out that since segmental organs are present in the Anamorpha but not in the Epimorpha, Fahlander's diagnosis of function implies that nitrogenous excretion might be carried out quite differently in the two groups of centipedes. The head is not a major site of ammonia excretion. A possible function of the nephridia is the production of grooming fluids (Bennett & Manton, 1963).

17

The reproductive system and reproduction

Geophilomorpha

Secondary sexual characters

In most geophilomorphs the number of pairs of legs, which is always odd, varies and where this is so, the females tend to have a greater number of pairs of legs than the males, for example, in British *Schendyla nemorensis* there are 37–41 pairs of legs in males and 39–43 pairs in females. In the larger species *Haplophilus subterraneus*, there are 77–81 pairs of legs in males and 79–83 in females (Eason, 1964). In many species of the family Mecistocephalidae, however, the number of legs does not vary and is the same in both sexes, for example, 49 pairs in *Mecistocephalus insularis* (H. Lucas) from Africa and India and 51 pairs in *M. evansi* Brölemann from Iran. Analysis of broods of *Henia illyrica* (Meinert), *Pachymerium fernigineum*, *Strigamia acuminata* and *S. crassipes* has shown that female larvae have the same number of legs as the mother, male larvae two pairs less (Prunescu & Capuşe, 1972).

In many geophilids, for example species of *Chaetechelyne*, *Strigamia* and *Geophilus* the last pair of walking legs in males (Fig. 14) is tumescent and much more setose than that of the female which more closely resembles the normal walking legs. In some schendylids, for example *Hydroschendyla submarina* the last pair of legs are swollen and densely setose in both sexes. The setae are presumably sensory and the enlarged legs presumably contain glands but these have yet to be described. It is tempting to suggest that modifications of the last pair of legs are associated with mating behaviour in which male and female come together head to tail and antennae tap the last pair of legs. Possibly the legs of the male produce pheromones.

In *Strigamia maritima* juveniles of both sexes are of a similar size but by the time they have reached the stadium maturus junior (see Chapter 18) females are considerably larger than males (mean weights: females 20 mg, males 15 mg) (Lewis, 1961).

In the mecistocephalid genus *Tygarrup* from the Himalayas the females lack sternital pore fields but the males possess them. This unusual case of sexual dimorphism resulted in the male being placed in a new genus *Brahamaputrus* Verhoef a matter rectified by Crabill (1968).

Reproductive organs

Fabre (1855) described the anatomy of the male reproductive organs of *Himantarium gabrielis* and two species of doubtful identity: '*Geophilus Ilicis*' and '*G. convolvens*'. Descriptions of the organs of other species by Schaufler (1889), Tuzet & Manier (1953), Prunesco (1968) and Breucker (1970) are essentially the same as Fabre's.

In all species there is a pair of fusiform testes, each opening by two vasa efferentia, one leaving each end. The vasa efferentia fuse to form a much coiled vas deferens which consists in *Clinopodes linearis* and *Strigamia maritima*, and perhaps other species, of a thin distal and thick proximal region, the latter dividing to pass round the gut, fusing again to open ventrally on the 'penis'. Two pairs of tubular accessory glands open into the genital atrium (Fig. 155*a*). Most of the published figures of male reproductive organs are highly diagrammatic: an accurate figure was given by Breuker (1970) for *Clinopodes linearis* (Fig. 155*b*). In this species the vasa efferentia and vas deferens consist of a simple epithelium with an extraordinarily thick basement membrane and a very thin layer of striated muscles. The vasa efferentia are characterised by deep regular folding of the basal cell membrane, abundant mitochondria and a border of microvilli. The narrow region of the vas deferens is characterised by lateral interdigitations between the cells, stereocilia and cytoplasmic protrusions suggesting secretion (Fig. 155*c*). The thick part of the vas deferens is characterised by an abundance of cell organelles, especially Golgi complexes and ergastoplasm. There is a large quantity of secretory material in its lumen.

Fabre (1855) described the female reproductive system of *Himantarium gabrielis*, *Geophilus electricus*, and '*G. Ilicis*' and '*G. con-*

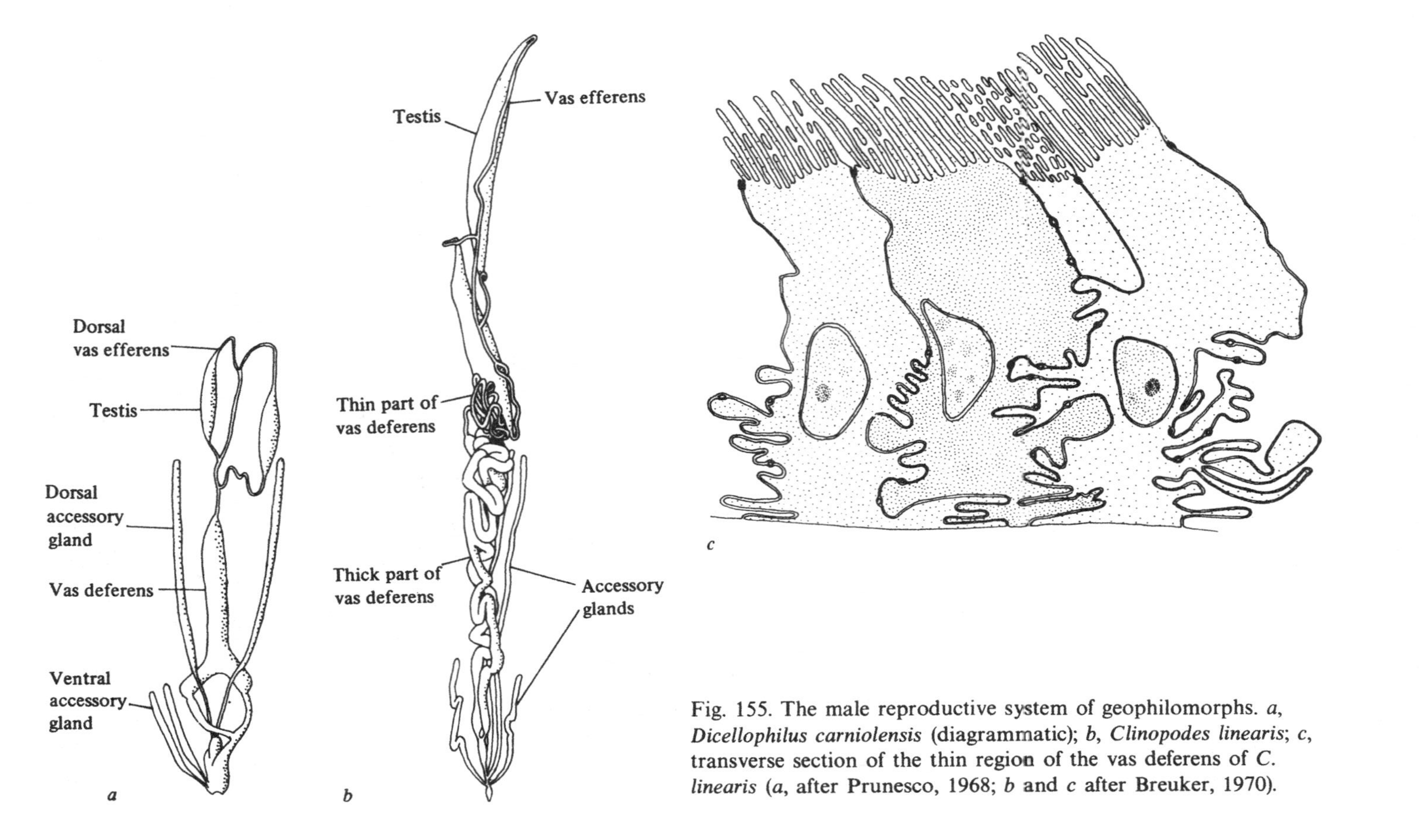

Fig. 155. The male reproductive system of geophilomorphs. *a*, *Dicellophilus carniolensis* (diagrammatic); *b*, *Clinopodes linearis*; *c*, transverse section of the thin region of the vas deferens of *C. linearis* (*a*, after Prunesco, 1968; *b* and *c* after Breuker, 1970).

volvens'. Schaufler (1889) described the system in *Clinopodes flavidus* C. L. Koch and Prunesco (1967*a*) those of *Dicellophilus carniolensis*, *Pachymerium ferrugineum* and *P. tristanicum* Attems, *Strigamia acuminata* and *H. gabrielis*. In all species there is a single median dorsal ovary leading to a short oviduct which divides to pass round the gut, the two branches fusing again before opening into the genital atrium (Fig. 156*a*). Fabre recorded one pair of accessory glands in female geophilomorphs, the glands being very small in *Geophilus* spp. In *Clinopodes flavidus* there are large ventral and very small dorsal accessory glands (Schaufler, 1889) whilst in

Fig. 156. Diagrammatic representation of the female reproductive system of geophilomorphs. *a*, *Dicellophilus carniolensis*; *b*, *Himantarium gabrielis* (after Prunesco, 1967*a*).

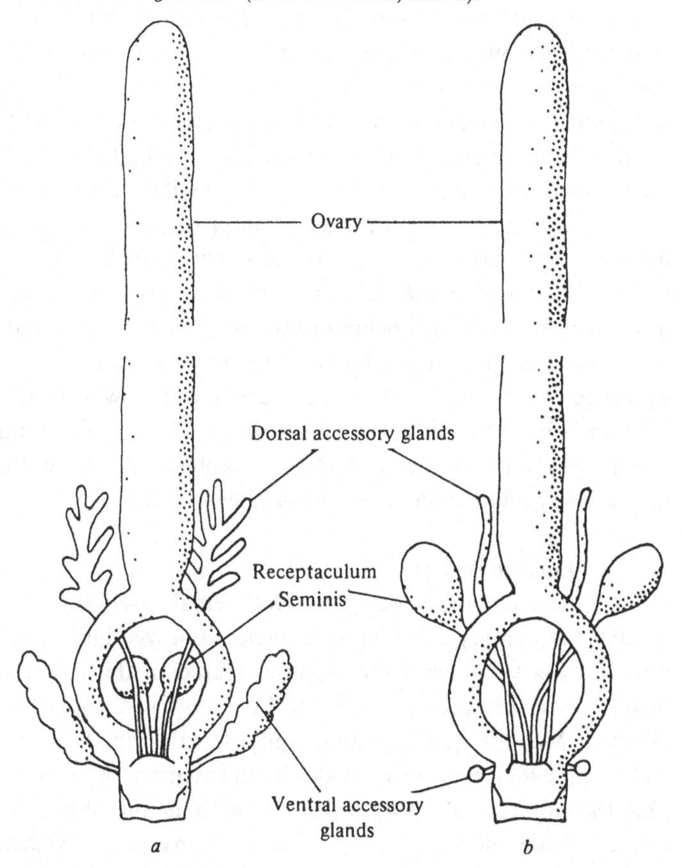

Dicellophilus carniolensis there is, in addition, a pair of very small 'supplementary' glands 40–60 μm long also opening into the genital atrium. The dorsal glands of this species are branched and the ventral glands relatively large (Fig. 156*a*). In *H. gabrielis* the dorsal glands are simple and the ventral glands very small (Fig. 156*b*) (Prunesco, 1967*a*). There is confusion in the nomenclature of the female accessory glands: those described as dorsal glands by Prunesco were termed *untere Drüsen* by Schaufler. The development of the gonads of *Pachymerium ferrugineum* was described by Prunescu & Capuşe (1972).

The variation in the development of the glands in different species may be due to seasonal changes. Alternatively it may reflect their greater importance in some species than in others: Weil (1958) suggested that these glands produce the secretion that sticks the eggs together after they are laid and not all geophilomorphs glue their eggs together (see below).

A pair of spherical or oval seminal receptacles is situated in the region of the antepenultimate or penultimate leg-bearing segment. Each opens by a narrow duct into the genital atrium and stores sperm prior to fertilisation. The seminal receptacles arise as epidermal invaginations and so are lined with cuticle. The cuticular lining is shed at each moult but in *Strigamia maritima* and probably other species, it remains behind in the receptacle because the lumen of the duct is too narrow for it to be withdrawn. In sections and squashes the lining of the receptacle and the whorls of sperm contained by them are easily visible (Figs. 157*a*, *b*). The number of sperm whorls in the squashes of the receptacles has been used as a means of ageing female *S. maritima* (Lewis, 1961).

Parthenogenesis

Sograff (1882) observed that female *Geophilus proximus* always had empty seminal receptacles and yet laid eggs which developed without fertilisation when reared in the laboratory. In Scandinavia the species appears to be almost invariably parthenogenetic: Meinert (1870) found only females in Denmark and Palmen (1948) found only females in an extensive survey of Finland and the adjacent Russian territory north of the White Sea and quoted Lohmander as having found no males in Sweden. Both sexes occur, however, in Latvia (Trauberg, 1932).

Jeekel (1964) suggested that *Strigamia maritima* might occur as a parthenogenetic race on the coasts of the Baltic sea. This supposition was based on Bergsøe & Meinert's (1866) data. Enghoff (1976) re-examined this material and showed that the authors misidentified several juvenile males as females.

Sperm transfer

Fabre (1855) observed simple webs, each bearing a white spherical globule which proved to contain sperm of his '*Geophilus convolvens*' and suggested that these were spermatophores deposited in the soil by males to be found later by females. This observation was largely overlooked: Schaufler (1889) considered that the structure of the genitalia in centipedes allowed copulation and Heymons (1901) and Attems (1930*a*) considered that copulation might take place.

Demange (1956) observed the web and spermatophores of the lithobiomorph *Lithobius piceus gracilitarsus* Brölemann. Klingel (1959) observed sperm transfer in *Necrophloeophagus longicornis*: individuals ready to 'pair' show lateral 'wiping' movements of the

Fig. 157. The seminal receptacle of a maturus senior female *Strigamia maritima*. *a*, drawing of a squash preparation; *b*, stereogram (from Lewis, 1961).

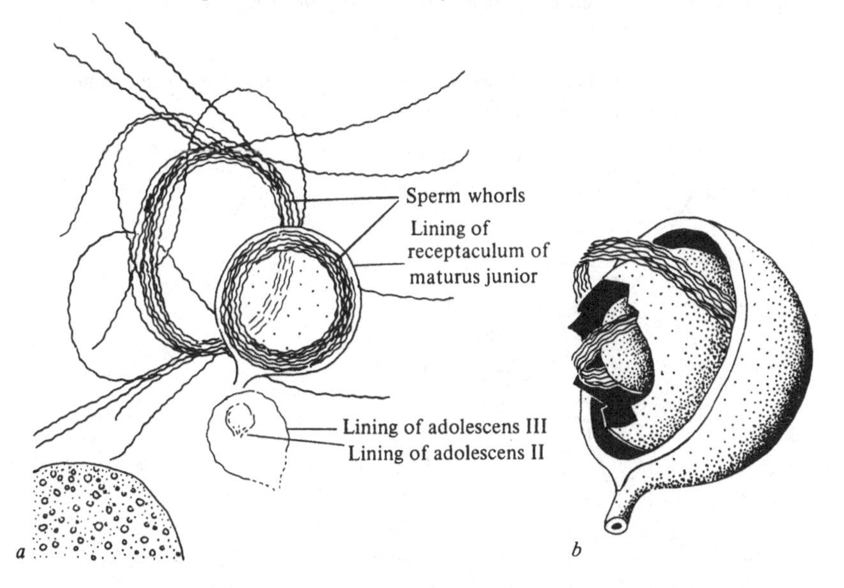

last segments and the male and female usually come together head to tail and antennae and anal legs are tapped. The individuals then separate and the male spins a simple web of zigzag threads across the burrow (Fig. 158) on which he deposits a spermatophore or, more accurately, an unencased sperm droplet. The female may not return for 3 or 4 h. When she does, she locates the spermatophore with her antennae and moves over it waving her posterior end rhythmically in a wiping motion thus picking up the sperm.

Egg-laying and brooding

Fabre (1855) appears to have been the first to observe brooding in centipedes. This process, in which the female remains curled round the eggs until they have hatched and the young moulted several times, occurs in both geophilomorphs and scolopendromorphs. Brooding has been observed by many workers: only the most detailed accounts will be mentioned here.

In America, *Geophilus rubens* commonly oviposits under bark. Clutch sizes vary from 17 to 73 (mean, 39). When ovipositing, the female holds her body in a wide coil so that the genital segments are near the antennae. Eggs appear at the rate of about one every 6 min. They are held briefly in the poison claws and then transferred to a heap about which the body is coiled. When about a dozen eggs have been deposited the posterior part of the body is swung round

Fig. 158. Web and spermatophore of *Necrophloeophagus longicornis* (after Klingel, 1959).

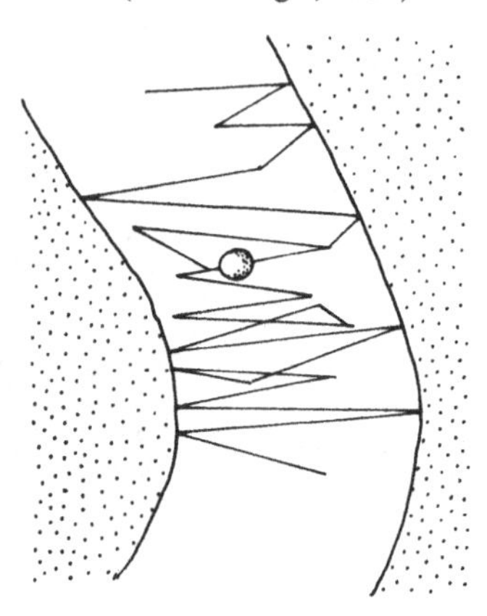

so that the anal segment rests on the pile of eggs: eggs subsequently laid slip into position without manipulation by the poison claws (Johnson, 1952). The female observed laid about 30 eggs in three hours. The tergites not the sternites are in contact with the eggs. Manton (1965) noted that many species of geophilomorph take up a coiled resting position with the head end well covered and the sternal surfaces exposed so that the sternal glands can discharge their noxious fluids in all directions. *G. rubens* explores its eggs periodically with its antennae: they are frequently held and turned with the poison claws and brushed with the maxillae.

In Germany *Necrophloeophagus longicornis* oviposits in soil at a depth of 5–20 mm. Between 12 and 73 eggs are laid (mean, 34). Brooding females were found from 8 June to 14 August (Weil, 1958). During oviposition the female forms a loop with the mid-region of her body such that the segments at each end of the loop are in contact and form a platform on which the eggs are deposited (Fig. 159). The last three to five segments are held off the ground and regular contractions of the posterior segments force the eggs out. Each egg, covered by a sticky secretion is swung forward to the

Fig. 159. *Necrophloeophagus longicornis* laying eggs (after Weil, 1958).

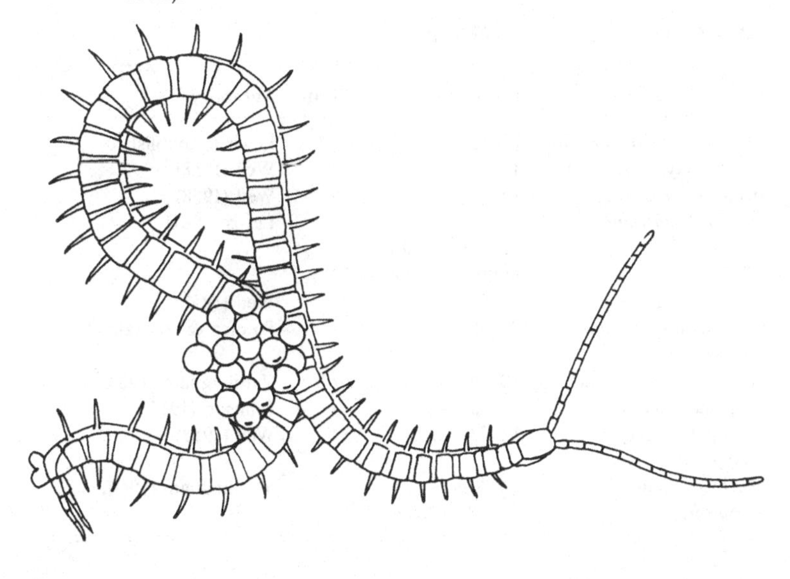

pile in the mid-region of the body. The last pair of legs are held vertically out of the way. The specimen observed by Weil laid 47 eggs in 5–6 h. When oviposition is completed the female winds about the eggs with her sternites turned outwards.

In Finland, *Pachymerium ferrugineum* lays between 20 and 55 eggs in brood cavities 0.8–1.2 cm in diameter in beds of the moss *Racomitrium*. As in other geophilomorphs, the female does not feed while brooding which lasts 40 to 50 days in *Pachymerium* (Palmen & Rantala, 1954). Palmen & Rantala discuss brooding at some length: they observed, as did Sograff (1882) and Schaufler (1889) amongst others, that the female usually ate her brood when disturbed but noted that she was less likely to attack larvae than eggs. Sograff (1882) observed that the eggs of *P. ferrugineum* became infected with fungi when separated from the female; Palmen & Rantala repeated this observation and suggested that brooding was necessary to prevent the brood becoming infected by fungi or micro-organisms. They suggested that a fungicidal secretion was produced by the 'oral glands' or coxal glands. Johnson (1952) considered that the fungicidal secretion must be produced by head glands as the sternites and coxae are never in contact with the brood. Weil (1958) succeeded in keeping eggs isolated from the

Table 6. *Brood size in Geophilomorpha*

	Brood size	Mean	Authority
Hydroschendyla submarina	14–15	14.5	Lewis (unpublished)
Chaetechelyne vesuviana	17	17	Weil (1958)
Strigamia crassipes	15	15	Weil (1958)
Strigamia maritima	3–44		Lewis (1961)
	maturus junior	13	
	maturus senior	27	
Clinopodes linearis	13–51	25	Weil (1958)
Pachymerium ferrugineum	20–55	—	Palmen & Rantala (1954)
Geophilus carpophagus	10–14	—	Vaitilingham (1960)
Geophilus rubens	17–73	39	Johnson (1952)
Necrophloeophagus longicornis	12–73	34	Weil (1958)
Brachygeophilus truncorum	6–20	—	Vaitilingham (1960)

female alive and infection free until they hatched but believed that secretions from the female 'had a definite effect'.

It is probable that most geophilomorphs lay their eggs in soil. In addition to the brooding sites mentioned above the following have been recorded: *Geophilus proximus* under bark, *P. ferrugneum* in sand (Sograff, 1882), *Schendyla nemorensis* in soil of an ant hill (Lewis, unpublished data). Jones, Conner, Meinwald, Eisner & Eisner (1976) recorded *Orphnaeus brasilianus* (Humbert & Saussure) brooding in any empty cocoon of the moth *Megalopyge opercularis* in Florida. The littoral centipede *Strigamia maritima* lays its eggs in sand and fine shingle behind the shingle of the main beach (Lewis, 1961) or in grit and decaying organic matter between flat rocks at the storm line (Lewis, 1962). *Hydroschendyla submarina* broods its eggs and young in rock crevices in the *Verrucaria* and *Chthamalus* zones at Plymouth (Lewis, 1962).

A brooding female *Necrophloeophagus longicornis* will defend her brood against another female of the same species: during the combat the posterior two-thirds of her body remain curled round the eggs. *Clinopodes linearis* females sometimes moult while brooding their eggs (Weil, 1958).

Data on brood size and time of brooding are shown in Table 6.

Latzel (1880) found a nest of *Dicellophilus carniolensis* (C. L. Koch) which in addition to 20 young contained 12 young of *Geophilus pygmaeus* Latzel.

Scolopendromorpha

Secondary sexual characters

There is no sexual dimorphism in leg number in the Scolopendromorpha but in the males the last pair of legs sometimes shows specialisation. In *Scolopendra morsitans* the dorsal side of the prefemur, femur and sometimes the tibia is flattened and has a prominent ridge along each lateral border (Fig. 160). In the subgenus *Parotostigmus* of *Otostigmus* from South America and Africa there is frequently a cylindrical process on the inner side of the prefemur of the last pair of legs in males (Fig. 161*a*) and in *O.* (*Parotostigmus*) *insignis* Kraepelin males have, in addition, a large coxal process on the twentieth pair of legs (Fig. 161*b*). Male *O.* (*Parotostigmus*) *caudatus* Brölemann have an almost cylindrical finger-like process on tergite 21 (Fig. 161*c*).

Female *S. morsitans* tend to be larger than the male: Lewis (1968) recorded a maximum length of 81 mm for females and 69 mm for males. In male *S. morsitans* from Zaria the last five to seven tergites are marginate, in females only the last four or five.

Fig. 160. Dorsal view of the prefemur and femur of the last leg of a male *Scolopendra morsitans* (after Attems, 1930*a*).

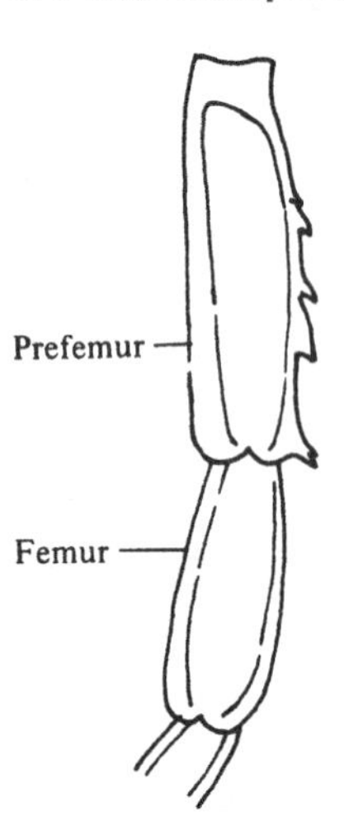

Fig. 161. Secondary sexual characters in male *Otostigmus* spp. *a*, dorsal view of tergite 21 and the terminal prefemora of *O.* (*Parotostigmus*) *insignis*. *b*, ventral view of the terminal segments of the same; *c*, tergite 21 of *O. caudatus* (after Attems, 1930*a*).

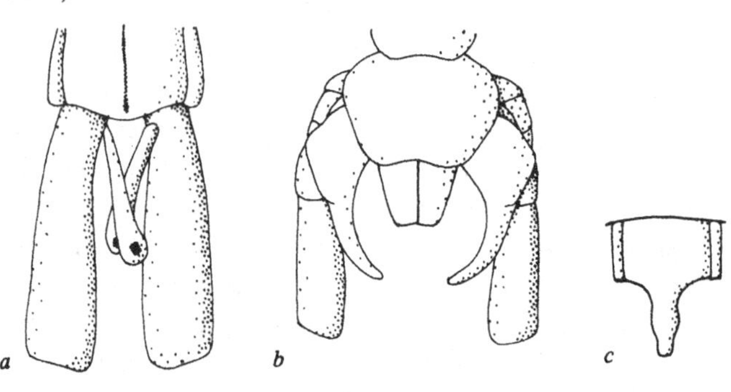

Male reproductive organs

Fabre (1855) described the anatomy of the male reproductive organs of *Scolopendra alternans*, and *Cryptops hortensis* and *C. savignyi*, and Heymons (1901) that of *S. cingulata*. Demange (1946) gave an account of the arrangement in four species of *Scolopendra* and *Otostigmus limbatus* Meinert, and Jangi (1956) gave details of the system in *S. morsitans*. Further data were presented by Demange & Richard (1969).

In Scolopendridae the testes are arranged in pairs. In *Scolopendra subspinipes* there are seven pairs, 10 in *S. valida* and *Cormocephalus punctiventris* (Newport), 12 in *S. viridicornis* Newport and 13 in *S. gigantea*. *Otostigmus orientalis* Porat has five pairs, *O. limbatus* Meinert 10. In the Cryptopidae the testes are arranged alternately along the vas deferens: there are four testes in *Cryptops hortensis*, 10 in *Newportia monticola* Pocock and 26 in *Otocryptops gracilis* (Wood).

Each testis discharges by an anterior and posterior vas efferens as in the Geophilomorpha. The vasa efferentia join the single vas deferens (*canale axiale*) which is continued anterior to the testes as a very thin filament. In most scolopendrids the posterior ducts of the testis pair join the vas deferens opposite the anterior ducts of the succeeding pair (Fig. 162*a*). The arrangement in cryptopids is

Fig. 162. The arrangement of the testes in scolopendromorphs. *a*, *Scolopendra valida*; *b*, *Cryptops hortensis* (after Demange & Richard, 1969).

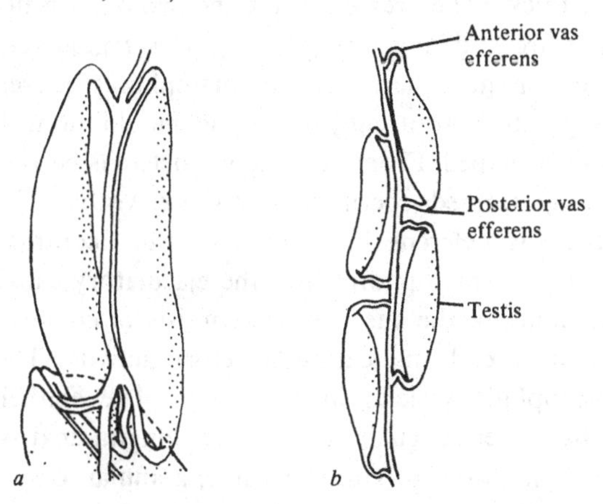

shown in Fig. 162*b*. Jangi (1956) pointed out that Heymons (1901) was in error in showing the vasa efferentia of *S. cingulata* converging to open on the vas deferens at the same point. The vas deferens runs dorsal to the gut and continues posterior to the testicular region as a coiled epididymis (Fig. 163*a*). The posterior part of the vas deferens is wider and contains dilations (*bourses à spermatophores*) in which spermatophores may be found. The vas deferens divides to pass round the gut before opening ventrally on the penis. The left duct is longer and more slender than the right, forming an M-shaped arch in the adult, the *arcus genitalis* of Heymons. The ducts join below the gut to form the terminal duct which opens dorsal to the common opening of the larger pair of accessory glands (Fig. 163*a*).

The most detailed account of the scolopendromorph vas deferens is Brunhuber's (1970*a*) for *Cormocephalus anceps anceps* (Porat). In this species sperm collect in the anterior part of the vas deferens where they aggregate into bundles and secretory granules are produced. The vas deferens then narrows to form the coiled 'epidyme' and then widens again. In this second wide region the sperm are seen to be coiled and a second secretion, that will form a granular mass at one end of the spermatophore, accumulates. A sphincter separates this region from a section which is sausage shaped at the time of active spermatophore production (January for *Cormocephalus* in South Africa) and which produces the inner envelope of the spermatophore. A sharp constriction separates this from a further thick-walled region where the two outer walls of the spermatophore are laid down (Fig. 163*b*). This region retains its shape whether or not there are spermatophores present and Brunhuber suggests that it may form the mould in which the spermatophore is shaped. From here the vas deferens becomes thin walled and fully formed spermatophores are visible. The final portion of the vas deferens has muscular walls, contraction of which forces the spermatophores into the ejaculatory canal. In *S. morsitans* the lining of the vas deferens consists of columnar cells with swollen inner ends indicating secretory activity. There are generally eosinophilic vesicles in the lumen. The epithelium is surrounded by inner circular and outer longitudinal striated muscles. The genital arch produces a white, granular, eosinophilic solid substance of unknown function (Jangi, 1956).

There are two pairs of accessory reproductive glands in scolopendromorphs. A dorsal pair whose ducts enter the penis where they fuse may extend as far forwards as the eighteenth segment in a full-grown *S. morsitans*, and a smaller pair whose ducts open into the genital atrium ventral to the penis. In *Scolopendra* the glands are lobulated, in *Cryptops* they are tubular and very elongated, one pair bearing 15

Fig. 163. The male reproductive organs of scolopendromorphs. *a*, *Scolopendra* (based largely on Jangi, 1956). *b*, the vas deferens of *Cormocephalus anceps anceps* (after Brunhuber, 1970*a*).

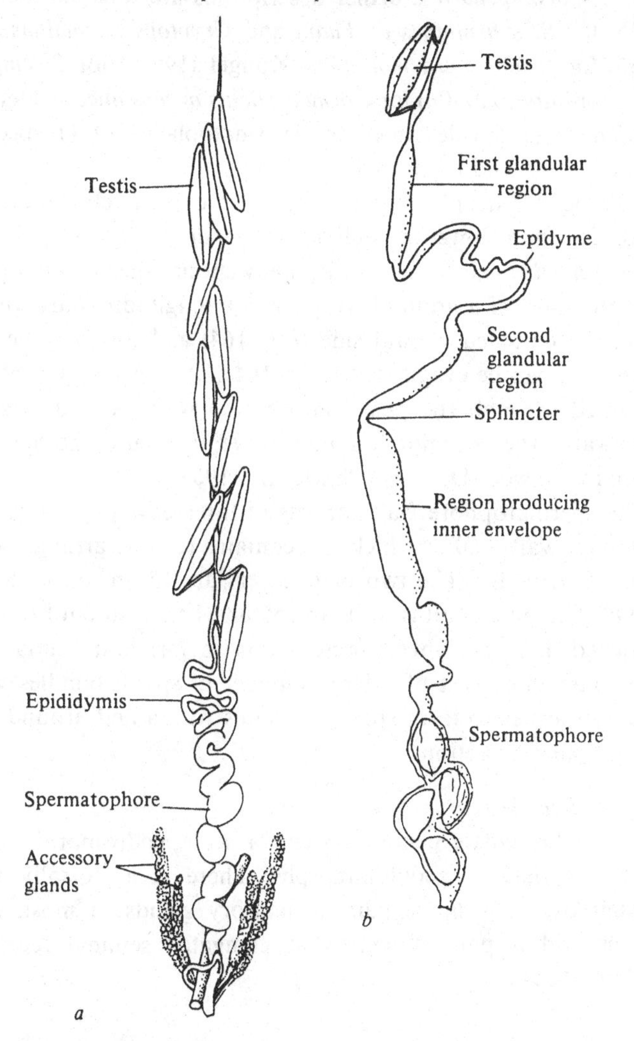

equally spaced vesicles identical to those of the female shown in Fig. 166*a* (Fabre, 1855).

Brunhuber & Hall (1970) suggested that the accessory glands in male scolopendromorphs secrete web material. This is supported by their observation that the glands are most active in *Cormocephalus* when the centipede is courting and web spinning.

The spermatophore

Fabre (1855) was the first worker to describe the spermatophore of *Scolopendra*. Further descriptions are those of Demange (1945) for *S. subspinipes de haani* and *Cryptops anomalans*, Jangi (1956) for *Scolopendra morsitans*, Klingel (1960*a*) for *S. cingulata* and Brunhuber (1970*a*) for *Cormocephalus anceps anceps*. Demange & Richard (1969) described the spermatophores of 11 species of scolopendrids and cryptopids.

Scolopendromorph spermatophores are relatively large structures, 2.5 mm long in Indian *S. morsitans*, 1.5 mm long in *Cormocephalus anceps anceps*. In many species the spermatophores have the form of a grain of wheat with a longitudinal invagination along the flattened ventral side (Fig. 164) and this may be much folded and can be evaginated (Figs. 164*d*, *g*). The spermatophore is deposited through the penis onto a web with its ventral surface downward; the invagination may serve to secure the spermatophore to the web (Demange & Richard, 1969).

The spermatophore wall consists of three envelopes (Fig. 164*h*). The outer wall is 60 μm thick and contains radially arranged canals 3 μm in diameter. The two inner layers are 15 μm thick (Klingel, 1960*a*). The outer wall is white in colour when fresh but brown and hardened after prolonged preservation in formalin (Jangi, 1956). The contents consist of a large number of sperm bundles wound into each other to form spring-like coils within and around which is a granular secretion.

Female reproductive organs

The female genital system of scolopendromorphs closely resembles that of geophilomorphs. There is a dorsally placed median ovary, a single pair of acessory glands in most, or all, species and a pair of somewhat elongated seminal receptacles (Fabre, 1855).

Fig. 164. The spermatophores of scolopendromorphs. *a* and *b* *Scolopendra valida*; *c*, *Rhysida nuda togoensis*; *d*, section of the same along the line *A–B*; *e* and *f*, *Cryptops hortensis*; *g*, evaginated spermatophore of *Scolopendra cingulata*; *h*, transverse section of the spermatophore of *S. cingulata* (*a–f* after Demange & Richard, 1969; *g*, *h* after Klingel, 1960*a*).

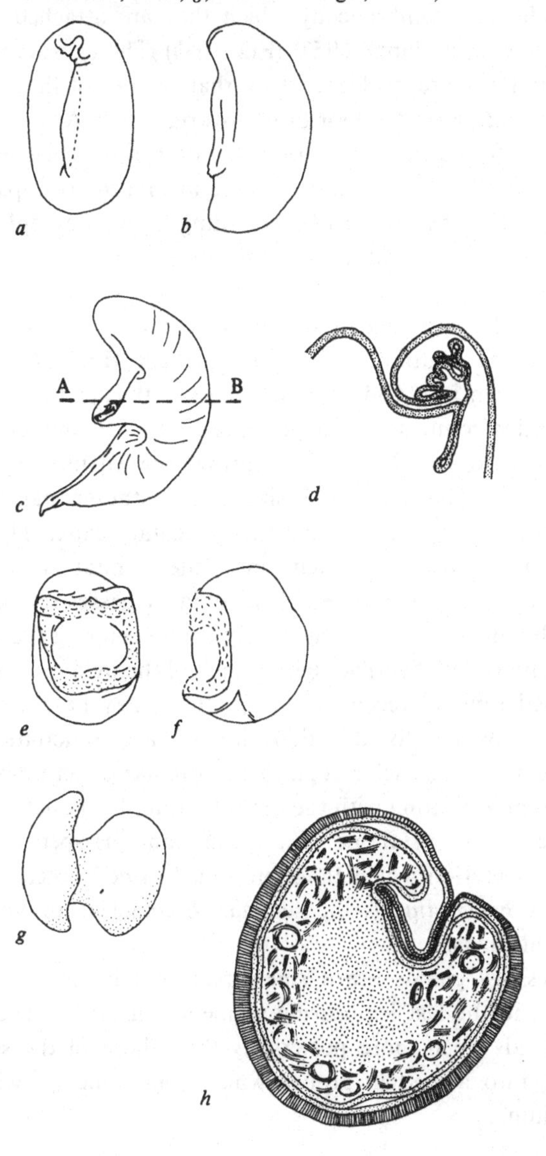

The ovary of *Scolopendra morsitans* is more or less uniform in width until the eighth segment where it narrows to form a thin terminal filament as far as the fifth segment (Fig. 165*a*). The wall of the ovary consists of an ovarian epithelium covered by longitudinal striated muscle fibres and connective tissue. The eggs are contained within a follicular membrane by which they are attached to the ventral ovarian wall (Jangi, 1957) (Fig. 165*b*). The eggs as well as the follicular layer are proliferated by that section of the ovarian epithelium which lines the floor of the ovarian tube. At the oocyte stage a thin, soft and fibrous chorion (the part of the shell secreted by the follicle cells) is present. A vitelline membrane is apparently lacking although a periplasm (the bounding layer of cytoplasm) is present surrounding a protoplasmic reticulum containing the yolk globules.

Posteriorly the egg tube passes over to the right side of the alimentary canal continuing into the oviduct (Fig. 165*a*). The oviduct is thrown into folds, the musularis is thicker than it is in the ovary and is composed of longitudinal and a few oblique fibres. The oviduct divides in the last leg-bearing segment into right and left genital ducts. The right one is shorter and stouter than the left which forms a genital arch around the alimentary canal. The right and left ducts open separately into the genital atrium. In *Ethmostigmus trigonopodus* (Leach) the oviduct passes to the right of the gut dividing into two immediately before entering the genital atrium (Prunesco, 1965*e*). The right branch of the oviduct is absent.

The paired seminal receptacles (Fig. 165*c*) may be asymmetrically placed owing to the difficulty of accommodation. In *Scolopendra morsitans* each receptacle is a tubular structure bent on itself and communicating with the genital atrium by a duct which is highly coiled at its distal end. The lumen contains sperm and an eosinophilic secretion which Jangi presumed to be 'protective' and nutritive. In *Ethmostigmus* the seminal receptacles are oval and have straight ducts.

The accessory glands are a single pair of lobular ectodermal organs sited in the last leg-bearing segment (Fig. 165*c*). The ducts of these glands open more posteriorly than those of the seminal receptacles, into a median groove which communicates with the genital atrium.

The accessory glands of *Cryptops hortensis* were described by Fabre (1855) as of considerable length, each bearing 15 equally spaced vesicles on its inner face (Fig. 166*a*). Prunesco (1965*b*) figured two pairs of glands in *C. anomalans*, the first resembling that described by Fabre and a second short pair opening into the common duct of the seminal receptacles. In a later publication (Prunesco, 1965*c*) he showed that there is only one pair of glands in *C. anomalans* and *C. parisi* (Fig. 165*b*). The differences between Fabre's and Prunesco's descriptions are surprising, further work on the anatomy of the glands is required. There is no trace of the left branch of the oviduct in *Cryptops* (Prunesco, 1965*c*).

Sperm transfer

Heymons (1901) reported seeing a male and female *Scolopendra* with their posterior ends in contact. He considered that copulation might take place.

Sperm transfer in scolopendromorphs was first described for *Scolopendra cingulata* by Klingel (1957, 1960*a*). Further details were added by Brunhuber (1969) for *Cormocephalus anceps anceps*. The courting ritual develops after a male and female have adopted the defence posture in which they grip each other along their entire

Fig. 165. The female reproductive organs of *Scolopendra morsitans*. *a*, ovary *in situ*; *b*, transverse section of ovary; *c*, seminal receptacles and accessory glands *in situ* (after Jangi, 1957).

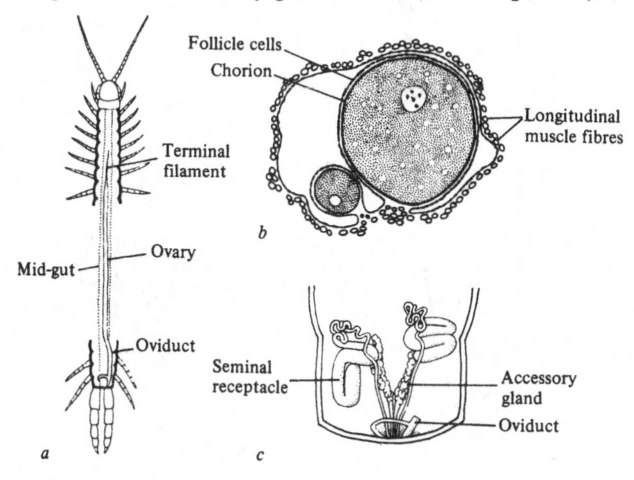

length, often lying head to tail. The male initiates the courtship. The antennae are used to tap the posterior part of the body of the partner, usually the legs, and in *Cormocephalus* this may last up to 14 h.

Tapping may be interrupted by the male moving forward but is resumed when the female follows. If she does not the male may reverse until he contacts the female again. The male spins a web in the tunnel or crevice with a gap in the threads through

Fig. 166. Female reproductive organs of *Cryptops*. *a*, *Cryptops hortensis*; *b*, *Cryptops anomalans* (*a* after Fabre, 1855, *b* after Prunesco, 1965*c*).

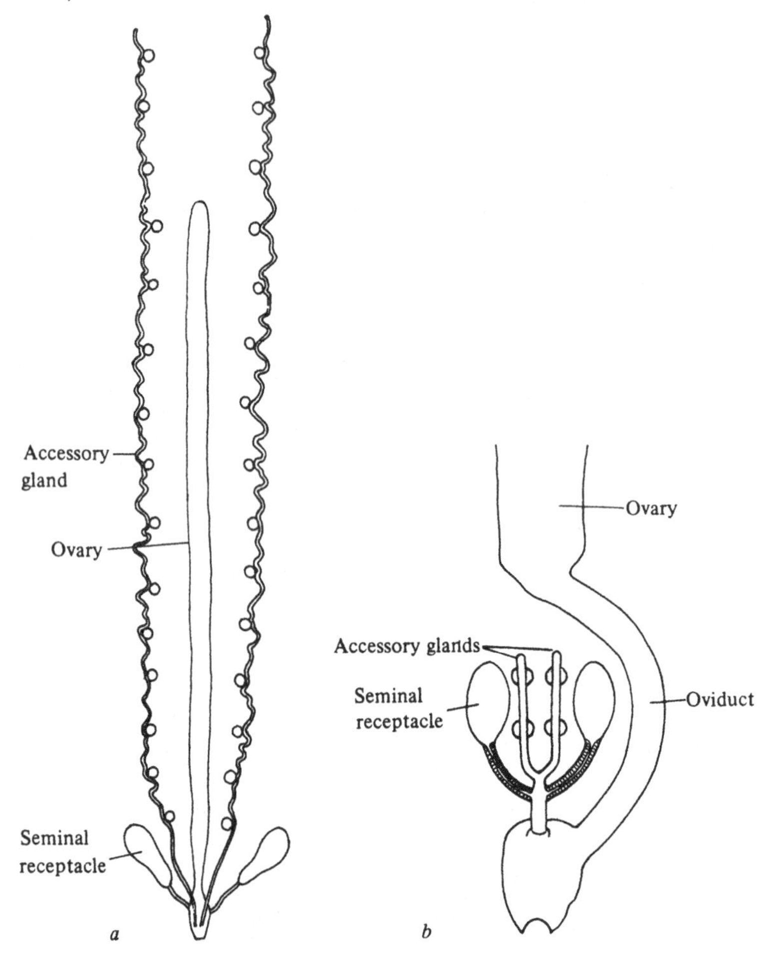

which the female will move. When the web is complete the male moves back onto it and deposits a single spermatophore. The female *S. cingulata* retains contact with the male while he spins the web, the female *C. anceps* loses contact and remains at the entrance to the tunnel until the male moves off, she then moves forward tapping the tunnel with her antennae until she contacts the web. As soon as a female *S. cingulata* touches the web with the posterior end of her body, the genital segment is stretched out and the spermatophore enveloped so that it lies horizontally across the genital opening. After several minutes the spermatophore bursts at its point of attachment to the body and the sperm enter the genital atrium. The rest of the spermatophore is then eaten.

Very different mating behaviour takes place in *Ethmostigmus platycephalus spinosus*. In this species the male seizes the female with his poison claws and works himself into a position where he overlies her, his ventral surface being in contact with her dorsum. A female ready to mate raises the last few segments of her body and the male then flexes his body, bringing the genital orifice up to that of the female and delivering a spermatophore into her genital atrium. The two then separate (Rajulu, 1970*f*). Rajulu made no mention of the preliminary tapping behaviour such as is seen in geophilomorphs and other scolopendromorphs.

Egg-laying and brooding

Scolopendromorphs resemble geophilomorphs in that females brood the eggs and larval stages (Fig. 167). Female *S. morsitans* lay a cluster of elliptical greenish-yellow eggs which are held together by a sticky secretion. The brood cavity is usually about 8 cm deep and hollowed out in the earth just below a stone (Jangi, 1966). In both *Cormocephalus anceps anceps* (Brunhuber, 1970*b*) and *C. multispinus* (Lawrence, 1947) the eggs and embryos are covered by a sticky secretion which causes them to adhere in a roughly spherical mass.

Female *C. anceps anceps* and *Scolopocryptops sexspinosus* have been observed to groom or mouth eggs (Auerbach, 1951), behaviour thought to protect the eggs by covering them with a fungicidal secretion. In *C. anceps anceps* the brooding female was normally coiled round the eggs with her ventral surface in contact with them, holding them off the ground. When grooming, the female uncurled and lying with her ventral surface uppermost, turned the egg ball with her legs, grooming the outer eggs for some ten minutes. When disturbed the female invariably ate or abandoned her eggs (Brunhuber, 1970*b*).

Data on the size of broods in Scolopendromorpha are given in Table 7.

Lithobiomorpha

Secondary sexual characters

There are no data on size differences between the sexes in lithobiomorphs but a number of species show structural differences

Table 7. *Brood size in Scolopendromorpha*

	Brood size	Mean	Authority
Scolopendra spp.	15–23	—	Heymons (1901)
Scolopendra morsitans	28–86	46	Lewis (1970)
Cormocephalus multispinus	20–40	—	Lawrence (1947)
Scolopocryptops sexspinosus	49–65	—	Auerbach (1951)
Cryptops hyalinus	9	9	Cornwell (1934)
Cryptops hyalinus	7	7	Johnson (1952)
Otostigmus tibialis *Otostigmus scabricauda* *Scolopocryptops ferrugineus*	15–30	—	Bücherl (1971)

Fig. 167. Female *Scolopendra* and brood (after Brehm, 1877).

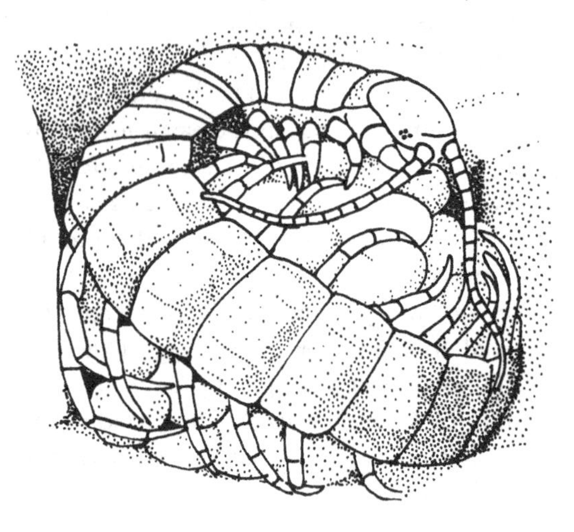

at the posterior end of the body associated with the sex of the individual. In *Lithobius calcaratus* the femur of the fifteenth pair of legs is particularly swollen along its posterior border and a wart-like process surmounted by a tuft of setae projects from the dorsal surface of the distal end of this swelling. In male *Lithobius aulacopus* Latzel there is a shallow channel on the dorsal surface of the femur and tibia of legs 14 and 15. Matic (1966) figured a number of Romanian species of *Lithobius* with similar hollows or grooves in the femur or tibia of the fourteenth and fifteenth pair of legs, sometimes, as in *L. mutabilis*, accompanied by a tuft of hairs and Eason (1973) gives clear figures of these structures in some central American species (Fig. 168). In the subgenus *Monotarsobius* the

Fig. 168. Modifications of the posterior legs of male *Lithobius* spp. *a*, posterior (internal) surface of fourteenth tibia of *Lithobius godmani* Pocock; *b*, posterior (internal) surface of fifteenth tibia of *L. godmani*; *c*, posterior (internal) view of fifteenth tibia of *Lithobius vulcani* Pocock; *d*, posterior (internal) view of fifteenth leg of *Lithobius humberti* Pocock (after Eason, 1973).

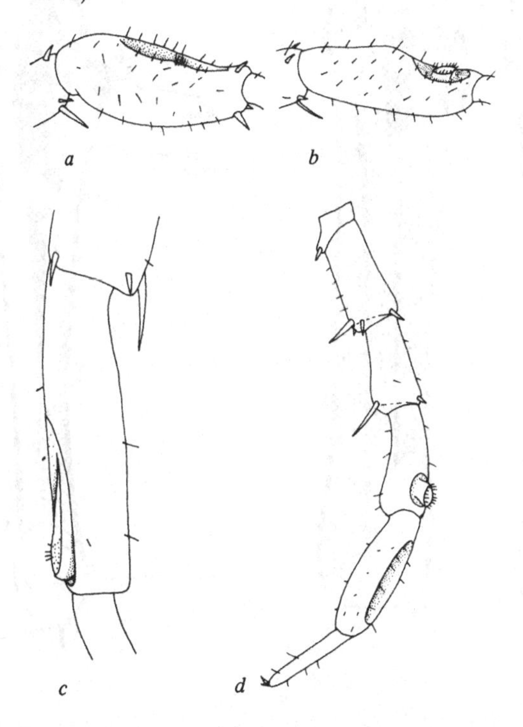

fourteenth and fifteenth pairs of legs are thickened in both sexes but more so in males than females. In male *Eupolybothrus* spp. much of the dorsum of the base of the femur of the fifteenth pair of legs is occupied by a pit, continuous with an internal dorsal sulcus. There is a pore-free area at the end of the femur which may bear special features such as a swelling, a concentration of setae or a small circumscribed group of very fine pores (Eason, 1970*b*) (Fig. 169).

In the *Lithobius* subgenus *Pleurolithobius* tergite 16 in males has curved posterior prolongations (Fig. 32) and in the subgenus *Dacolithobius* the fourteenth tergite of the male is very elongated with a fringe of setae along its posterior border. In the genus *Pterygotergitum* the male has tergites 10, 12 and 14 widened,

Fig. 169. Dorsal view of femur and distal part of prefemur of fifteenth leg of three species of *Eupolybothrus*. *a*, *E. fasciatus*; *b*, *E. grossipes*; *c*, *E. littoralis littoralis* (L. Koch) (from Eason, 1970*b*).

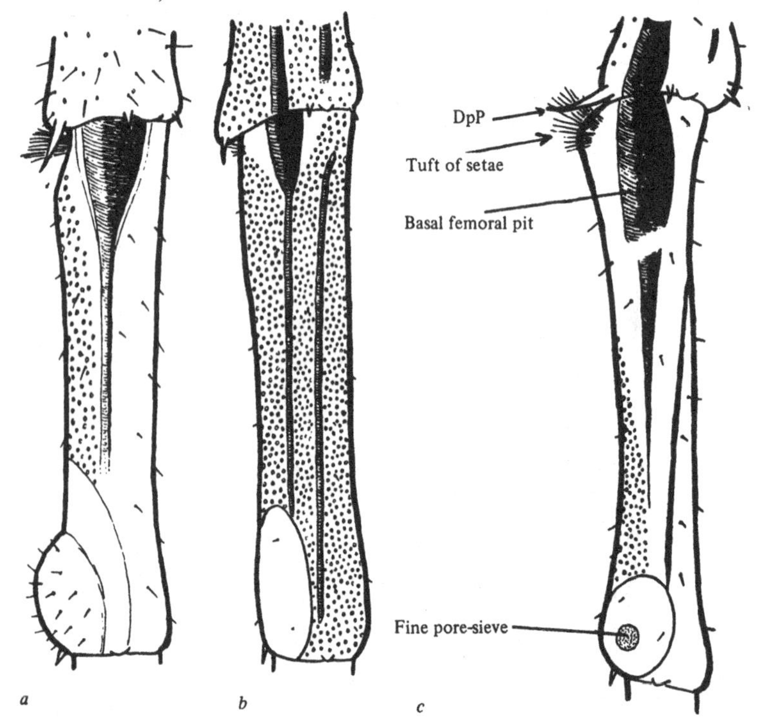

particularly tergite 12 which is produced into a large 'wing' on either side. These tergites are also toothed laterally (E. H. Eason, personal communication).

It may well be that the modifications of the posterior legs in some male lithobiomorphs and the modifications of the posterior tergites in others may be associated with mating and perhaps the production of sex pheromones as has been suggested above for the last pair of legs in some male geophilomorphs. Verhoeff (1902–25) believed that the tergital prolongations in *Pleurolithobius* were part of a coupling mechanism.

The American species *Paitobius zinus* is unique amongst centipedes in that there is a sexual dimorphism in the forcipules (Fig. 127), those of the female being short and robust as is typical for the order, whilst in males they are enormous with a very wide basal axis and projecting far forward (Crabill, 1960*b*).

Reproductive organs

The structure of the reproductive organs of lithobiomorphs was described by Fabre (1855), Fahlander (1938, briefly), Prunesco (1963, 1964, 1965*a*) and Rilling (1968).

In male *Lithobius forficatus* there is a single median tubular testis, normally reflected on itself twice (Fig. 170). The testis commences as a fine filament at about tergite 6 or 7, widens to form the main part of the testis and passes into a very narrow vas efferens which enters a very short vas deferens. This divides to pass round the gut, joining again below to open on the penis. Two very large seminal vesicles open into the vas deferens. These are absent in the Epimorpha. Three pairs of accessory glands open in the genital region. Their function is unknown: it is presumed that they produce material for the spermatophore. Prunesco (1964) reported that in addition to the three large pairs of glands a small pair of atrial glands is present.

Prunesco (1963) figured the male reproductive organs of *Lithobius forficatus*, *L. erythrocephalus*, *L.* (*Monotarsobius*) *burzenlandicus* (Verhoeff), *Harpolithobius banaticus* Matic, *Eupolybothrus leptopus* (Latzel) and *E. transsylvanicus* (Latzel). In general their arrangement resembles that of *L. forficatus* as described by Rilling (1968) but there is considerable variation in the extent to which the testis is coiled: six times in *L. erythrocephalus*, only once in *H. banaticus*. In *Anopsobius neozelandicus* Silvestri it is divided into two regions, a distal macrotestis in which large sperm develop and a thinner proximal microtestis in which small sperm develop. The macrospermatocytes are 25–50 μm in diameter, the microspermatocytes 2.5 μm in diameter. Elsewhere microtestes and macrotestes are only seen in the Scutigeromorpha (Prunescu & Johns, 1969).

In the tribe Lithobiini the dorsal accessory glands (Rilling's Gland 3) and the two pairs of ventral glands (Rilling's Glands 1 and 2) are very large with the exception of *Harpolithobius* where the dorsal glands are very small. The dorsal glands in the tribe Polybothrini are very small (Fig. 171*a*). The ventral glands are

Fig. 170. Male reproductive system of *Lithobius forficatus* (after Rilling, 1968).

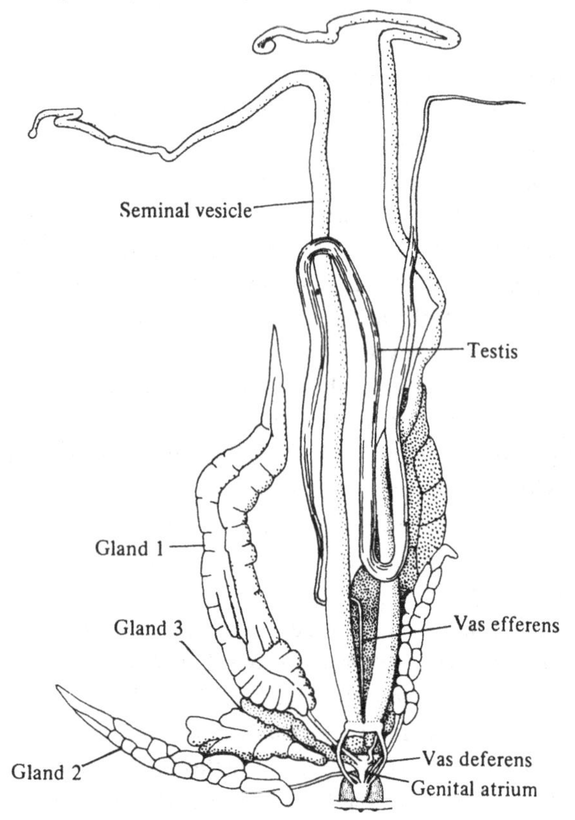

lobed, each lobe of the inferior ventral gland having its own duct leading to the main duct (Fig. 171*b*). There is a median ventral gland situated beneath the median canal leading from the ventral glands and a pair of atrial glands in the genital atrium (Prunesco, 1964).

The testis of *Lithobius* is formed from the fusion of paired testicular rudiments (Biegel, 1922). The wall consists of an outer epithelium lying on a thin basement membrane beneath which is a layer of longitudinal muscles, then a thick connective tissue layer containing large collagen fibres and circular muscles arranged as discontinuous rings. Finally there is an extremely 'interdigitated' inner epithelial layer (Camatini & Castellani, 1974).

Fig. 171. Diagrammatic representation of the male reproductive system of *Eupolybothrus*. *a*, complete system; *b*, detail of accessory glands (after Prunesco, 1964).

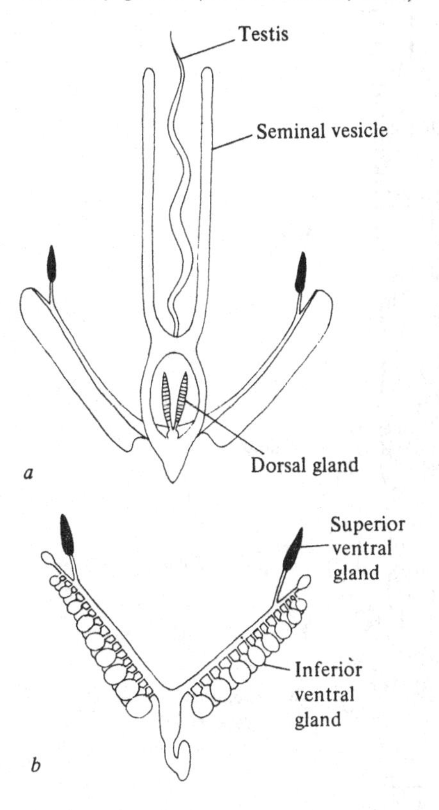

The female reproductive system of *L. forficatus* has been described by Fabre (1855), Schaufler (1889), Prunesco (1965*a*) and Rilling (1968). The single ovary which commences as a thin terminal filament is situated dorsal to the gut. In a mature female it reaches as far forward as segment 6. The oviduct divides to pass round the gut and joins to form the genital atrium. A pair of sausage-shaped seminal receptacles are each connected by a narrow duct with the genital atrium. There are two pairs of well-developed accessory glands which open into the genital atrium by long ducts (Fig. 172). The ova are produced from the ventral wall of the ovary and become surrounded by follicle cells which produce the primary oocyte membrane which is later replaced by the chorion which is

Fig. 172. Female reproductive organs of *Lithobius forficatus* (after Rilling, 1968).

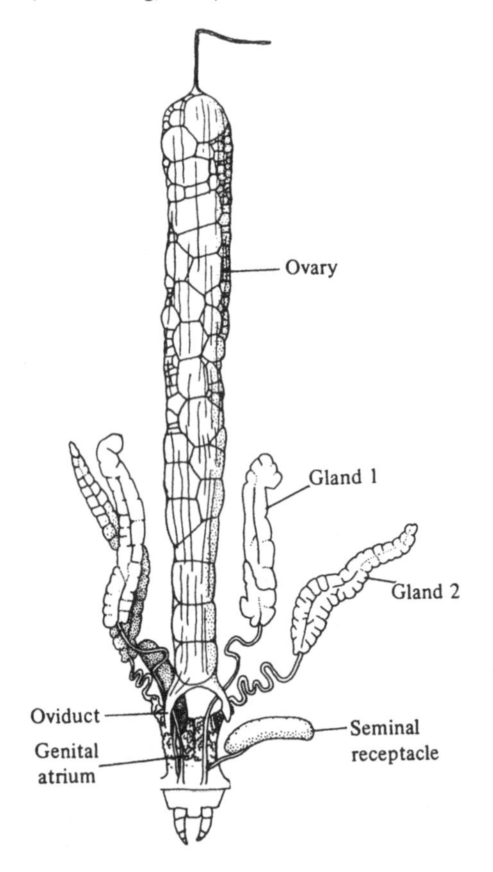

secreted by the ovarian epithelium (Herbaut, 1974, and see below).

Prunesco (1965*a*) described the female reproductive organs in *L. bulgaricus* Verhoeff, *L. muticus*, *L. burzenlandicus* Matic, *Harpolithobius banaticus*, *Eupolybothrus transsylvanicus* and *E. leptopus*. In some species the seminal receptacles are ovoid and the duct is coiled two or three times. A periatrial glandular mass surrounds the atrium into which it opens by several canals. In *Eupolybothrus* the seminal receptables are ovoid, the genital atrium larger than in *Lithobius* and the periatrial glands better developed.

Gynandromorphism

Matic (1958) reported a case of gynandromorphism in *Lithobius forficatus*, one gonopod being male, the other female. A similar case has been reported in *Lithobius* (*Monotarsobius*) *austriacus* Verhoeff: the left gonopod and terminal leg show female characteristics, the right male characteristics (Borek, 1969).

Descamps & Herbaut (1971) described a case of intersexuality in *L. forficatus*. The specimen which had male gonopods and reproductive system also possessed a small ovary. In *Anopsobius neozelandicus* Silv. an embryonic gonad persists alongside the testis in adult males (Prunescu & Johns, 1969) and *Dichelobius* also has two male gonads, one rudimentary (Prunesco, 1970*a*). Zerbib (1966) showed that when undifferentiated gonads from Larva II are grafted into larval or adult *L. forficatus* they always develop into ovaries suggesting that ovarian differentiation is probably determined by a sex hormone.

Parthenogenesis

Male *Lamyctes fulvicornis* have only been found in the Canary Islands and the Azores. European specimens may belong to a distinct parthenogenetic race (Palmen, 1948). Only female *Lithobius aulacopus* have been found in the British Isles and Scandinavia; Holland is the most northerly country from which males have been recorded (Eason, 1964). *Lamyctes coeculus* (Brölemann) may also be parthenogenetic: it occurs in greenhouses in Europe and also in Australia, Hawai, Mexico, Cuba and Tanzania. With the exception of the population in Cuba, males are very rare (Enghoff, 1975). The small size (3.5–5 mm) and soil dwelling habits of this species may have increased its chances of passive dispersal by man.

Sperm transfer

Although Fabre (1855) observed spermatophore production in '*Geophilus convolvens*' the process was not seen in lithobiomorphs until 1956 when Demange observed male *Lithobius piceus gracilitarsus* in captivity depositing spherical spermatophores on webs from which they were removed by females. The spermatophores are white, about 1 mm in diameter and extremely fragile. The envelope is probably formed by the drying of the liquid surrounding the sperm. The female picks up the spermatophore with her genital appendages and in the process covers it with a viscous liquid.

In *Lithobius forficatus* the ritual commences by partners touching each other with vibrating antennae. They frequently form a circle, the one stimulating the other's terminal legs. This behaviour may last for many hours and the male is the more active (Klingel, 1960*b*). There follows a period during which the male rocks the back half of his body up and down for as much as an hour after which sideways movements towards the female are accompanied by rocking and vibration of the antennae. The male then takes up a position 1.5–4 cm ahead of the female waiting for contact between his terminal legs and her vibrating antennae before moving forward. This last pattern is repeated many times in about half an hour.

Eventually the male spins a web measuring about 1 × 1 cm and consisting of about 120 strands, each of which takes 5 s to produce. Whilst the male is spinning the female remains behind the web tapping the terminal legs of the male with her antennae. On completion of the web, the male produces a small drop of transparent secretion followed 1 min later by a white translucent drop consisting of sperm which is pushed into the first drop. The spermatophore measures 2.5 mm in diameter, the sperm drop 0.6–0.8 mm.

Some 10–15 min after the production of the spermatophore the male creeps backwards and spins 200–240 threads 1 mm behind the spermatophore to form a 'slime strip'. A further one or two slime strips are produced behind the first (Fig. 173).

Finally the male, whilst vibrating his terminal legs, turns his trunk and touches the antennae of the female with his antennae

(Fig. 174). This is the signal for the female to move forward over the posterior half of the male. The male responds by moving forward 3–4 mm, his body pressed down to the ground and the spermatophore visible between his terminal legs. The female collects the spermatophore with her gonopods and runs off.

A minute or two later the male turns and eats the web. The female sometimes eats some of the spermatophore secretion. Some 20–30 min later the male shows sharp up and down jerky movements of his terminal legs. The female eats the rest of the spermatophore after an hour. Klingel pointed out that this behaviour

Fig. 173. Web of *Lithobius forficatus* with spermatophore and slime strips (after Klingel, 1960).

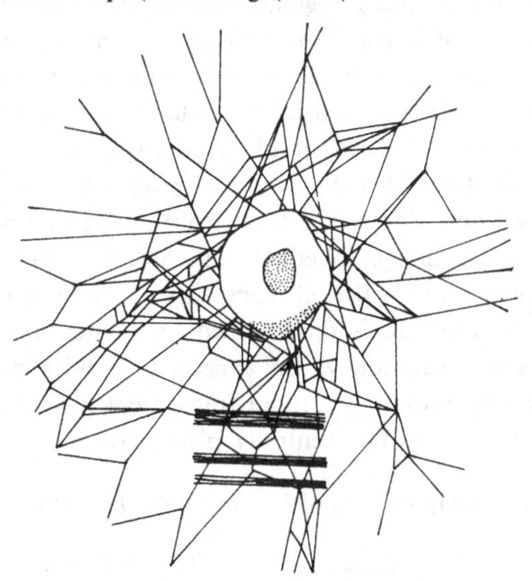

Fig. 174. Male *Lithobius forficatus* signalling to the female to collect the spermatophore (after Klingel, 1960).

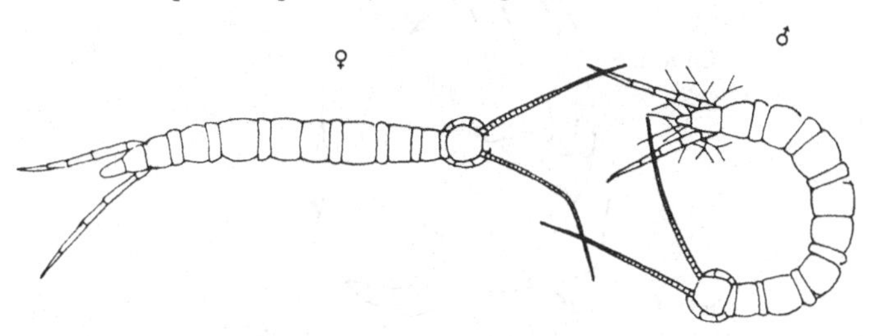

pattern is extremely long: the prespermatophore production period seldom lasting less than 2 h, the period after web-building 2–3 h.

Egg-laying

Lithobiomorphs and scutigeromorphs differ from geophilomorphs and scolopendromorphs in that their eggs are laid singly rather than in a single clutch. Brocher (1930) described the process in *L. forficatus* in detail. A female was observed to be carrying an egg between her vulvae at 09.00 h. An hour later the egg was being turned on itself continuously. The female scraped the soil with her gonopods detaching fine particles which were triturated and applied to the rotating egg. At the same time a secretion which cemented the soil particles together was poured onto the egg, presumably from the vulva. This process took about one hour. After two hours the egg resembles a piece of humus. Two or three hours later the female was seen at rest holding the egg which she then placed in the soil. Brocher observed egg-laying on 21 occasions from 25 December to 30 April after which he ceased making observations. Egg-laying took place in the morning. The egg of *Lithobius forficatus* is spherical and a little less than 1 mm in diameter: when covered with soil it measures 1.5 mm in diameter (Brocher, 1930). The egg is covered with an outer mass of an albumin-like white secretion which adheres to a thick protective hyaline sheath encapsulating a more transparent amber-coloured inner sheath (Fig. 175). The centrally placed ovum is yellowish-

Fig. 175. Newly laid egg of *Lithobius forficatus* (after Johnson, 1952).

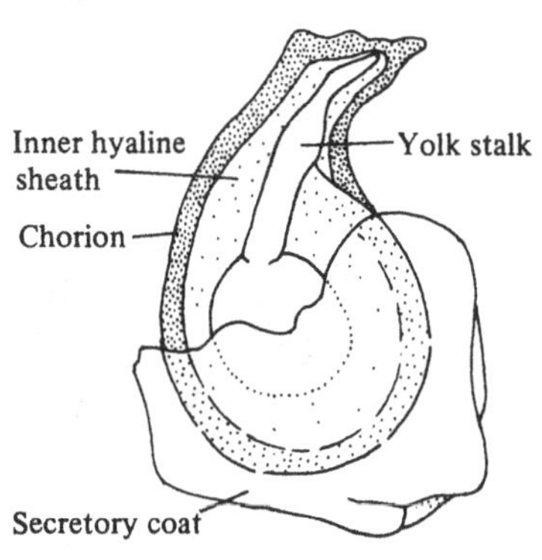

white and opaque. Recently extruded eggs are asymmetrical (Johnson, 1952).

The soil covered egg of *Lithobius piceus gracilitarsus* (ootheca) is lenticular in shape (Fig. 176) (Demange, 1956). In the laboratory, *Lithobius forficatus* and *L. piceus* scraped moist plaster of Paris from the bottom of their containers and covered the eggs with a thick layer of this before depositing them anywhere in the dish but *L. variegatus* and *L. duboscqui* did not coat their eggs with plaster; they laid them in sponge pads (Vaitilingham, 1960).

Scutigeromorpha

Secondary sexual characters

There appear to be no records of secondary sexual characters in scutigeromorphs but the order is poorly known taxonomically.

Reproductive organs

In male *Scutigera coleoptrata* there is a pair of oval testes each with a single coiled vas efferens which join to form a much coiled vas deferens; this divides into a straight median and two coiled lateral branches (Fig. 177). The median branch opens into a sinus formed by the fusion of the two elongated seminal vesicles, the lateral branches join the middle of the superior part of the ejaculatory canals of the seminal vesicles (Fabre, 1855). Bouin (1934) differentiated oval macrotestes which produce large sperm, from the spirally wound microtestes which produce small sperm.

Fig. 176. Soil covered egg (ootheca) of *Lithobius piceus gracilitarsus*. *a*, profile; *b*, from above (after Demange, 1956).

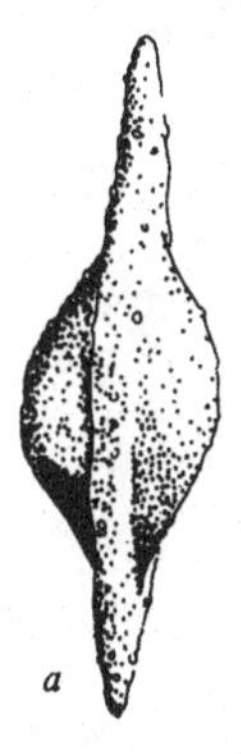

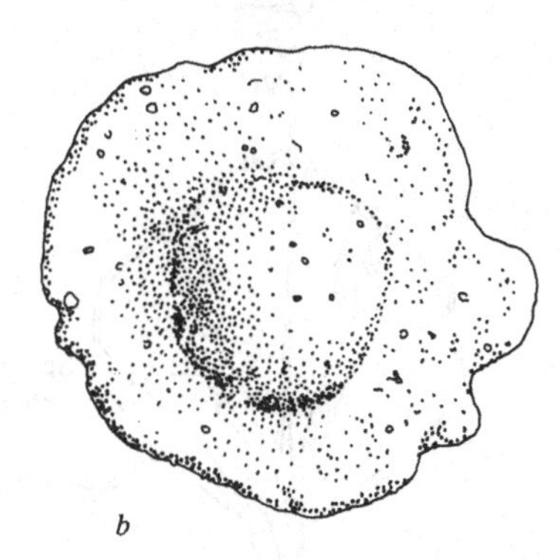

The microtestes correspond to the distal region of Fabre's vasa efferentia.

The arrangement is similar in *Thereuonema tuberculata* (Fig. 178). In this species the testes are U-shaped, consisting of a thicker distal macrotestis and a thinner proximal microtestis (Fahlander, 1938). The arrangement of the vas deferens is the same in *Scutigera* and *Thereuonema*. It is unusual as the median vas deferens does not open in the genital atrium as in other centipedes but via the lateral seminal vesicles.

The female reproductive system of *Scutigera* (Fig. 179) comprises a median ovary with a terminal filament, an oviduct which divides to pass ventral to the gut, a pair of club-shaped, bluish-white

Fig. 177. Diagram of the male reproductive system of *Scutigera coleoptrata* (after Fabre, 1855).

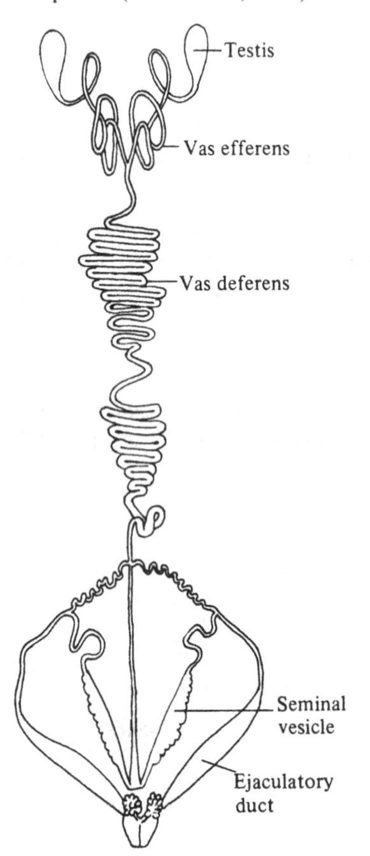

seminal receptacles and two pairs of accessory glands, the dorsal pair opening by a number of small ducts into the main one (Fabre, 1855; Prunesco, 1967*b*). The ova arise from paired dorso-lateral ridges in the ovary, suggesting that the latter was originally a paired structure (Knoll, 1974).

Sperm transfer

The process of sperm transfer in *Scutigera coleoptrata* commences, as it does in *Lithobius*, with the partners touching each

Fig. 178. The male reproductive system of *Thereuonema tuberculata* (after Fahlander, 1938).

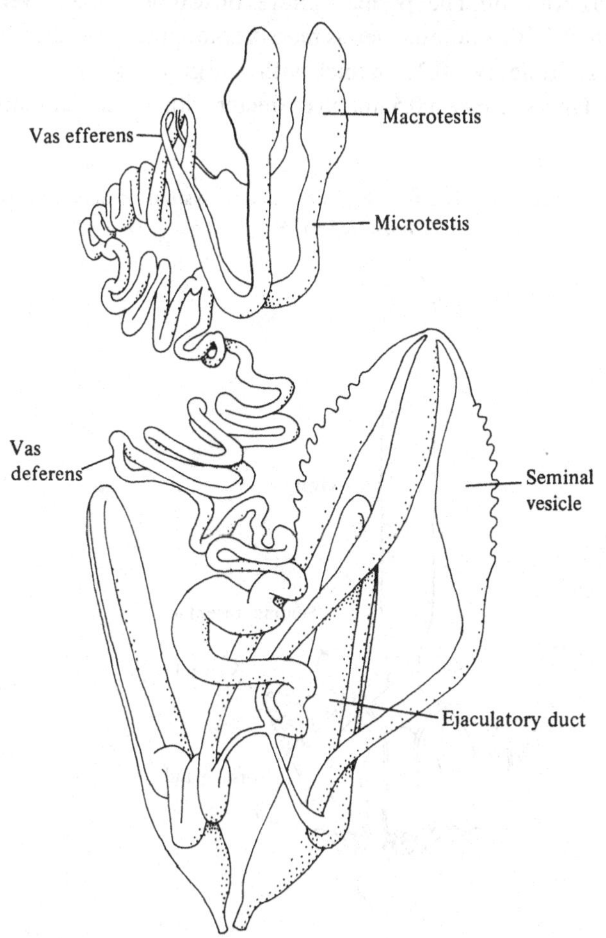

other with vibrating antennae and often forming a circle (Klingel, 1956, 1960*a*). In the second phase of the ritual, the male moves sideways and beneath the anterior part of the female with the front half of his body raised off the ground. Whilst in this position he makes a number of up and down movements with the front half of his body. In between these bouts of rocking he either rests or moves away. Rocking is followed by the deposition of the spermatophore which measures 4.5 × 2.3 mm and is lemon shaped. Immediately after spermatophore deposition the male turns backwards, touching the flanks of the female with his anterior legs (Fig. 180) and leads the female over the spermatophore from which she removes the sperm with her posterior end. The spermatophore consists of an outer layer 0.2–0.3 mm thick with a homogeneous matrix containing granules 10–15 μm long, a middle layer 0.2 mm thick with oval granules 2–8 μm long and a central mass of sperm 0.5 mm in diameter with a granular matrix. The

Fig. 179. The female reproductive system of *Scutigera coleoptrata* (after Prunesco, 1967*b*).

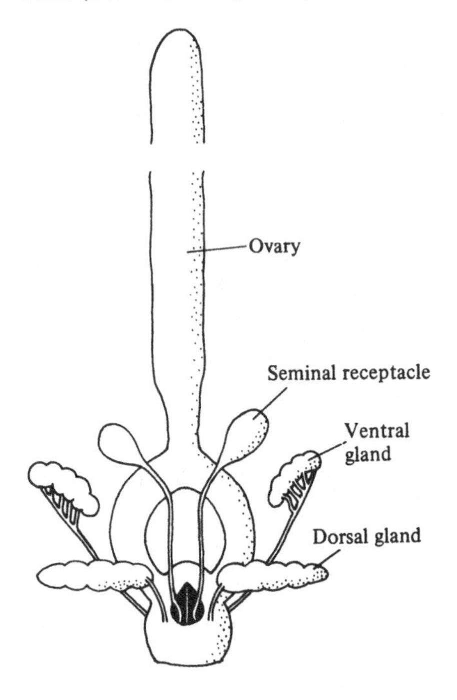

outer layer is incomplete, leaving the middle layer exposed at the pointed end of the spermatophore (Fig. 181).

In *Thereuopoda decipiens cavernicola* Verhoeff pairing commences with the partners mutually vibrating their antennae and the first four pairs of legs. The male then pushes sideways under the

Fig. 180. Position of male and female *Scutigera coleoptrata* just before the uptake of the spermatophore (after Klingel, 1960*a*).

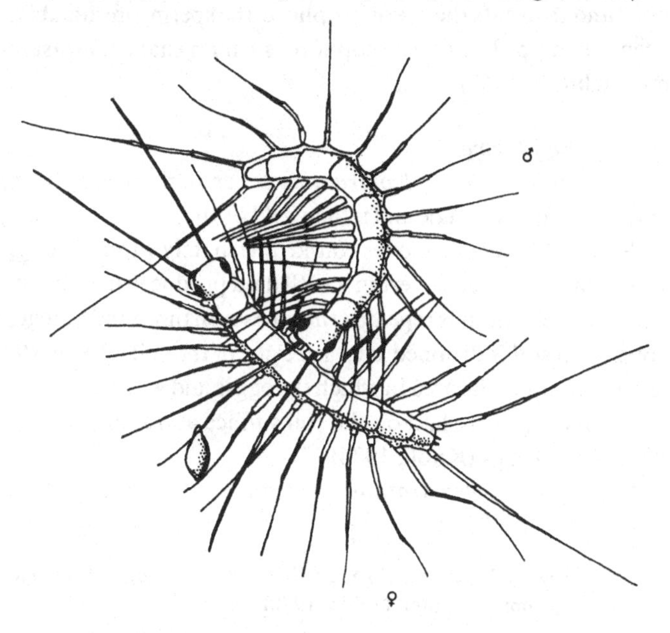

Fig. 181. The spermatophore of *Scutigera coleoptrata*. *a*, side view; *b*, posterior view; *c*, dorsal view; *d*, transverse section through *A–B* in *a* (after Klingel, 1960*a*).

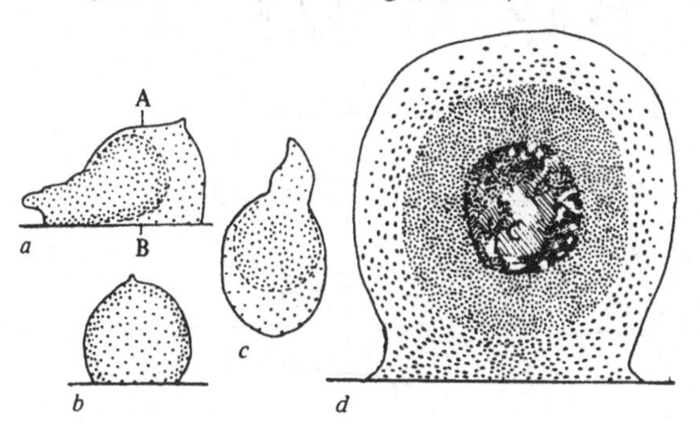

raised anterior part of the female's body and raises and lowers his body three to five times in a jerky manner and finally places a spermatophore on the ground. He then pushes the female forwards and sideways so that the spermatophore lies between them. The male picks up the spermatophore with his forcipules and pushes it into the genital opening of the female: the partners then separate. The female carries the spermatophore around at her hind end for an hour and then eats the spermatophore, the sperm, presumably, having been taken up. The spermatophore is lemon shaped, measure 7 × 3.5 mm (Klingel, 1962).

Egg-laying

In *Scutigera coleoptrata* each egg first appears between the anal valves of the female and is then held by the genital appendages as in *Lithobius*. The female moves round, alternately wiping the egg on the soil surface and then covering it with fluid produced from hèr posterior end. This behaviour is repeated three to seven times and the egg, by now covered in soil is dropped into a crevice in the soil (Dohle, 1970). The female tramples the area in which the egg is laid with her posterior legs. Sometimes eggs are laid on the soil surface and sometimes as a loose nest of 7–10 eggs (Knoll, 1974).

The eggs of *S. coleoptrata* are oval in shape, measuring 1.1 × 1.2–1.25

Fig. 182. Section through the shell of the egg of *Scutigera colcoptrata* (after Dohle, 1970).

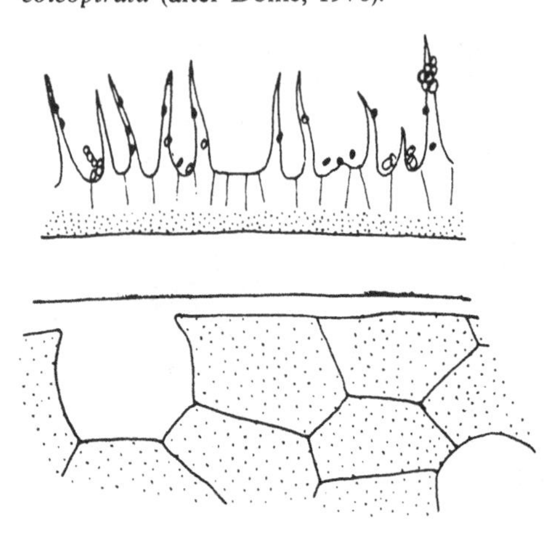

mm. They are covered by two membranes, the inner membrane is 7 μm thick, homogeneous and transparent, the outer, also 7 μm thick, is granular and bears hair-like processes to which soil particles adhere (Fig. 182) (Dohle, 1970).

In the laboratory females laid about four eggs per day: the maximum number laid by a single female in 24 h was 20. The egg-laying season in the south of France is early May to late June (Dohle, 1970). Knoll (1974) found that the egg-laying period lasted from the end of April, or earlier, until early July and was interrupted by a moult. He found that the average number of eggs laid by 24 females was 63. The highest number was 151. On average 53 eggs were laid before and 9 after the moult.

Craterostigmus

The female reproductive system of *Craterostigmus* consists of a median ovary, a pair of seminal receptacles and two pairs of accessory glands (Fig. 183). Prunesco (1965*d*) considered the structure to be intermediate between that of the Lithobiomorpha and the Scolopendromorpha.

In Tasmania the breeding season lasts from November to

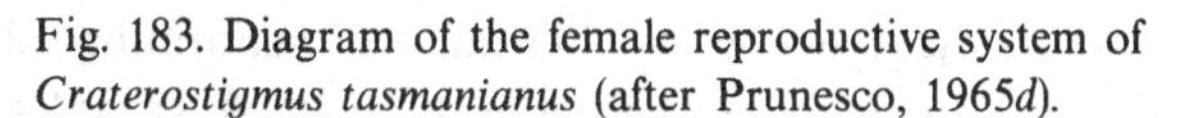

Fig. 183. Diagram of the female reproductive system of *Craterostigmus tasmanianus* (after Prunesco, 1965*d*).

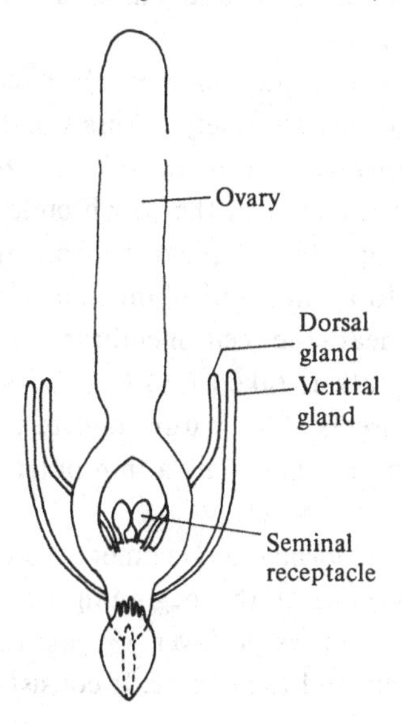

February. Females lay 'egg capsules' in clusters in small crevices or chambers in damp decaying logs and hump the body over the egg mass as do scolopendromorphs (Manton, 1965).

Gametogenesis

Oogenesis

Oogenesis may be divided into four phases: premeiosis in which the nuclei are at the diplotene stage of meiosis; previtellogenesis in which the nuclear and cytoplasmic volumes increase rapidly; vitellogenesis in which vitelline reserves accumulate and maturation, during which the chorion is deposited around the oocyte.

During the phase of previtellogenesis in *Lithobius forficatus* the nucleolar material breaks up into numerous masses and laminae and nuclear extrusions are very frequent. The number of ribosomes and mitochondria increases markedly, the mitochondria show pronounced morphological modifications: stacking and honeycomb-like figures. Lysosomal formations are also numerous. During vitellogenesis, the nucleus shows little activity and yolk accumulates in the cytoplasm. The number of ribosomes and mitochondria in the oocytes decreases. During the maturation phase of oogenesis the Golgi apparatus seems to bud off vacuoles in the cortical cytoplasm (Herbaut, 1972*a*).

The three classical types of vitelline reserves glycogen, lipid droplets and protein yolk, appear successively during the development of the oocyte of *L. forficatus* (Herbaut, 1972*b*). Glycogen synthesis appears to occur in contact with the rough endoplasmic reticulum. Lipid droplets appear to be elaborated from material which passes through the endoplasmic reticulum and Golgi apparatus. Yolk first appears near the cell membrane which is produced into microvilli. The yolk is taken into the oocytes from without by pinocytosis as in insects. The mature oocyte is almost completely filled with yolk, there remains only a thin outer layer of cytoplasm containing very few organelles.

The developing oocytes of *L. forficatus* become covered by a primary membrane of glycoprotein at the beginning of the previtellogenetic phase. This appears as a layer of homogeneous material between the follicle cells and the oocyte. It consists of two layers, the inner being of fibrillar nature. It is this primary membrane that is concerned with the uptake of yolk during vitellogenesis (Herbaut, 1974).

The youngest oocytes are found on the ventral wall of the ovary, older oocytes are attached to this wall by a cellular cordon, their primary membranes are in contact with the basement membrane of the ovarian epithelium (Fig. 184). Towards the end of maturation the oocyte migrates towards the posterior region of the ovary and its primary membrane disappears to be replaced by a secondary membrane, the chorion, 10–15 μm thick and formed from a homogeneous ground substance in which particles of glycoprotein fuse together. The chorion is secreted by the ovarian epithelium which forms cup-like evaginations, the cells releasing vacuoles which fuse to form the ground substance of the chorion. The glycoprotein component is elaborated by the cells between the cupules.

Fig. 184. The development of the oocytes of *Lithobius forficatus* (after Herbaut, 1974).

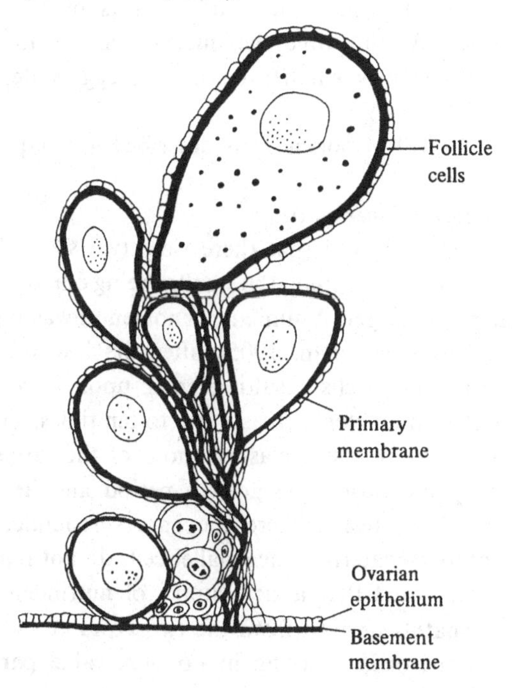

The ovary of *L. forficatus* always contains some degenerating oocytes, the *Nährzellen* of Tönniges (1902). Cytoplasmic lysis takes place before nuclear lysis and the vitelline reserves resist lysis the longest. Degenerating previtellogenetic and poorly vitellogenetic cells are essentially recovered by the follicular cells. If large vitellogenetic or mature oocytes are affected, their vitelline reserves are released into the ovarian lumen and included in the ovarian epithelial cells. The number of regenerating oocytes reaches a maximum at the end of the spring (Herbaut, 1976*b*).

The oocytes of *Scutigera coleoptrata* are produced from a pair of ridges on the dorsal wall of the ovary. This contrasts with the situation in scolopendromorphs and lithobiomorphs where the germinal epithelium is ventral. The oogonia become covered with a layer of small follicle cells and as the oocyte moves away from the ovary wall the follicle cells form a cellular cordon similar to that found in *Lithobius*. Older oocytes become detached from the cord and the surrounding cells degenerate and secretion is taken over by the folded epithelial cells (Fig. 185*a*). A chorion is deposited around the oocyte. It is about 7 μm thick and consists of 10–15 fine lamellae (Fig. 185*b*). At this time the nucleus has an amoeboid appearance but it becomes smaller before the egg is deposited (Knoll, 1974).

The endocrine control of oogenesis is described in Chapter 7.

The structure of spermatozoa

Bouin (1903) observed that there were two sizes of spermatocytes in *Scolopendra morsitans*, the smaller being one-quarter to one-fifth the size of the larger. A similar phenomenon was reported in *Scolopendra heros* (Blackman, 1905): after the last spermatogonial division the small cells resulting enter upon a period of growth in which their diameter increases five to ten times. The cells are early divisible into two size classes: those of the larger type remain united in pairs during the growth period and lie in uncrowded regions of the testis where they are surrounded by a plentiful supply of food material. The smaller cells do not remain in pairs and are so crowded that a rich supply of nutriment is not possible. The spermatids derived from the two types of spermatocytes produce spermatozoa differing in no observable particular

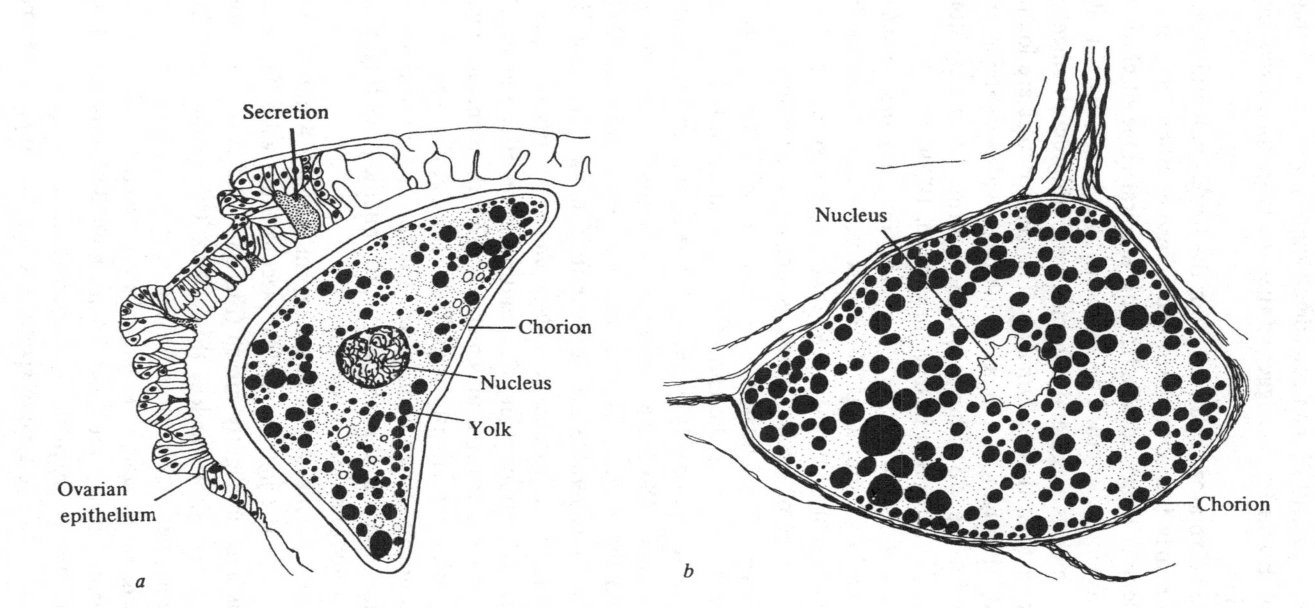

Fig. 185. Development of the oocyte of *Scutigera coleoptrata*. *a* and *b*, successive stages in the development of the chorion (after Knoll, 1974).

except their size. Blackman considered that they were probably all functional. Bouin (1920) suggested that the large sperm was female determining.

Mature spermatozoa derived from the large spermatids reach a length of nearly 1 mm, the head and acrosome being 260 μm long. The nucleus is long and slender with a spiral ridge of chromatin running its entire length. The acrosome is a long slender pointed filament which is large and conspicuous during development and much larger than that of *Lithobius*. The centrosomes are found at the point of origin of the axial filament which contains spiral mitochondria as do other centipede sperm. A similar 'double spermatogenesis' occurs in *Cryptops* (Aron, 1920).

There is only one size of sperm in *Cryptops trisulcatus* Brölemann, it is 3 mm long. The acrosome is curved and the head in which the chromatin is spirally wound, measures 10–11 × 1–2 μm. At the base of the head there is a centrosome followed by the middle piece containing two interlaced mitochondria. There is a second centrosome at the end of the middle piece followed by a short terminal filament (Fig. 186*a*, *b*). The mature sperm are rolled up in twos (Fig. 186*c*): they unroll in the seminal receptacle of the female and the middle piece fragments (Tuzet & Manier, 1953).

There is considerable detailed variation in the structure of the sperm in geophilomorphs. In *Schendyla nemorensis* the sperm is 4 mm long and 10 μm wide. The head measures 50 μm in length. It consists of clear cytoplasm which contains the compact nucleus and is prolonged anteriorly as two horns (Fig. 186*d*, *e*), there follows a normal long middle piece. The sperm are rolled up singly in the male and unroll in the female, the middle piece fragmenting. There is only one type of sperm.

In *Chaetechelyne vesuviana* the sperm measure 4 mm in length, the head 20 × 1–2 μm, the middle piece is 3 μm wide. The narrow acrosome is often slightly curved (Fig. 186*f*, *g*). Some sperm appear to have a very short nucleus (Fig. 186*h*). The sperm unroll in the female and the middle piece fragments in the seminal receptacle. The sperm of *Himantarium gabrielis* are several millimetres long; the acrosome 8 μm, the head 12 μm long. They are rolled up in groups of five or six in the vas deferens.

There is only one type of sperm in *Geophilus osquidatum* and *G.*

carpophagus. In both species the sperm are 2 mm long but the heads are very small, being 3 or 4 μm long. The sperm of *G. osquidatum* unroll in the seminal receptacle of the female and the middle piece degenerates.

Fig. 186. Centipede spermatozoa. *a*, anterior and *b*, posterior end of sperm of *Cryptops trisulcatus*; *c*, two sperm of *C. trisulcatus*; *d* and *e*, head of sperm of *Schendyla nemorensis*; *f*, head and *g*, posterior end of sperm of *Chaetechelyne vesuviana*; *h*, sperm of *C. vesuviana* with short nucleus; *j*, anterior part of sperm of *Lithobius duboscqui*; *k*, anterior part of sperm of *Lithobius forficatus* (after Tuzet & Manier, 1953).

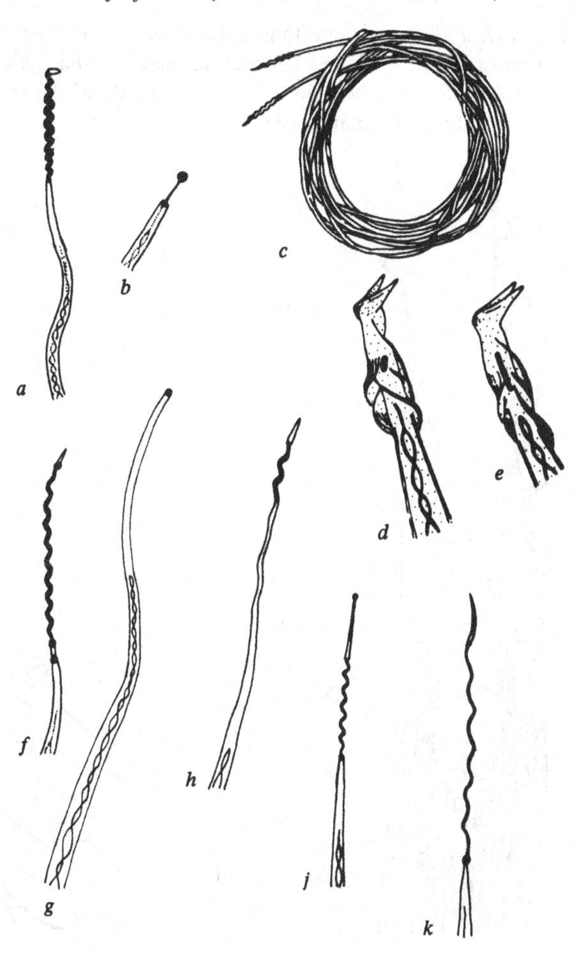

The ultrastructure of the sperm of *Clinopodes linearis* was investigated by Horstmann (1968). They are about 3 mm long with an acrosome which resembles a rod with delicate striations, 7 μm long and 0.3 μm wide. The head is 27 μm long and the elongated nucleus is surrounded by a spiral ridge of 50–60 turns. There follows a connecting piece 6 μm long which provides the complicated attachment for the nucleus (Fig. 187*a*) and the region of origin of the axial filament. The axial filament is surrounded by a striated cylinder which consists of two layers. The axial filament and striated cylinder are continuous with the principal or middle

Fig. 187. *a*, diagrammatic longitudinal section of the posterior region of the nucleus and connecting piece of *Clinopodes linearis*; *b*, diagram of a portion of the mantle of the sperm of *C. linearis* (after Horstmann, 1968).

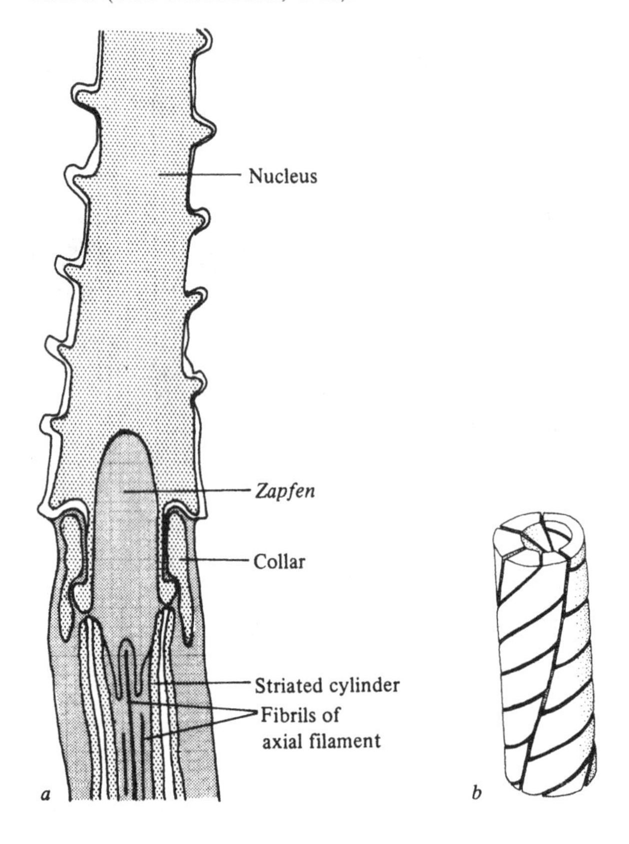

piece of the sperm. Between the axial filament and the outer membrane of the middle piece there is a large submembranous cleft containing a complex organelle developed from the mitochondria, the mantle. This consists of three spiral septa running longitudinally and about 10 000 oblique septa so that the mantle is subdivided into about 30 000 sections of very complex structure (Fig. 187*b*).

In mature spermatozoa the axis fibre terminates in the end piece, whereas in immature sperm it continues to form a long tail.

There are two sizes of sperm in *Lithobius forficatus*, *L. calcaratus* and probably in *L. duboscqui* var. *exarmatus* Brölemann. In the last species the sperm is only 600 μm long and the head 11–12 μm long with a spiral acrosome (Fig. 186*J*). In *L. calcaratus* the head is either 87 or 172 μm long. As in other centipedes there is a long middle piece with two spirally wound mitochondria, two centrosomes and a short terminal flagellum (end piece). In this species the intermediate segment fragments in the seminal receptacle of the female (Tuzet & Manier, 1953).

The spermatozoon of *Lithobius forficatus* is about 2 mm long. The acrosome, surrounded by exogenous fibrillar material is about 4 μm long and 0.2–0.3 μm wide. The spiral nucleus is 300–400 μm in length, there is a complex connecting piece. The middle piece is very long (1.5 mm), the end piece only 6–7 μm long (Descamps, 1972*b*). The mitochondria are highly differentiated and similar to those observed in *Clinopodes linearis* by Horstmann (Camatini, Saita & Cotelli, 1974). According to Tuzet & Manier (1953) the head of *L. forficatus* is either 57 or 114 μm long (Fig. 185*k*). These figures are at odds with those of Descamps.

In *Scutigera coleoptrata* large sperm are produced by the macrotestis, small sperm by the microtestis. In both types of sperm the haploid chromosome number is 18, comprising 17 autosomes and an X chromosome. The macrosperm have a large X chromosome, the small sperm a small X. There are more microsperm than macrosperm (Bouin, 1934).

Mature sperm have a rod-shaped acrosome 4 μm long, a helical nucleus 150 μm long, which is joined to the nucleus by a short cone-shaped connecting piece. In the middle piece, four mitochondria surround the axoneme (Camatini, Franchi, Saita & Bellone, 1977).

Spermatogenesis

The spermatogonia of *Lithobius forficatus* are about 20–25 μm in diameter with large nuclei. The primary spermatocytes are produced after a period of growth and an increase in the number of mitochondria and Golgi elements. These undergo two mitotic divisions. Two meiotic divisions follow to produce spermatids which have perinuclear filaments (Descamps, 1969*a*).

During the growth of the spermatocyte, RNA and basic proteins are synthesised: spermiogenesis is characterised by a gradual decrease in the amounts of RNA and basic proteins (Descamps, 1969*b*). The spermatogonia have nuclei with dispersed chromatin and a compact nucleolus. The cytoplasm contains free ribosomes and polysomes but mitochondria and dictyosomes are not abundant. During the growth of the spermatocyte the chromatin becomes more dispersed and the nucleolus shows frequent budding. Ribosomes, mitochondria, dictyosomes and endoplasmic reticulum are very numerous (Descamps, 1971*c*).

Degenerating spermatocytes and the more rarely degenerating spermatids in the ovary are phagocytosed by the spermatocytes (Descamps, 1971*b*). Tönniges (1902) described *Nährzellen* in the testis of *L. forficatus* which degenerate to form a nutritive syncytium.

Spermatogenesis in *Scutigera coleoptrata* was described by Camatini *et al.* (1977).

18

Post-embryonic development and life history

The embryonic development of centipedes will not be described here: it has been reviewed by Johannsen & Butt (1941). The most important work is that of Heymons (1901) on the embryology of *Scolopendra* spp. Verhoeff (1902–25) quoted at length from Heymons and added further data. Dohle (1970) and Knoll (1974) described the embryological development of *Scutigera coleoptrata.*

The Chilopoda exhibit two distinct patterns of post-embryonic development. In the epimorphic orders Geophilomorpha and Scolopendromorpha, the young hatch with a full, or almost full, complement of legs and the eggs and early post-embryonic stadia are brooded by the female. In the anamorphic orders Lithobiomorpha and Scutigeromorpha, the eggs are laid singly and are not brooded by the female. The young hatch with less than the adult number of legs and the number gradually increases during the early moults. Data for the *Craterostigmus* are currently very fragmentary.

Geophilomorpha

Larval stadia and brooding

Latzel (1880) gave brief notes on the larval stages of a large number of geophilomorphs. More detailed accounts have been given for *Geophilus proximus* and *Pachymerium ferrugineum* by Sograff (1883), for *Dicellophilus carniolensis* (C. L. Koch) together with notes on some other species by Verhoff (1902–25), for the American species *Geophilus rubens* by Johnson (1952), for *Strigamia maritima* by Lewis (1960) and for *Necrophloeophagus longicornis* and *Clinopodes linearis* by Weil (1958).

The process of hatching in geophilomorphs is a gradual process, the egg splitting into two halves to reveal the 'last embryonic phase'. This is doubled over to form a horseshoe shape (Fig. 188*a*,

190*b*). The anterior third of the body is much inflated with yolk. The thin cuticle shows some sign of segmentation but the antennae are unsegmented and the mouthparts, poison claws and limbs are represented by simple buds. An egg tooth (*ruptor ovi*) is present on the second maxilla in a *Geophilus* sp. (Metschnikoff, 1875), *G. carpophagus* (Verhoeff, 1902–25) and *N. longicornis* (Weil, 1958) (Fig. 190*a*). Johnson (1952) stated, however, that in *G. rubens* it was a sharp dorsal ridge of the embryonic head that splits the embryonic membranes.

The last embryonic stage is succeeded by the peripatoid stadium which is characterised by having short, thick antennae with 14 indistinctly differentiated segments, no spiracles, no apparent tracheal system and a weakly developed musculature (Figs. 188*b*, 189*a*, 190*c*). Setae are absent and neither tergites nor pleurites are clearly differentiated. The genital zone is embryonic and sternal and coxal glands are lacking (Fig. 189*b*). In *Geophilus rubens* the peripatoid stadium is milk-white in colour and 3–4 mm long when fully extended. The peripatoid stadium of *Strigamia maritima* is pinkish-violet. The pigment is located in the gut, the rest of the body being whitish and translucent.

The peripatoid stadium is followed by the foetus or foetoid stadium. This has distinctly segmented appendages with claws present but weakly developed. The 14 antennal segments are clearly

Fig. 188. *a*, the last embryonic stadium of *Strigamia maritima* (after Lewis, 1960); *b*, the peripatoid stadium of *Dicellophilus carniolensis* (after Verhoeff, 1902–25).

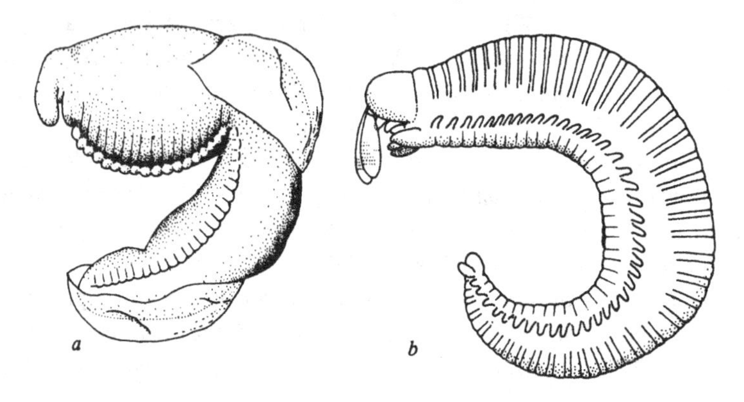

distinguishable and sparsely setose, the mouthparts and poison claws are segmented but not functional (Fig. 190*d*) Johnson described the tracheal system as 'nascent', the spiracles either lacking or occasionally developed subcutaneously. The yolk mass in the anterior portion of the intestine is diminished; the tergites, pleurites and coxal sclerites apparent; sternal and coxal glands nascent but no external pores are present. The foetus stadium of *G. rubens* is yellowish-cream in colour, 8–10 mm long and quite active. Localised movements are in evidence, particularly in the appendages. The foetus stadium of *Strigamia maritima* is of similar colour to the peripatoid, dorso-ventrally flattened and its average length is 7.0 mm. It is capable of making 'writhing' movements.

The last stadium to be brooded by the mother is the first adolescens stadium. In the adolescens I the appendages are

Fig. 189. The peripatoid stadium of *Dicellophilus carniolensis*. *a*, the head from below; *b*, the posterior end from below (after Verhoeff, 1902–25).

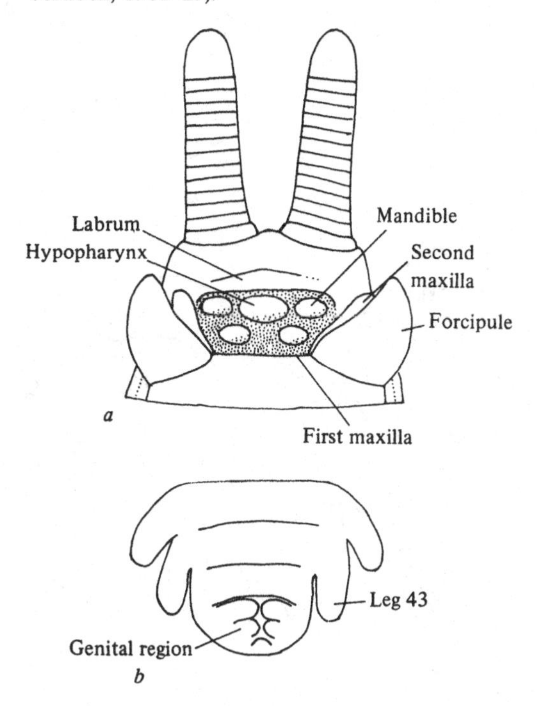

functional, the tracheal system complete, the spiracles are open and the coxal and sternal glands are prominent and their pores are present. The adolescens I of *G. rubens* is 12–15 mm long and yellowish-cream in colour. The adolescens I of *S. maritima* is 10.5 mm long and does not show the pinkish-violet colour that charac-

Fig. 190. The development of *Necrophloeophagus longicornis*. *a*, the egg shortly before hatching showing the egg-teeth; *b*, the hatching embryo; *c*, the peripatoid stadium; *d*, the second embryonic stadium (after Weil, 1958).

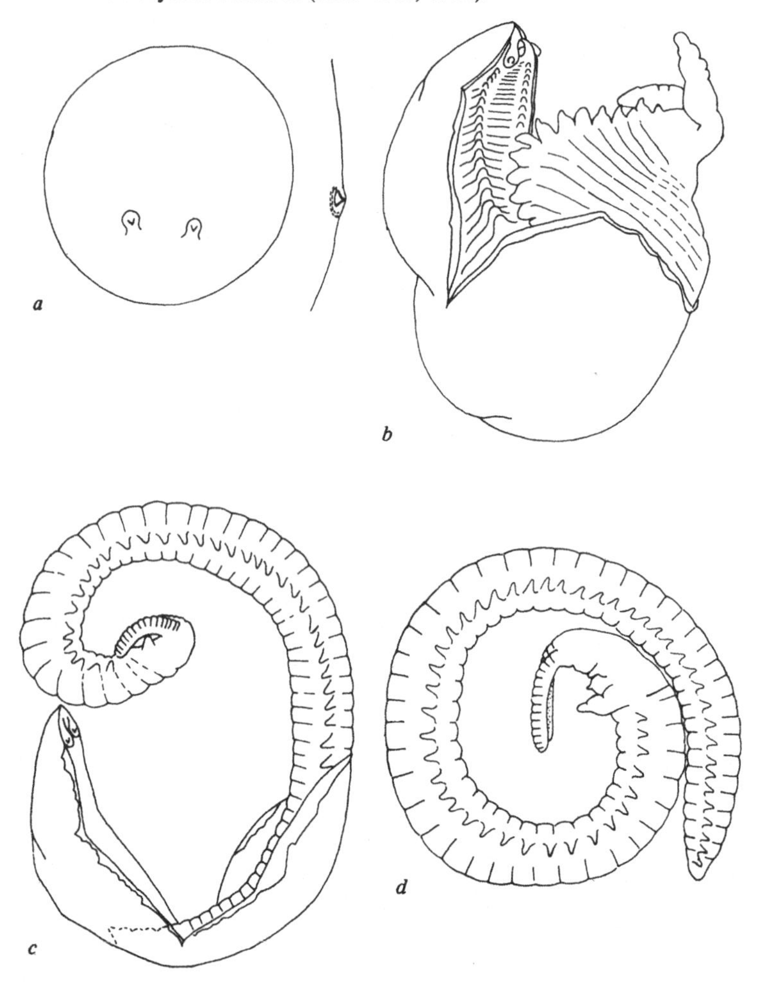

terises the earlier stadia. The behaviour of the adolescens I when first moulted is similar to that of the foetus stadium. Specimens from older broods can crawl but tend to move in circles. Later, however, they can crawl in a straight line (Lewis, 1960). The young centipedes leave the brood chamber in the first adolescens stadium.

In *Necrophloeophagus longicornis* and *Clinopodes linearis* three embryonic stadia are followd by a transitional stage (*Übergangsstadium*) which moults to the first free-living stadium. The latter two stadia seem to be equivalent to the young and older adolescens I stadium of other authors. It could be that they overlooked a moult. Further observations would appear to be required on these early stadia.

In *Geophilus rubens* the leg number varies from 47 to 51 pairs in peripatoid larvae whereas it is 49–51 in adult males and 51–53 in adult females (Johnson, 1952). In German *Necrophloeophagus longicornis* the average number of segments is 53.7 compared to 59.4 in adults. Corresponding figures for *Geophilus electricus* are 56.8 and 61.1, suggesting a small post-embryonic increase in segment number (Misioch, 1978).

In temperate regions egg-laying and brooding take place in the summer months. Thus Verhoeff (1902–25) found *Dicellophilus carniolensis* brooding adolescens I at the beginning of September in the Istrian mountains and *Geophilus carpophagus* broods in June and July in Brandenburg. Palmen & Rantala (1954) found the eggs of *Pachymerium ferrugineum* from 22 May to 2 September in Finland whilst Johnson (1952) found the eggs of *Geophilus rubens* from 14 June to 10 August in Michigan. Weil (1958) reported the time of brooding to be 3 June to 30 October in *N. longicornis* and 8 June to 21 November in *C. linearis*, but these dates were based on laboratory observations. Vaitilingham (1960) reported that the egg-laying period for *Geophilus carpophagus* and *Brachygeophilus truncorum* in southern England lasted from the beginning of May to the end of August.

Lewis (1961) found the eggs of the littoral centipede *Strigamia maritima* from the end of May to mid-June in southern England. Foetus and peripatoid larvae occurred from the end of June to the end of July. Broods at the first adolescens stage were found from late July to early August. Eggs of another littoral species, *Hydro-*

schendyla submarina, were found at Plymouth, England, in late June (Lewis, 1962).

All the geophilomorphs for which adequate data are available, with the exception of *S. maritima* have an egg-laying period of two or more months. The egg-laying and brooding period is much shorter in *S. maritima* and Lewis suggested that this might be necessary to prevent the eggs and larval stadia from being exposed to immersion by the high autumn tides.

Distinction of post-larval stadia

Centipede life histories have been studied either by close observation of laboratory cultures or by regular sampling in the field. If the latter technique is to be successful reliable methods of distinguishing post-larval stadia must be devised.

Verhoeff's (1902–25) discussion of methods of distinguishing adolescens stadia in *Dicellophilus carniolensis* forms a basis for life history studies. The characters he used were the number of coxal gland openings on the coxopleura of the last pair of legs (1, 13, 25 respectively in the first three adolescens stadia of *D. carniolensis*) and the formation of the genital zone in both male and female, together with the number of hairs on the sternite of the intermediate segment which Verhoeff called the genital sternite (0, 2, 12 in the first three adolescens stadia). Other characters considered were the number of clypeal setae, the number of whorls of hairs on the antennal segments, the degree of development of the teeth on the poison claw and the number of setae on the coxopleuron of the last pair of legs. Unfortunately these characters vary in degree only, differences being insufficient to separate successive stadia with certainty. Verhoeff pointed out that many regions of the body show a gradual increase in the number of setae citing the hemi-intercalar segment (pretergite) in the region of segments 24–28. On this there are 6–7 setae on each side in the adolescens I, 7–8 in the adolescens II, 9 in the adolescens III and 15–16 in the adult.

The characters used to distinguish stadia of one species may not be applicable to another: Lewis (1961) found the number of coxal pores on the last pair of legs to be very useful in *Strigamia maritima* (see below) but this character is of little use in *Brachygeophilus truncorum* where there is a single pore on each coxa in the adolescens I, and two in all successive stadia or in *Haplophilus subterraneus* where there are numerous pores on each coxa.

Demange (1943) described small setae (*microchètes*) from segments 5, 9 and 13 of the antennae of *Hydroschendyla submarina* which varied in number with the developmental stage. He suggested that they might be useful in life history studies. They also are present in *Strigamia maritima* but do not change with increasing age.

The number or whorls of setae on the antennal segments and on the segments of the last pair of legs in females shows an increase with age in *Strigamia*. The former from three or four in the smallest individuals to four or five in mature animals, the latter from one to three or four but these differences are not sufficient to characterise successive stadia. There is, however, a marked increase in the number of setae on the presternites, the smallest individuals having three to four and mature animals 10 to 15 but difficulties in counting render the character unsuitable for categorising large numbers of individuals. Although there are marked differences in the structure of the genital region in the younger stadia, the structure of this region is of no value in separating older stadia.

The most useful characters for separating instars are the number of coxal glands on the last pair of legs in both sexes and the number of setae on the sterhite of the intermediate segment in males which Lewis called the genital sternite (Table 8). Body length

Table 8. *Characteristics of the post-embryonic stadia of* Strigamia maritima *from Lewis* (1961)

Stadium	Body length (mm)	Average number of coxal glands	Number of hairs on intermediate sternite in male
Adolescens I	11–15.	1	0
Adolescens II	14.5–20	3.5–5	2
Adolescens III	♂17–24	6.5–10.5	5–11
	♀21–27	6.5–10	
Maturus junior	♂24–29.5	7.5–11.5	11–21
	♀26.5–33	7.5–13	
Maturus senior	♂29–36.5	7.5–13	22–27
	♀31–43	12–17	

and weight are of little value as both increase during the intermoult period. Head width, however, gives a fairly good separation of stadia: plots of head widths for a population from Cuckmere Haven in the south of England showing a clearly quadrimodal distribution for both sexes (Lewis, 1961). The best separation of stadia is achieved when head width is plotted against the number of coxal glands in females and against the number of setae on the intermediate sternite in males. These plots suggest that there are five post-larval stadia in each sex. Lewis termed the first three adolescens I, II and III and the last two stadia the maturus junior and maturus senior as both contained developing gametes.

Further data on the number of instars in females was provided by a study of squashes of the seminal receptacles. These are lined with cuticle and since the duct of the organ is very narrow the lining remains in the receptaculum after a moult and in older females squashes show that a series of concentric cast linings are present. In adolescens III females only one small cast lining is visible. Two linings are seen in maturus juniors. The linings contain whorls of sperm so that maturus juniors contain one whorl of sperm and maturus seniors two but some specimens have three or four whorls of sperm (Fig. 157) and have presumably, therefore, moulted as maturus seniors.

Manton (1965) pointed out that the cuticular lining of the median atrium of the tracheal system in geophilomorphs could not be pulled out of the body when the animals moulted. In a specimen of *Geophilus carpophagus*, sections showed five concentric cast cuticles lying in the atrium. (Fig. 129*a*). If the first free-living stadium had utilised the smallest central cuticle, this specimen must have been in its sixth stadium.

Number of post-larval stadia and life span

There is very little information on the number of stadia and life span in geophilomorphs. Palmen & Rantala (1954) concluded from the size of brooding female *Pachymerium ferrugineum* that the largest were in their third summer or 'possibly even still older'. Weil (1958) reared one *Necrophloeophagus longicornis* and one *Clinopodes linearis* in the laboratory. Both passed through five adolescens stadia. Both species reached maturity in 21 months.

There were three maturus stadia in *N. longicornis* and two in *C. linearis* (Fig. 191).

In *Strigamia maritima* there are only three adolescens stadia. The maturus junior stadium is reached in just over two years, the maturus senior in just over three. Mature animals probably continue to moult at yearly intervals. Lewis (1961), using the number of whorls of sperm in the seminal receptable as an indicator of age in females, concluded that those with four whorls were between five and six years old.

The pattern of the succession of stadia in *N. longicornis* and *C. linearis* is remarkably similar: both have an adolescens I stadium of very short duration. In contrast the adolescens I stadium in *S. maritima* lasts for about 10 months and the adolescens II is of relatively short duration (Fig. 191). It is possible that the slower rate of succession of stadia in *S. maritima* as compared with the two terrestrial species may be related to a paucity of suitable food on the sea shore. It cannot, however, be assumed that the pattern of life history of *N. longicornis* and *C. linearis* is the same in the field as it is in the laboratory.

Research workers have paid little attention to geophilomorph life

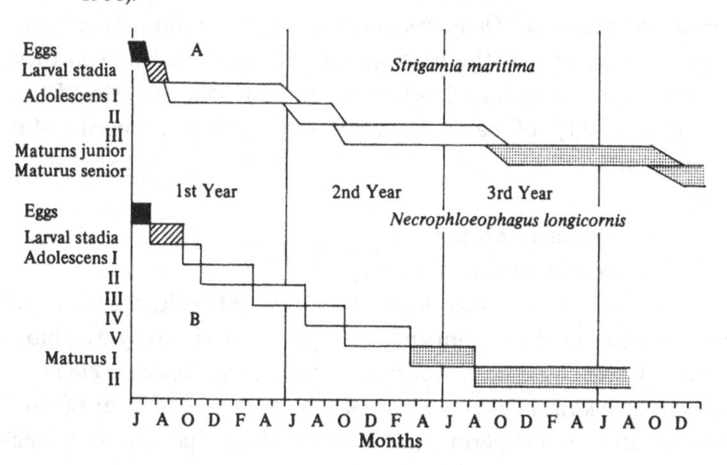

Fig. 191. Diagrammatic representation of the life-histories of geophilomorphs. *a*, *Strigamia maritima* (after Lewis, 1961); *b*, *Necrophloeophagus longicornis* (based on data from Weil, 1958).

histories, an area of investigation which might provide interesting comparative data.

Development of gonads and onset of maturity

In *Strigamia maritima* sperm first appear in the vas deferens of mature males in August. Small groups of sperm, which are about 2 mm long, pass down the vas deferens becoming coiled as they do so. By January the vas deferens is much distended with sperm and remains in this condition until May when the quantity of sperm begins to diminish. The vas deferens is empty by mid-June.

In females there is a slow increase in the diameter of the oocytes in the ovary during the first three adolescens stadia. There is an acceleration in the growth rate of the oocytes in maturus junior and by October the ovary contains oocytes of two sizes: the larger to be laid the following summer, the smaller to be laid in subsequent years. Occasionally large specimens at stadium adolescens III have large oocytes in the ovary suggesting that they will lay eggs before moulting to the maturus junior stadium. One adolescens III was found with a clutch of three eggs (Lewis, 1961). Examination of the seminal receptacles of females and a consideration of the migratory behaviour of males (see Chapter 21) prompted Lewis to suggest that maturus females are inseminated in May before egg-laying, whereas newly recruited maturus juniors are inseminated in August. Palmen & Rantala (1954) showed that female *Pachymerium ferrugineum* collected in the autumn can successfully raise broods the next spring and concluded that in these cases fertilisation occurred before hibernation but they stated that in the majority of cases fertilisation appeared to occur after hibernation.

Scolopendromorpha

Larval stadia and brooding

The first detailed account of the embryology and larval development of the Scolopendromorpha was given by Heymons (1901). It was based on a laboratory study of *Scolopendra cingulata* and *Scolopendra dalmatica* C. L. Koch. A period of cleavage and the formation of the germ band is followed by a period of segmentation and the establishment of the major organs. The further development of organs takes place during two embryonic stadia and a foetus stadium. There follows a first adolescens stadium which is the last stage to be brooded by the female.

The germ band of the developing *Scolopendra* embryo is a broad

Fig. 192. The development of *Scolopendra cingulata a*, the germ band of a developing larva seen from the side; *b*, diagrammatic transverse section through an embryonic segment (after Heymons, 1901).

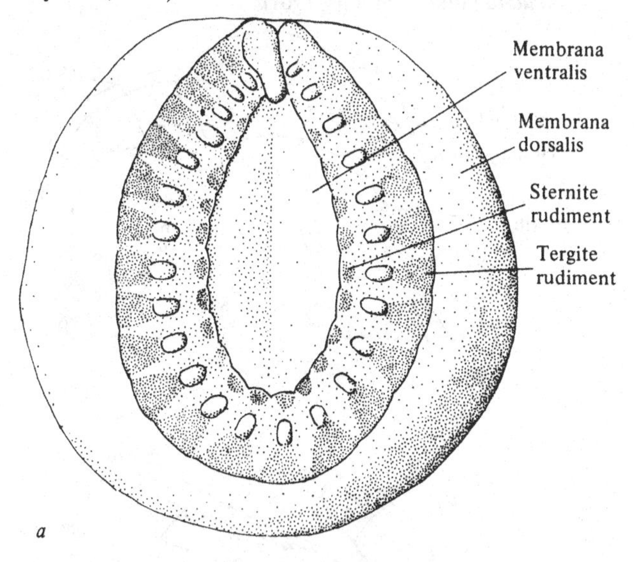

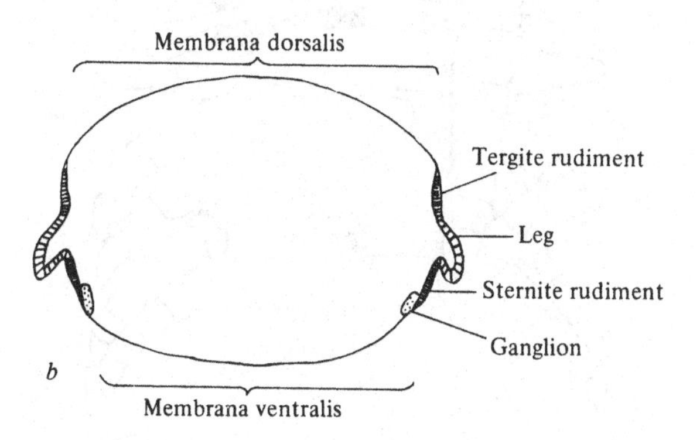

strip bent into a circle so that the head and posterior end are almost touching (Fig. 192*a*), the limb rudiments are bud-like: immediately above them are the rudiments of the tergites, immediately below, the rudiments of the sternites. The former are separated by a wide *lamina dorsalis*, the latter by a *lamina ventralis* (Fig. 192*b*). The first embryonic stadium takes place within the egg

Fig. 193. Ventral view of the anterior end of *a*, the first embryonic stadium and *b*, the foetus stadium of *Scolopendra cingulata* (after Heymons, 1901).

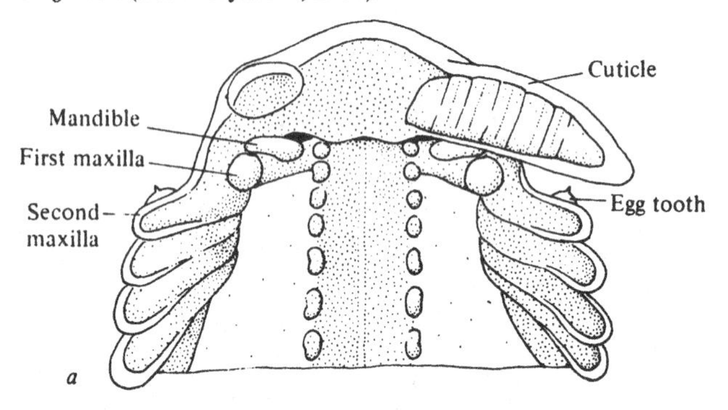

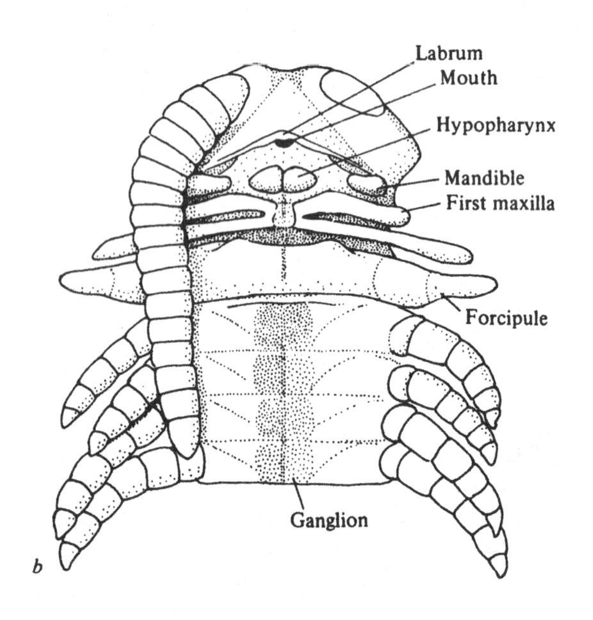

shell. The only sclerotised regions of its thin cuticle are the edge of the telson and egg teeth, one on each second maxillary rudiment (Fig. 193*a*). The only sign of segmentation of appendages is seen in the indistinctly four-segmented antennae.

The egg teeth of the first embryonic stadium split the egg shell around its equator and the cuticle is moulted, the egg teeth being lost, to reveal the second embryonic stadium. This stadium assumes a horseshoe shape: it is approximately cylindrical in cross section. Spiracles are present but it is not clear whether or not they are open. The body segments are distinct, the walking legs have eight clear segments, the antennae seventeen (Fig. 194*a*).

The second moult produces the foetus or intermediate stadium. This has a cuticle sculptured with polygonal areas and is more elongated than the second embryonic stage, with the anterior and posterior ends of the body dorso-ventrally flattened, the mid-part remaining cylindrical. The first signs of movement are seen in this stage and after some time the young animals creep over each other. The foetus is colourless except for four brown ocelli on each side of the head capsule (Fig. 194*b*). The antennae are seventeen segmented, the mandible bears a row of five teeth in *S. cingulata* and four in *S. dalmatica*. The maxillae and prehensors are segmented but the forcipular coxopleural tooth plate lacks teeth. Generally, this stadium lacks accessory claw spines and spines on the terminal legs and their coxopleural processes.

The third moult, observed only in *S. dalmatica*, produces the adolescens stadium which is yellowish in colour due to the sclerotisation of the cuticle. The spines on the legs are well developed and the eighth segment of the walking legs forms a claw. The young leave the brood chamber at this stage having become darker in colour, the legs being greenish, the trunk brownish. After 15 or 20 days a fourth moult takes place. The resulting second adolescens stadium is 2.0–2.5 cm long and has virtually attained the olive green coloration of the adult.

The eggs, embryonic and foetus stages are covered by a slime sheet. Heymons observed numerous injuries and scars on the embryos and foetus stadium and believed that these were probably caused by the sharp claws of the female.

Lawrence (1947) described the early post-embryonic develop-

Fig. 194. *Scolopendra cingulata*. *a*, newly moulted second embryonic stadium; *b*, the foetus stadium (after Heymons, 1901).

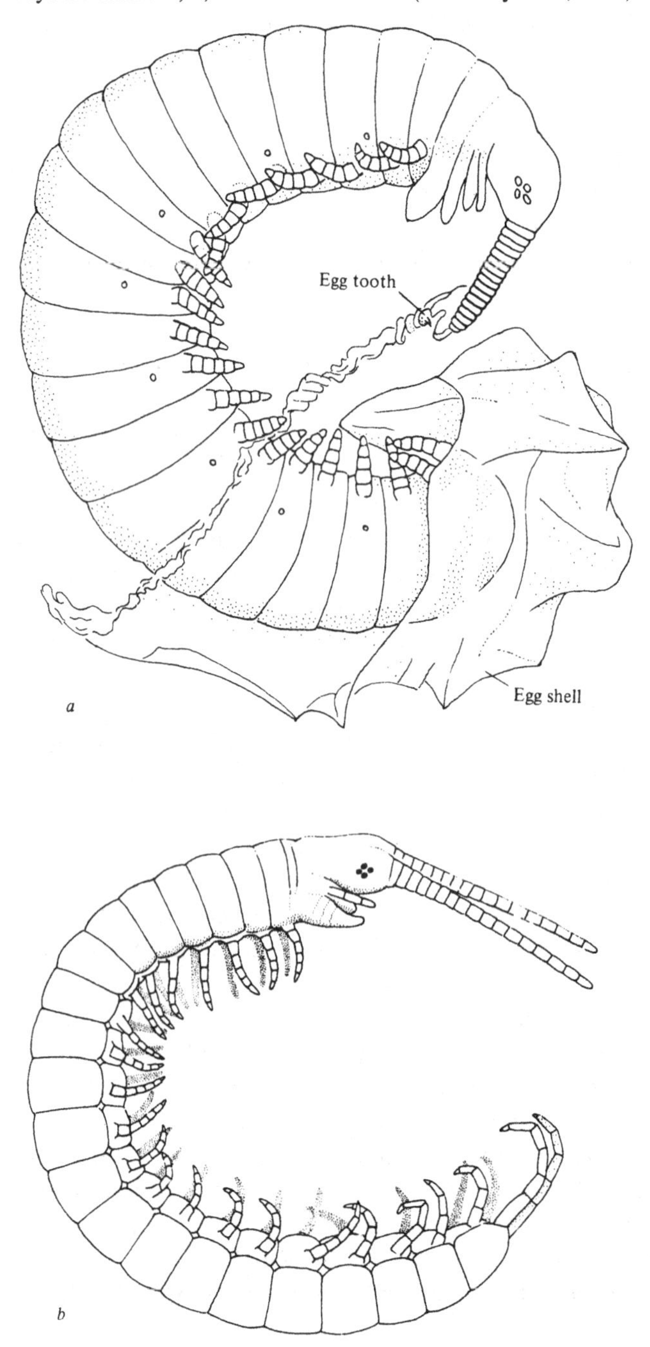

ment of the South African scolopendromorph *Cormocephalus multispinus* (Kraepelin). This description differs in a number of respects from that of Heymons. Lawrence distinguished three embryonic and three early adolescens stadia in *Cormocephalus*. In the embryonic phase of development the body is bent on itself in a horseshoe shape and is either incapable of movement or movement is very simple and restricted. The body is cylindrical or slightly flattened laterally and in the earlier stages the limbs are bud-like. The antennae are folded backwards under the head. The embryonic stadia are covered with a sticky secretion.

The first two embryonic stages distinguished by Lawrence are passed within the egg. In the first embryonic stadium there are no sutures defining the body segments and the legs, mouthparts and antennae are unsegmented. Egg teeth are present and these and the telson are the only sclerotised structures (Fig. 195*a*). In the second embryonic stage the segmentation of the body is fairly clear although not completely defined, the sutures not passing right across the body. The antennae are 'ringed but not segmented' and still relatively short. The walking legs are more pointed and have a large wide-based rudimentary claw. The mouthparts are still bud-like. Lawrence states that this stage is not figured by Heymons but resembles the peripatoid stage of the geophilomorph *Dicellophilus* (Fig. 195*b*).

The third embryonic stage has long antennae and lacks egg-teeth. In some individuals the latter are seen to be attached to the newly moulted cuticle at the anterior end of the body. Early third embryonic stage larvae (form A of Lawrence) have ill-defined tergital sutures which do not pass right across the body, the antennae are not completely divided into segments neither are the legs fully segmented. The mouthparts are still unsegmented. At a later stage (form B of Lawrence) the antennae and legs are clearly segmented with a very long blunt claw (Fig. 195*c*). The second maxillae are segmented and leg-like but the first maxillae and mandibles are still incompletely developed. Lawrence points out that this corresponds to the first stage of *S. cingulata* on its emergence from the egg, as illustrated by Heymons (the second embryonic stadium), with the exception that neither ocelli nor spiracles are present. The difference between forms A and B

Fig. 195. The development of *Cormocephalus multispinus*. *a*, a_1, The first embryonic stage; *b*, b_1, the second embryonic stadium; *c*, c_1, the third embryonic stadium. (after Lawrence, 1947).

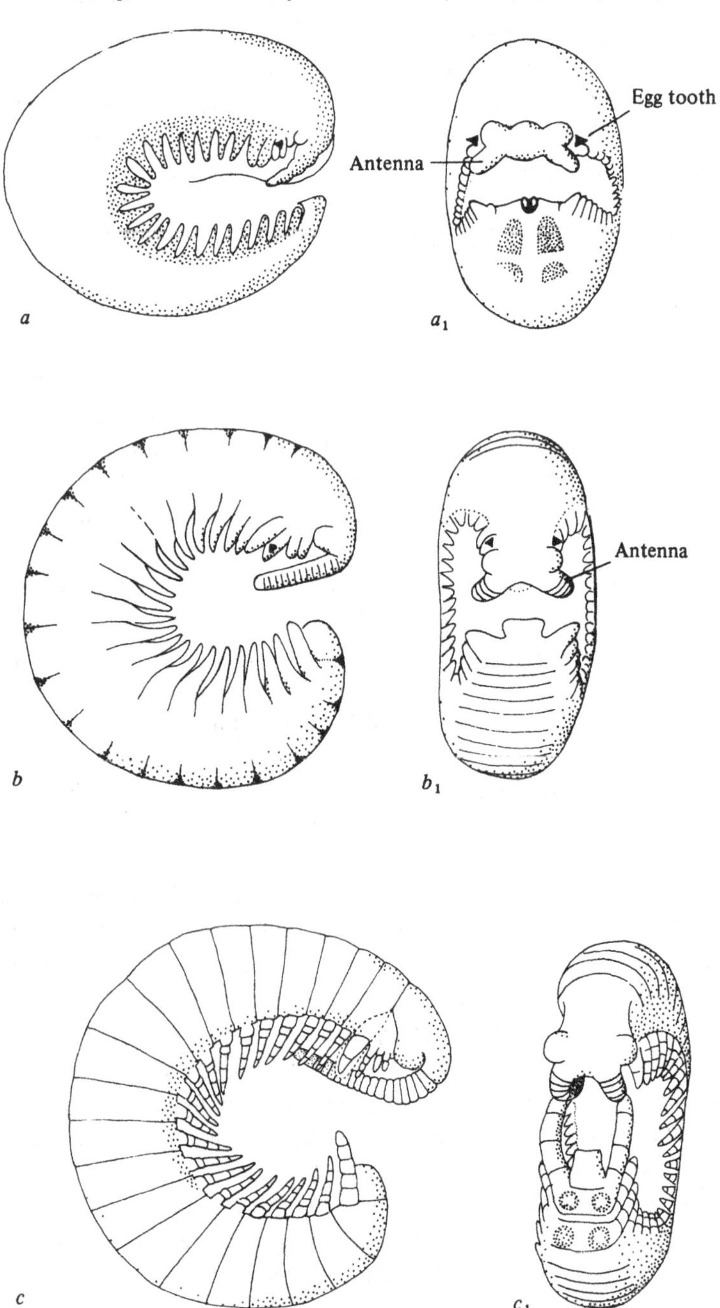

probably represents differences between embryos just before and just after leaving the egg.

The adolescent phase which follows the embryonic phase is characterised by Lawrence as having the body fully extended with the antennae and legs in the positions they retain in the adult. The body is dorso-ventrally flattened like that of the adult but lacks the adult pigmentation and, in the first adolescens stage, the sclerotisation of the various epidermal structures. In the first adolescens stage the young are white in colour except for the ocelli which are represented by four reddish or orange pigment spots under the skin. No lenses can be distinguished. The antennae have a number of minute setae and the tarsi of the legs are provided with a minute chitinous claw and two rudimentary claw spurs may or may not be present: the legs bear a few minute setae. The forcipule is still blunt and imperfectly developed, the tarsus of the second maxilla is club-shaped and clawless, or with a very rudimentary claw and the telopodites of both the first and second maxillae are without hair pads. The mandibles are unsclerotised with four rudimentary triangular teeth distally. The last pair of legs have rudimentary teeth on the prefemora but these are unsclerotised. There are no pores on the coxopleura of the last pair of legs and no spines on the coxopleural processes. The spiracles are represented by small, blunt conical processes but appear to be without true openings. The yolk occupies half the length of the body.

In the second adolescens stage the head is darker than the unpigmented body. The ocelli have colourless lenses. The mouth-parts and forcipules are fully developed, the coxopleura of the last pair of legs bear pores and the coxopleural processes bear spines. The spiracles are open and longitudinal sutures are visible on the tergites and sternites. The yolk forms a small patch about one-seventh of the total length of the body and is situated near the posterior end. There is, in addition, some dark matter in the anterior part of the intestine which may represent food particles; it is probable that feeding begins at this stage.

The young begin to leave the brood chamber in the third adolescens stage. They differ from the previous stage in their larger size and darker coloration.

Differences between Heymons' description of the development of

Scolopendra and Lawrence's of *Cormocephalus* reflect not only differences between the animals but also differences of opinion and terminology. Lawrence says of the three embryonic stages in *Cormocephalus* that 'each is a forward leap unconnected with other stages by a graduated series of intermediate forms: such development is brought about by a series of moults, and though the exuviae can only be clearly seen in the third stage, it is assumed that they also take place within the egg during the first two stages. It is not known how the moulted embryonic cuticles are disposed of in the case of these stages.' Heymons clearly felt that this stage of development was gradual and did not involve moulting, the first moult coinciding with hatching. Lawrence's first and second embryonic stages therefore correspond with Heymons' first embryonic stage. Lawrence noted the similarity between his third embryonic stage and Heymons' second embryonic stage. In most respects Lawrence's first adolescens stage of *Cormocephalus* corresponds with the foetus stage of *Scolopendra*. Coxopleural pores are absent in this stage in *Cormocephalus* but present in *Scolopendra*; prefemoral spines are present on the last pair of legs of *Cormocephalus* but do not appear until the next stage in *Scolopendra*. The first adolescens stage of *Scolopendra* clearly corresponds to the second in *Cormocephalus*. The young *Scolopendra* leave the brood chamber at this stage but in *Cormocephalus* a further moult takes place before they disperse.

The post-embryonic development of a second South African species, *Cormocephalus anceps anceps* has been investigated by Brunhuber (1970*b*). She distinguished the same embryonic and adolescens phases as Lawrence. Noting the increase in the size of the larvae while they are still within the egg shell she concluded that this can only be accounted for by an increase in body water and suggests that this is absorbed from the atmosphere. Brunhuber found, as did Lawrence, that the second adolescens stage larvae (Heymons' first adolescens stage) contained dark brown material, probably food, in the gut. In this stage the animals explored the brood chamber but returned to the female after a short time.

The humidity conditions under which the young were reared was found to be of the utmost importance. Many young died of fungal attack as a result of the culture boxes being kept at the relative humidity preferred by adults. It was found that the young should be kept under drier conditions until the end of their first year.

Cormocephalus multispinus lays eggs from September to December (usually in October or November) and the young leave the brood chamber in late February or early March; thus the period of development from egg to free living adolescens stage is about four months. In *C. anceps anceps* adolescens from eggs laid in late September disperse in January.

Lawrence (1947) gave a brief account of the post-embryonic development of *Cryptops australis* Newport a common centipede in Natal. He distinguished three embryonic stages which correspond well with the same stages in *Cormocephalus*. The second and third stages are relatively longer and more slender than the same stages in *Cormocephalus* and instead of being bent into a horseshoe shape, they are coiled into a spiral. The egg tooth is similar to that of *Cormocephalus*: it may be either present or absent in the third stage. The first adolescens stage is rather less developed than the same stage in *Cormocephalus*. The body is curved in the shape of a comma, with the antennae bent backwards as are the legs. All appendages are clawless except the last few pairs of walking legs, which have rudimentary claws. The sutures defining the segments of the legs are clear but not complete: the body and legs are without hairs, spines or teeth.

In the second adolescens stage the segments and appendages are covered with numerous spines and hairs and the legs have well-developed claws; this stage resembles the adult in general appearance. As in scolopendrids, the eggs and larvae are covered by a sticky secretion.

Verhoeff (1902–25) discussed the larval stages of *Cryptops hortensis* and described the characteristics of the post-embryonic stadia. The more important characteristics are summarised in Table 9.

Post-larval stadia and life span

No precise study has yet been made of the number of post-larval stadia and their characteristics in the Scolopendromorpha, but a number of workers have estimated life spans in members of the order. Brunhuber (1970*b*) estimated that *Cormocephalus anceps anceps* reached sexual maturity in one and a half to two years and

Table 9. *Some characteristics of the post-embryonic stadia of* Cryptops hortensis *selected from Verhoeff* (1902–25)

Stadium	Adolescens I	Adolescens II	Adolescens III	Maturus
Body length (mm)	6	9–10	13	20–22
Mandible lamellae	5(6)	6	7–8	9–10
Setae on anterior wall of forcipular coxosternum	1 + 0–1	1 + 3–4	1 + 4–5	1 + 4–5
Coxpleural pores of terminal legs	11–12	27–30	56–57	65–90
Saw-teeth on tibia and tarsus 1 of terminal legs	3 + 1(2)	—	6–7 + 3–4	8–11 + 5–8

suggested from the size of some animals kept in the laboratory for two years after capture that their life span was at least six years.

On the basis of measurements of head width, body length and antennal segment number Lewis (1966) suggested that there were three free-living adolescens stadia and a maturus stadium in *Scolopendra morsitans* from Khartoum, Sudan. The conclusions were based on a sample of only 50 specimens collected in a period of less than four months and were unjustified as was the assumption that each stadium lasted for a year. In Zaria which is situated in the guinea savanna zone of northern Nigeria and where there is a dry season from October to March, *Scolopendra morsitans* produces two generations a year. Young individuals appear in March just before the beginning of the rains. These specimens reach full size by October at which time further small individuals appear (Lewis, 1970). Lewis considered that the species matured in one year or possibly six months in some cases. The species is unusual in that it is active throughout the dry season in Zaria, finding high humidities and abundant food in cow dung. Attempts to find characters to separate the stadia proved unsuccessful. None of the characters which change with age, analysed by Lewis (1968), namely body length, weight, number of antennal segments, number of tergites with complete paramedian sutures, number of marginate tergites or coxopleural teeth showed sufficiently abrupt changes for them to be of use in this respect. A plot of head widths in 520 females showed five clear peaks in animals up to 45 mm long (maximum length of females 81.5 mm) and in a plot of 396 males four clear peaks were visible in animals less than 37 mm (maximum length of males 69.5 mm). It is obvious that there is a large number of free-living stadia. Lewis (1970) quoted a personal communication from Jerrard who had found that *Scolopendra subspinipes* moulted nine or possibly ten times in captivity in two and a half years.

Three other scolopendrids have been investigated in northern Nigeria, namely *Rhysida nuda togoensis*, *Ethmostigmus trigonopodus* (Lewis, 1972*a*) and *Asanada sokotrana* (Lewis, 1973). Unlike *S. morsitans* they do not appear in surface habitats until the first rains have fallen. In *Rhysida*, new recruits, 12.5–14.5 mm long, appear in May about a month after the first rains have fallen. Growth rates are rapid and by September these specimens have reached 33–40 mm. They can be found in damp habitats for some time after the rains have ended but become rare in December. By the following April specimens are 46–53 mm long and fully grown. This species reproduces at the end of its first year. Small *Ethmostigmus* 22–30 mm in length also first appear in April. As with *Rhysida*, growth rates are very rapid and specimens measure 45–61 mm by September. They become rare after the end of October. By the beginning of their second year, specimens measure 69–89 mm so growth evidently continues through the dry season. It was thought probable that the animals reproduced at this size. Two very large specimens, females of body length 109 and 119 mm would have been in their third year.

Asanada sokotrana, which is invariably found in deserted *Trinervetermes* mounds, lays eggs sometime between January and April in Zaria and the young appear in March. The smallest specimens are 10 mm long. Growth rates are rapid in the early part of the rainy season and by July the mean length is about 30 mm but there seems to be little increase in size during the remainder of the rains. In April and May the testes are very small in males but by June or July they are functional and there is a gradual increase in the number of spermatophores in the vas deferens which continues until at least January. Spermatophore transfer must take place between January and March after which the males die, being about a year old. Data suggest that some females, however, live for more than one year.

These three species continue to grow through the dry season although they are absent from surface habitats. Growth rates of young scolopendrids are high during the rainy season so that they reach a large size by September or October presumably increasing their resistance to dessication. The mobility of *Ethmostigmus* and *Rhysida* enables these species to seek favourable habitats where they can continue to feed after the rains have ended but they become less common on the soil surface as the dry season progresses. Manton (1952*b*) maintained that scolopendromorphs do not burrow actively or deeply. If so, they must enter the soil through the crevices that appear as it dries out. The soil probably contains a substantial fauna through the dry season: Lewis (1972*b*) reported that spiders and crickets were flushed in considerable numbers from open thorn savanna near Lake Chad, Nigeria, during irrigation in December at the height of the dry season.

Lithobiomorpha

There is no maternal care of the brood in the anamorphic orders Lithobiomorpha and Scutigeromorpha, the eggs being laid singly. In these orders the young become active as soon as they hatch from the egg which they do at a more advanced stage than the young of the epimorphic orders. In the Anamorpha the development is in two phases. During the anamorphic phase extra pairs of legs are added at successive moults: the stadia with an incomplete number of legs are known as larvae. The larvae are small and pale coloured but the legs are well formed and the tracheal system is functional. In the second, epimorphic phase, the post-larval stadia show a progressive development of the genital region and an increase in the number of ocelli, antennal segments and in Lithobiomorpha the number of pores on the coxae of the last four pairs of legs.

There is a considerable volume of literature on the post-embryonic development of Lithobiomorpha and much of the work has been duplicated. The early work of Fabre (1855), and Latzel (1880) has been reviewed by Verhoeff (1902–25) who added a detailed account of the development of *Lithobius forificatus*. Verhoeff's terminology has been adopted by a number of authors and is used here.

Anamorphic (larval) stadia

Verhoeff distinguished a foetus stage and five larval stages in *L. forficatus*. The term foetus is used to describe the stage which has left the egg but is still covered by the embryonic cuticle. It has seven antennal segments, seven pairs of legs and a pair of limb buds (Fig. 196*a*).

The foetus is followed by four larval stages and a 'larva media'. Their characteristics are shown in Table 10. The posterior ends of larva I and II are shown in Fig. 196*b*, *c*. Spiracles appear on segments 3 and 5 in Larva I, on segment 8 in Larva III and on segment 10 in Larva IV (Demange, 1967). The larva media is produced very rarely (Joly, 1966*a*; Andersson, 1976) and is not typical of the development of *L. forficatus*. Larva II is the first to feed (Scheffel, 1969).

Scheffel (1969) repeated Verhoeff's work on the anamorphic stadia of *L. forficatus* but used a different terminology calling the foetus larva stadium I and larval stadia I to IV stadium II to V. His findings differ in detail from Verhoeff's: Scheffel found 11 antennal segments in the Larva I (Verhoeff gave 12), and 17 in 90 per cent of larva III rather than 18. He recorded 20–23 segments in larva IV,

Table 10. *Characteristics of the larval stadia of* Lithobius forficatus *from Verhoeff* (1902–25)

Stadium	Legs[a]	Limb buds[a]	Antennal segments[b]	Ocelli[b]	Forcipular coxopleural teeth[b]
Foetus	7	1	7		0
Larva I	7+1[c]	2	12	2	2
Larva II	8	2	14	2	2
Larva III	10	2	18	2 or 3	3
Larva IV	12	3	19–21	4	4
Larva media	12 or 13	2 or 3	21	5	4

[a]Number of pairs.
[b]Number on one side.
[c]Half-developed leg.

Fig. 196. Dorsal view of the posterior end of *a*, the foetus stadium; *b*, larva I and *c*, larva II of *Lithobius forficatus* (after Scheffel, 1969).

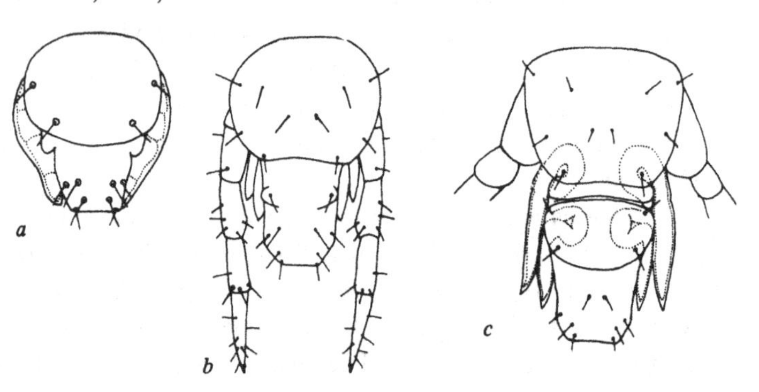

Verhoeff recorded 19–21. There are similar minor differences in the number of ocelli.

Andersson (1976) investigated the variation in size, number of coxal pores, ocelli, antennal segments, teeth on the forcipular coxosternum, spinulation, projections on tergites 9, 11 and 13, the accessory apical claw on the fifteenth pair of legs and genitalia in both anamorphic stadia and epimorphic stadia.

Variation in characters observed by different authors may reflect geographical variation in the species, or differences in rearing conditions in the laboratory.

The post-embryonic development of a large number of lithobiids has been described, namely, the European *Lithobius variegatus* (Eason, 1964) and *L. crassipes* (Wignarajah, 1968), and the Japanese species *Lithobius* (*Archilithobius*) *pachypedatus* Takakuwa (Murakami, 1960*a–c*), *L.* (*Archilithobius*) *canaliculatus* Murakami (Murakam, 1963), *Monotarsobius nihamensis* Murakami (Murakami, 1961*a*) and *Bothropolys asperatus* (Murakami, 1958*b*). In these species there is a foetus stage and four larval stadia as in *Lithobius forficatus*. In the Japanese species *Esastigmatobius longitarsis* Verhoeff (Family Henicopidae) the early stages differ from those described for lithobiids, there being eight pairs of legs and two pairs of limb buds in the foetus (seven and one in lithobiids) and eight pairs of legs and two pairs of limb buds in the larva I (seven and three in lithobiids). The number of pairs of legs and limb buds in the succeeding stadia is the same as in lithobiids, (Murakami, 1961*b*).

Epimorphic (*post-larval*) *stadia*

Verhoeff distinguished seven post-larval stadia in *L. forficatus*: their characteristics are shown in Table 11.

Joly (1966*a*) gave data for the number of antennal segments, forcipular coxopleural teeth and coxal pores for French *L. forficatus*. These are similar but not identical to those given by Verhoeff. Andersson (1976) gave a detailed account of the variation of the epimorphic stadia based on Swedish material. He distinguished nine post-larval stadia and found that the best character for separating the different post-larval stadia is the number of coxal pores on each of the last four pairs of legs. At each moult no more

Table 11. *Characteristics of the post-larval stadia of* Lithobius forficatus *from Verhoeff* (1902–25)

Stadium	Length (mm)	Coxo-pleural teeth[a]	Antennal segments[a]	Ocelli[a]	Coxal pores[b]	Setae on genital sternite[a]
Agenitalis	7.5–9	4	25–26	7–10	1.1.1.2 or 2.2.2.2	0
Immaturus	10–10.5	4	32	—	3.3.3.3	3
Praematurus	11.5–13.5	4–5	34	12–16	—	♂ 6–8 ♀ 3–5
Pseudomaturus primus	13.5–14	5	36–38	21	4.5.5.6	♂ 12–13
Pseudomaturus secundus	17–17.5	5	—	28	5.6.6.6 or 6.7.7.7	♂ 20–21
Maturus junior	♂ 18	5	41–42	28–29	5.7.7.7 or 4.6.6.7	25–30
	♀ 17.5–19.5	5	37–40	—	5.7.7.7 or 6.6.6.6	—
Maturus senior	♂ 21.5–23.5	6	38–44	35–37	5.6.6.6 or 5.7.8.8	30+
	♀ 22–23	6	36–40+	24–31	5.8.8.7 or 6.7.7.8	

[a]Number on one side.
[b]Successive figures are for twelfth, thirteenth, fourteenth and fifteenth legs.

than one pore is added to each coxa; thus there are never more coxal pores on each coxa on the thirteenth to fifteenth pairs of legs than the stadium number but on the twelfth pair of legs there can be one more because one pore is formed in Larva IV. Verhoeff's agenitalis stadium thus contains two instars, a fact realised by Brölemann (1930) and Joly (1966*a*) who distinguished an Agenitalis I stadium with one coxal pore and an Agenitalis II with two coxal pores on each coxa. Clearly, there is considerable variation in the number of pores in the older stadia but Andersson suggested that Brölemann's (1930) reference to specimens with 10 or 11 coxal pores suggest that there may be more than nine post-larval stadia.

Eason (1964) described the post-embryonic development of *Lithobius variegatus* in detail, including an account of the changes in the spinulation of the legs. He distinguished post-larval stadia 1 to 4, corresponding to agenitalis, immaturus, praematurus and pseudomaturus. The increase in the number of coxal pores is more regular than in *L. forficatus* being 2, 1, 1, 1; 3, 2, 2, 2; 4, 3, 3, 3 and 5, 4, 4, 4 in successive stadia. The maturus has 6, 5, 5, 5 coxal pores. Eason found it difficult to define the maturus senior stadium on any character other than coxal pores (7, 6, 6, 6). He reported some variation in the number of coxal pores in each stadium although the arrangement is more constant than in most species studied.

Wignarajah (1968) working in northern England described four post-larval stadia and a maturus stadium in *Lithobius crassipes* and found that some characters stabilise in the third stadium (praematurus): both stadium 3 and 4 have 20 antennal segments and 2, 3, 3, 3 coxal pores.

Murakami (1960*b*, *c*) distinguished six epimorphic stadia in the Japanese species *Lithobius pachypedatus*, namely agenitalis I and II, immaturus, praematurus, one or two pseudomaturus stadia and a maturus. The numbers of prehensorial coxopleural teeth (2+2) is constant from the first larval stadium, the number of antennal segments (usually 20) stabilises at the agenitalis I, the number of ocelli on each side (1+5) at the immaturus and the number of basal spines on the female gonopod (3+3) at the pseudomaturus. An analysis of the numbers of coxal pores in each stadium showed that these do not always correlate with the number of moults that have taken place, thus in the agenitalis I the coxal gland formula is

2, 1, 1, 1 or 1, 1, 1, 1 but in the agenitalis II it varies from 2, 2, 2, 2 to 3, 3, 3, 3 and in the pseudomaturus from 4, 5, 5, 4 to 5, 6, 5, 5.

In *Monotarsobius nihamensis* and *Lithobius* (*Archilithobius*) *canaliculatus* there are four and five post-larval stadia respectively before the maturus stadium (Murakami, 1961*a*, 1963) and, as is the case with *L. pachypedatus*, the number of ocelli and antennal segments stabilise early. Murakami gave a detailed account of the spinulation of the legs in *M. nihamensis*. In another Japanese species, *Bothropolys asperatus* Koch there are two agenitalis stadia, an immaturus, praematurus, pseudomaturus and two stadia which may either be pseudomaturus or maturus, before the definitive maturus (Murakami, 1958*b*). A female reared in captivity passed through two pseudomaturus stadia and then moulted five times as an adult. In *Bothropolys* the number of antennal segments (mean = 20) stabilises at the agenitalis II stadium but the number of prehensorial coxopleural teeth (up to 9 + 9) and the number of coxal pores (about 30) increase until the maturus stadium.

In all the species which he has investigated, Murakami has distinguished two agenitalis stadia, whereas most European workers have described only one. This is probably due to differences in interpretation and suggests that the stadia are not as clearly characterised as the general adoption of Verhoeff's terminology suggests. As characters such as antennal segment number, coxal gland number and the number of ocelli vary so much in incremental pattern from species to species, the distinction of the epimorphic stadia must rest on the characters of the genital segments. Data are insufficient for this to be done in all cases. In any event, it is probable that different species do not moult at precisely the same stage of development.

Demange (1967) regarded several characteristics of the anamorphic development of lithobiomorphs as indicating the inhibition of development of a segment between segments 7 and 8. He also cited anatomical evidence for this (see Chapter 4).

He argued that the non-appearance of spiracles in Larva II and the appearance of three pairs of limb buds in Larva I rather than four was due to the inhibition of a segment (true segment 8). The same process, he suggested in the Larva IV where three segments, 13, 14 and 15 appear together; the supposed missing fourth segment could be the genital segment which lacks appendages. Demange's conclusions derive from the assumption that the primitive pattern in anamorphic development is for two segments to be added at each moult; some might question this.

Life history and duration of stadia

Table 12 summarises the data on the duration of larval and post-larval stadia in lithobiomorphs. Scheffel's (1969) data for *L. forficatus* came from specimens reared in the laboratory at 25 °C, Murakami's (1958*b*) for *B. asperatus* from specimens reared at 28–31 °C. Roberts' (1956) data was based partly on field observations. Joly (1966*a*) raised *L. forficatus* at 18–22 °C. He found the duration of stadia to be longer than those of Scheffel's specimens and suggested that this might be due to the existence of different races of *L. forficatus* or to climatic differences.

Table 12. *Duration of stadia in days in various lithobiomorphs*

	L. forficatus (Scheffel, 1969)	*L. forficatus* (Joly, 1966)	*L. variegatus* (Roberts, 1957)	*Bothropolys asperatus* (Murakami, 1958*b*)
Stage				
Egg	—	35–40	25	26
Foetus	15.2	2–4	0.5	2
Larva I	3	4–6	7	2–3
Larva II	8	9–28	15	7–9
Larva III	10	14–40	31	8–10
Larva IV	14	14–40	28	10–14
Agenitalis I	10–25	29[a]	35	150
Agenitalis II	8–21	22[a]		14
Immaturus	17–30	31[a]	46–152	28
Praematurus	17–18	31[a]	42	37
Pseudomaturus	—	39[a]	54	24
Maturus junior	—	46[a]	—	—
Maturus senior	—	55[a]	—	—

[a] Mean of number of days.

The most detailed account of the life-history of a lithobiid is that of Murakami for *B. asperatus* in Japan. Eggs are produced over a prolonged period, from late March to late October. Eggs laid between March and July hatch to produce larvae that complete their anamorphic development in 32–56 days depending on the season. The young overwinter in stadium 5, 6 or 7 and reach the maturus stadium (stadium 11) the following November and become mature the following March when they are about two years old. Specimens that hatch from eggs later in the year overwinter in stadium 2, 3 or 4. They reach stadium 10 by the following November, overwinter at this stadium and mature the following spring when they are rather less than two years old. *B. asperatus* continues to moult after reaching the maturus stadium: a female kept in captivity moulted five times at six monthly intervals attaining a life span of just over three years.

The life-history of *Lithobius variegatus* appears to resemble that of *B. asperatus*. The main period of oviposition is early summer (May) and the anamorphic larval stadia are completed during the first summer and the species overwinters in the early epimorphic stadia. The maturus stadium is reached in the second summer about 15 months after oviposition (Roberts, 1956). The length of life is approximately two years. Mating occurs in the animal's second autumn, earlier in the maturus senior than in the maturus junior, larval stages are most abundant in the F_x layer (the previous year's litter) while immature and mature stadia are particularly common at the base of the F_0 layer (recent litter). At times of climatic extreme in summer and winter they move deeper into the litter and into logs.

Verhoeff (1902–25) suggested that *Lithobius forficatus* lived for five to six years. Joly (1966*a*) suggested that the adult phase might last three to five years.

Egg-laying and sperm transfer

According to Sinclair (1895) *Lithobius forficatus* oviposits from June to August in England: Brocher (1930) recorded oviposition from December to April in Switzerland. Verhoeff believed that it occurred throughout the 'warm season' in Germany as young centipedes could be found throughout the year. In Michigan females have been seen carrying eggs from May to September with a peak in July but in the laboratory oviposition takes place throughout the year although the peak still occurs in July (Johnson, 1952).

In England large ova are present throughout the year in the ovaries of mature *L. forficatus.* There is an increase in size from March to June but very few large ova are present in July and August, so presumably the bulk of eggs are laid before July. The average egg size increases rapidly in August and September and falls progressively from September to January (Fig. 197*a*). This decrease could be due to the addition of young maturus individuals to the population or to further egg-laying during the autumn and winter (Lewis, 1965). One female was found carrying one egg in April, two in May and one in June.

Herbaut & Joly (1972) showed that egg-laying in *L. forficatus*

Fig. 197. Seasonal variation in the size of ova in the ovaries of *a*, mature female *Lithobius forficatus*; *b*, mature *L. variegatus* (from Lewis, 1965).

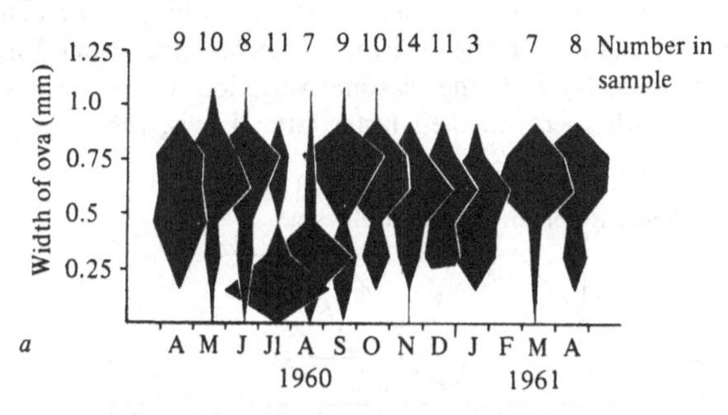

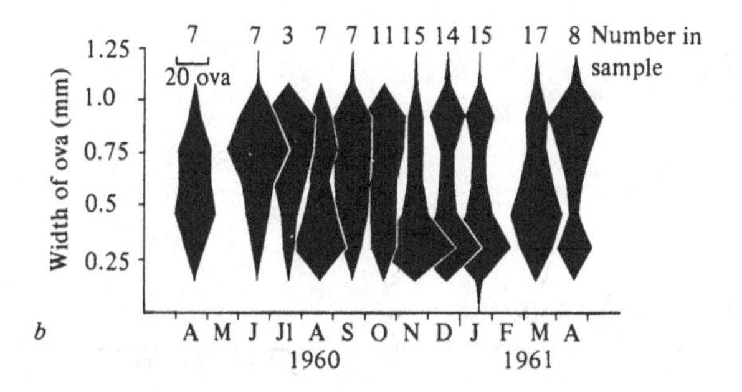

could take place throughout the year with peaks in spring and autumn: egg-laying is rare in the summer months. The most favourable conditions for oogenesis are temperatures of 12–15 °C, it is disturbed or ceases at 18–20 °C (Herbaut, 1975*b*). There is an antagonism between somatic growth and oogenesis. Moulting takes place in spring, oocyte growth in the autumn (Herbaut, 1977*b*).

Seasonal changes in the size of ova in *L. variegatus* are less marked than they are in *L. forficatus*: small ova are most abundant from November to January but large eggs are always present (Fig. 197*b*) so that oviposition could take place at any time of the year (Lewis, 1965). Females were found carrying eggs in the field, twice in April and once in November. Roberts (1956) observed females carrying eggs from the second half of April to the first half of June but his observations only lasted from March to June.

Lewis (1965) pointed out that the number of large oocytes in the ovaries of *L. forficatus* and *L. variegatus* is low, averaging five, and

Fig. 198. Sperm production in *Lithobius forficatus*. *a*, spermatogenesis in the male (continuous line) and the development of the seminal receptacles in females (pecked line) (after Joly & Descamps, 1969); *b*, the seasonal variation in the percentage of males with sperm in their testes (after Lewis, 1965).

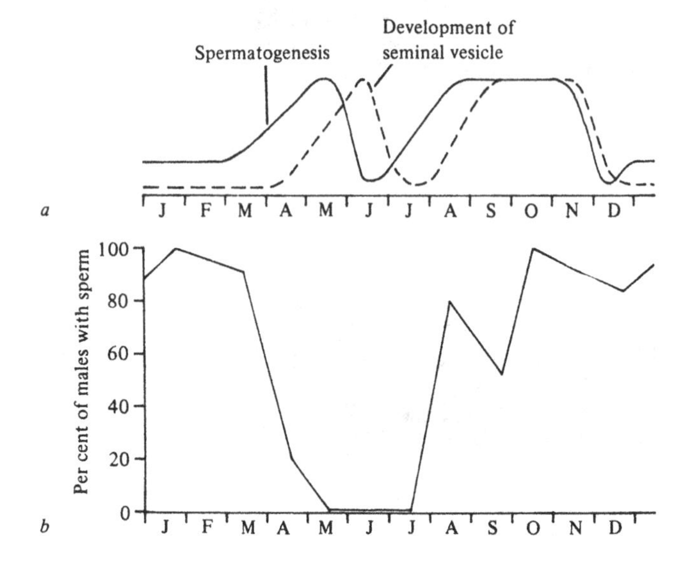

can give no indication of the number of eggs laid in a year. Brocher (1930) reported that two female *L. forficatus* laid 20 and 21 eggs respectively over a period of four months, Vaitilingham (1960) found the mean number of eggs laid by this species to be 20. Roberts (1956) gave the mean number for *L. variegatus* as six, Vaitilingham gave 15. Murakami (1958*b*) found that *Bothropolys asperatus* lays 130–190 eggs from the early part of March to the latter part of September in Japan. Lewis (1965) found a female *L. forficatus* with 83 fully developed oocytes in her body cavity. There was no trace of an ovary wall: presumably the oocytes had been released into the body cavity instead of being laid.

In *L. variegatus* there is a rapid fall in the number of mature males with sperm in their testes in April and May and sperm are absent in June (Lewis, 1965). Examination of the seminal receptacles of females from January to July showed a rapid fall off in the number containing sperm in June and July (Roberts, 1956). The author suggested that mating probably occurred in the autumn, Lewis suggested that it took place in spring. Joly & Descamps (1969) demonstrated two periods of spermatogenesis in *L. forficatus* during the year, from March to the beginning of May and from July to the beginning of August. Sperm appear in the seminal receptacles of the female in May and June (Fig. 198*a*). The number of sperm in the testes falls again in November: there was no indication of this second fall in Lewis's English material (Fig. 198*b*).

Spermatogenesis is inhibited at 5 °C and commences earlier at 24 °C than at 18–20 °C (Descamps, 1971*d*). Somatic growth and moulting delay spermatogenesis (Descamps, 1977*b*).

There appears to be no seasonality in the life cycle of lithobiomorphs in Romanian caves (Negrea, 1969).

Scutigeromorpha

Larval and post-larval stadia

Verhoeff (1902–25) reviewed the older literature on the post-embryonic development of the Scutigeromorpha and described the post-larval stadia of *Scutigera coleoptrata* in considerable detail together with information on *Podothereua insularum* Verhoeff.

The development of scutigeromorphs resembles that of lithobiomorphs, there being a series of larval stadia showing a progressive increase in leg number followed by a number of post-larval stadia. There are six larval stadia with 4, 5, 7, 9, 11 and 13 pairs of legs respectively. The first five stadia each have four pairs of limb buds, the sixth two (Fig. 199) (Murakami, 1959*a*). The late embryo of *S. coleoptrata* has a pair of anterior pigment spots (*Stirnocellen*) which are simple ocelli and become neurosecretory structures of unknown function (Fig. 109). The pigment disappears after the first moult (Knoll, 1974).

Fig. 199. The posterior end of larval *Thereuonema hilgendorfi*. *a*, late third instar; *b*, late sixth instar (after Murakami, 1959*a*).

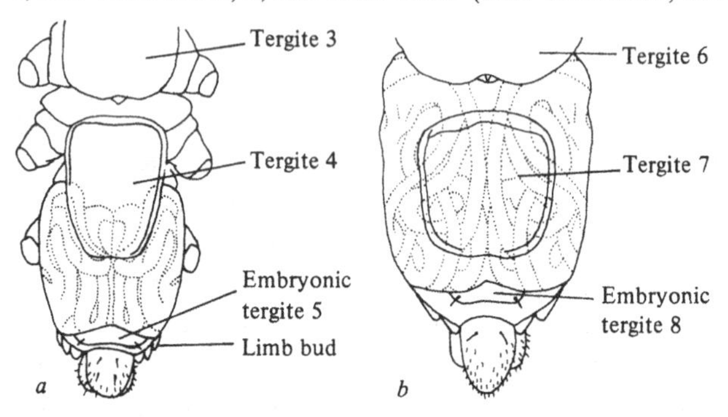

Fig. 200. The first instar larva of *Scutigera coleoptrata* (after Knoll, 1974).

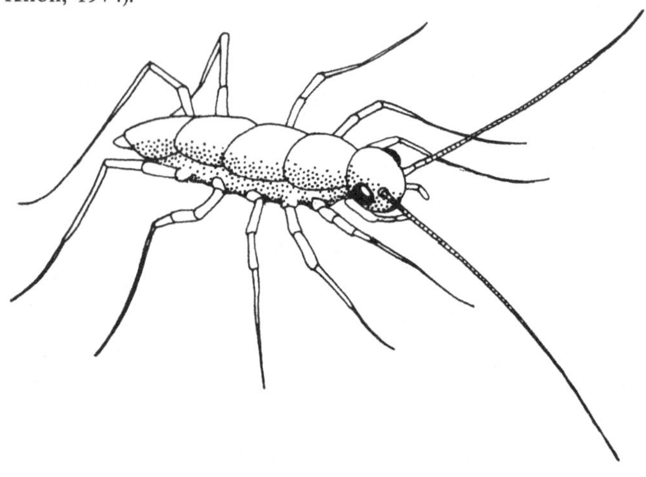

The Larva I (Fig. 200) corresponds to the foetus stage of lithobiomorphs although it has fewer pairs of legs, so the numbering of the stadia is different in these orders: larva II in *Scutigera* corresponding to larva I in *Lithobius*. The first two larval stadia contain yolk and lack tracheae and spiracles. Larva III is the first stadium to feed; it takes Collembola and small spiders (Dohle, 1970). Larva III is also the first feeding stadium in *Thereuonema hilgendorfi* Verhoeff (Murakami, 1958*a*). Hatching in *S. coleoptrata* is facilitated by an egg tooth on the second maxilla (Knoll, 1974).

The post-larval stadia in scutigeromorphs are the agenitalis I and II, immaturus, praematurus, pseudomaturus and maturus. They are characterised by the number of antennal segments, the number of tarsal segments and the spinulation of the legs. Verhoeff (1902–25) gave a very detailed account of these characters in *S. coleoptrata* together with some data for *Podothereua insularum* Verhoeff. Data for *Thereuopoda ferox* and *Thereuonema hilgendorfi* were given by Murakami (1959*b*) and Murakami (1956*a*) respectively. Some of these characteristics are shown in Table 13.

Life history and duration of stadia

Laboratory observations have shown that *Thereuonema hilgendorfi* lays 130–290 eggs from the latter part of May to September (Murakami, 1956*b*). The eggs hatch in 14–21 days. First instar larvae occur from late June to early November, reaching a maximum in August. Animals were reared at 16–32 °C and the stadia therefore showed very considerable variation in duration: Larva I, 1–15 days; Larva III, 5–30 days; Larva VI, 6–83 days and Pseudomaturus 13–67 days.

The eggs of *S. coleoptrata* hatch in 30–38 days at 20–21 °C (Knoll, 1974). Verhoeff (1938) found that the duration of the first three instars was 4, 7 and 8 days, Dohle (1970) gave figures of 7–9, 12–13 and 13–17 days for the same stadia reared at 20 °C. There are three pseudomaturus stadia in *Thereuonema hilgendrofi* and the adults moult five times. In its first year the species overwinters at stadia between larva IV and pseudomaturus. Females lay eggs in their third year (Fig. 201). A female *S. coleoptrata* which hatched in spring 1966 lived until January 1969 (Dohle, 1970). In southern

Table 13. *Some characteristics of the developmental stadia of* (*a*) Scutigera coleoptrata (*from Verhoeff*, 1902–25), (*b*) Thereupoda, (*from Murakami*, 1959) *and* (*c*) Thereuonema hilgendorfi (*from Murakami*, 1956*a*)

	Number of pairs of legs	Body length (mm)			Number of antennal segments			Number of segments on first and second tarsus of first pair of legs		
		a	*b*	*c*	*a*	*b*	*c*	*a*	*b*	*c*
Larva I	4	—	4–4.5	2.5	—	unseg.	unseg.	—	2+14	unseg.
Larva II	5	—	4–4.5	2.5	—	3–4	unseg.	—	3–4+17	unseg.
Larva III	7	—	5–6.5	2.4–3.3	—	9–12	5	—	5+25	4+14
Larva IV	9	—	6–7.5	3.8–4.6	—	8–21	(8)9	—	5–6+25–27	4+14–15
Larva V	11	—	7–8	5.6–6.5	—	28–32	13(11–14)	—	7–9+26–28	5+15
Larva VI	13	—	10–12	6.5–7.5	—	34–40	17(15–23)	—	10+33	5+16–17
Agenitalis	15	6–7.5	14–15		25–28	45–53		8+20	9–11+30–34	
Immaturus	15	8.5–11	15–19		35–41	55–61		14+24	14–18+38–43	
Praematurus	15	12–14	22–26		56	62–71			20–22+50–51	
Pseudomaturus	15	16–20	25–30		65	72–75		14+32–34	21–22+52–54	
Maturus	15	24–26	30–46		73–76	72–85		14+35	21–26+48–57	

Europe eggs laid in the autumn could reach stadium 5 in about five months (Knoll, 1974).

Scutigera mohamedanica lays one egg a day for 35 days. The eggs hatch in 12 days, the sixth moult takes place about 58 days after hatching and the animals live in the laboratory for more than 15 months (Kaestner, 1967). Data for animals reared in the laboratory should be treated with considerable caution as figures may bear little relationship to those obtaining in the field.

Craterostigmus

The only observations on the life cycle of *Craterostigmus* are those of Manton (1965). The breeding season on Mount Wellington, Tasmania lasts from November to February (the southern summer). Mature females penetrate damp decaying logs and lay their eggs in clusters in small crevices or chambers in the wood. The female humps her body over the egg mass, as do the scolopendromorphs, and remains with the young for some time. The young hatch with 12 pairs of legs and soon reach the adult number of 15 pairs. The adults and young then leave the logs and are difficult to find during the rest of the year.

Fig. 201. Diagrammatic representation of the life-history of *Thereuonema hilgendorfi* (based on data from Murakami, 1959).

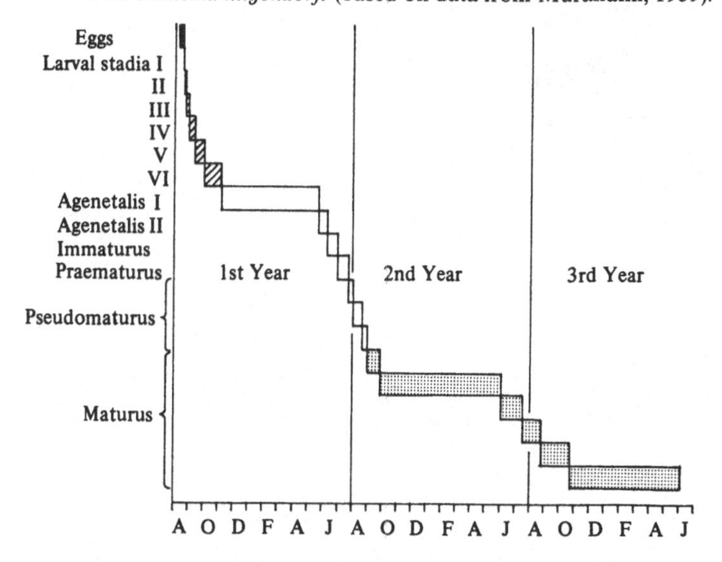

Abnormal development

Verhoeff (1902–25) quoted Brölemann's reports of two cases of abnormal segmentation in *Himantarium garielis.* Fig. 202 illustrates one such case.

Fig. 202. Part of the trunk of *Himantarium gabrielis* showing abnormal development of the segments (ventral view) (after Brölemann).

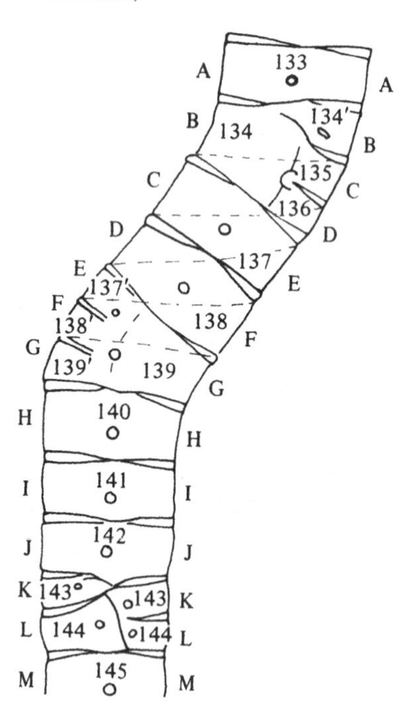

19

Epidermal glands and their function, defence and predators

Epidermis and gland cells

In centipedes, epidermal glands occur both singly and in groups. The head glands are dealt with in Chapter 15, the poison glands in Chapter 9. The remaining epidermal glands are dealt with in this chapter.

Geophilomorpha

There are no detailed accounts of the epidermis and single epidermal glands in geophilomorphs. Blower (1951) showed that there is a continuous layer of gland cells in *Haplophilus subterraneus* interspersed with small epidermal cells set in a very thick basement membrane (Fig. 48).

In many geophilomorphs the metasternites often bear pores, either singly or in groups (Fig. 19). These are the openings of unicellular glands that frequently produce luminous secretions (see below).

The inflated coxae of the terminal legs frequently bear pores, the openings of the coxal glands (Fig. 14). An electron microscope study of the coxal glands of *Clinopodes linearis*, *Necrophloeophagus longicornis* and *Haplophilus subterraneus* shows that the canal leading inwards from each pore is slightly dilated at its base like a drum stick and the intima is here somewhat thicker and of different texture from the cuticle of the body surface. Epidermal cells are radially arranged around the proximal dilation: they represent a typical transport epithelium. The nuclei as well as the poorly developed Golgi zones are located in the basal third of the cells.

Deep infoldings of the cell apex extend as far as the nuclear region. The extracellular clefts are usually narrow and filled with a finely granular material resembling the procuticle at the base of the canal in structure and density. Parallel bundles of longitudinally arranged microtubules are characteristically present in the cytoplasm. Vesicles of low electron density are also found. Large mitochondria occupy almost the entire width of the apical extensions approximately 0.8 μm from the cuticle. Other organelles are absent from the cells.

The basal region is also characterised by infoldings which are less regular than the proximal ones and usually wider. The stilt-like cell bases rest on the basement membrane. The cytoplasm contains numerous small transport vesicles 400–500 A in diameter which are predominantly spherical but sometimes rod-shaped and resemble smooth endoplasmic reticulum. Smooth as well as rough endoplasmic reticulum is also found at the cell base. Free ribosomes are frequently present. There are indications of vesicle formation (Rosenberg & Seifert, 1977). The authors assumed that this epithelium is involved in diuresis in animals living in moist environments. When animals are kept extremely dry, however, the epithelium appears to function as an organ for water uptake: the behaviour of animals (unspecified) and changes at the basal and apical infoldings were interpreted as signs of reversal in the transport direction. Single neurosecretory axons with typical synaptoid areas indicating exocytosis of a secretory product suggest a reversal of function which is probably under neuroendocrine control.

Paired anal glands occur in many geophilomorphs (Fig. 20). They resemble the coxal glands of lithobiomorphs in structure.

Mention has been made of the possible glandular function of the inflated terminal legs present in some species. Their histology has not been investigated.

Scolopendromorpha

The glandular epithelial cells of *Scolopendra cingulata* are pear-shaped with basal nuclei and open by fine chitinous intracellular ducts. Each duct is swollen at its base to form an ampulla. The gland cells are surrounded by an envelope of epithelial cells (Duboscq, 1898). Similar gland cells form a continuous layer immediately under the antennal cuticle (Fig. 203*a*) (Fuhrmann, 1922).

As in Geophilomorpha, coxal glands are frequently present on the terminal legs of scolopendromorphs. The duct of each gland is club-shaped and the chitinous intima shows widely spaced windings (Fig. 203*b*) (Herbst, 1891). They resemble the coxal glands of *Lithobius.*

Lithobiomorpha

In *Lithobius* the epithelial cells of the sclerite region are 25 μm high, in the articular regions they are only 5 μm high. The

Fig. 203. The epidermal glands of *Scolopendra cingulata. a*, glandular cells from the antenna (after Fuhrmann, 1922); *b*, transverse section of coxal glands (after Herbst, 1891).

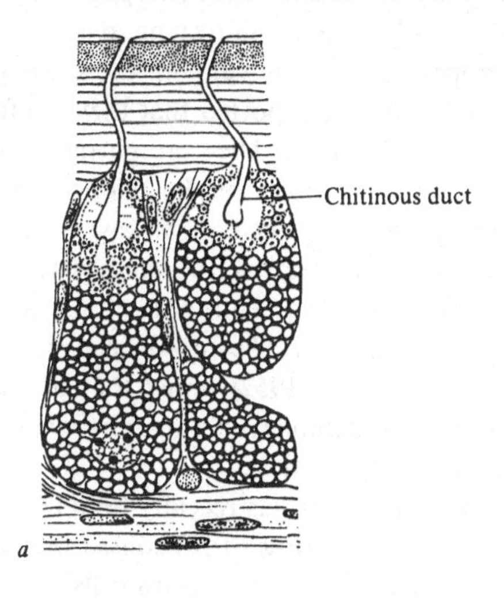

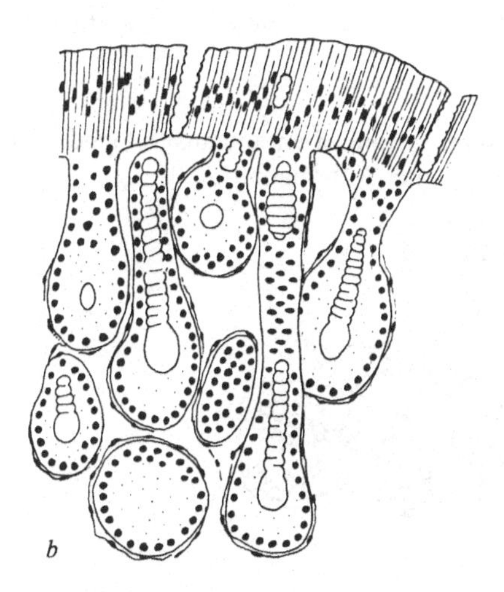

epithelium is very high in the vestibule of the spiracle (Duboscq, 1898).

In the antennae of *L. forficatus* a layer of more or less cuboidal cells lies beneath the shallow epidermal cells (Fig. 204). There are more cells than ducts, several cells opening into each pore canal. The pore canal opens through a chitinous ampulla into a pit in the exocuticle (Fuhrmann, 1922). The author suggested that the cells produce a defensive secretion.

Rilling (1968) recognised two types of gland cells in *L. forficatus*. The simpler type are large pear-shaped cells (telopodal glands), the narrowed apical ends of which open into a pit by means of an ampulla. These cells appear to be identical with those described in the antenna by Fuhrmann. Rilling reported that they are found in sclerotised regions all over the body. They are particularly abundant on the inner (posterior) borders of the telopodites on the fourteenth and fifteenth pairs of legs in *Lithobius* where they open on the femur, tibia, tarsus and metatarsus. In some species there are a few glands on the twelfth and thirteenth pair of legs and, more rarely on all other legs (Eason, 1964). They produce a sticky, string-like secretion (see below). Blower (1952) stated that the secretion appears to be a lipoid-protein complex or a mixture similar to the sclerotin and prosclerotin of the cuticle.

The ultrastructure of these glands has recently been described by Keil (1977). The telopodal glands (Fig. 205*a*) consist of a complex opening, one or two gland cells and two sheath cells. A pit in the cuticle leads to a bell-shaped cavity lined by exocuticle from the

Fig. 204. Transverse section of an antennal telopodal gland of *Lithobius forficatus* (after Fuhrmann, 1922).

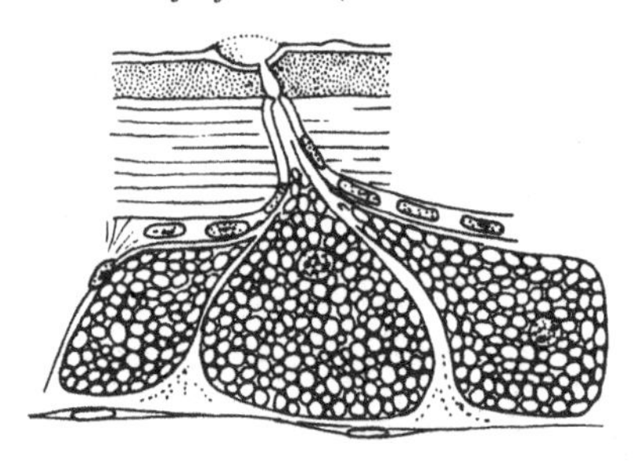

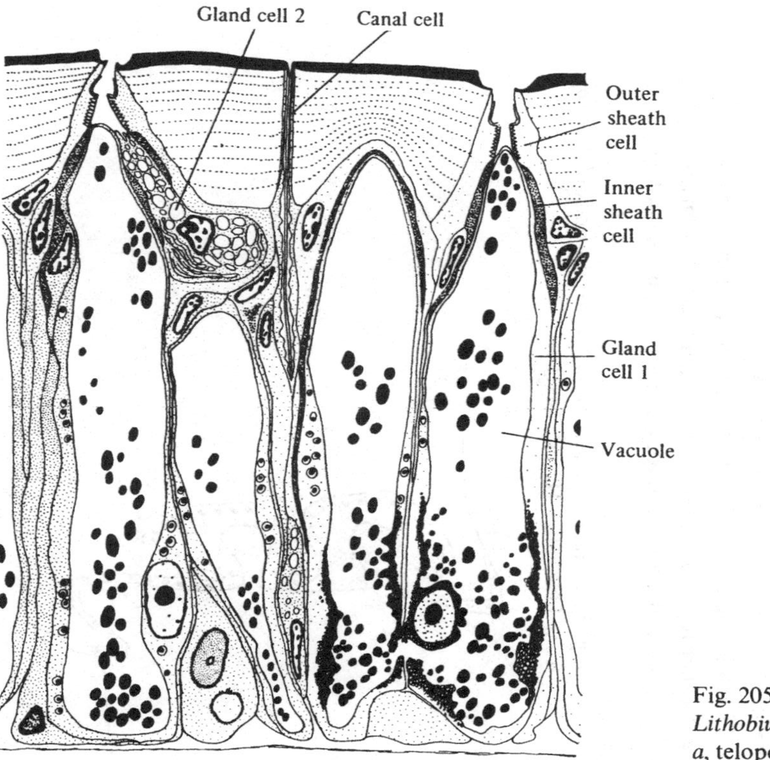

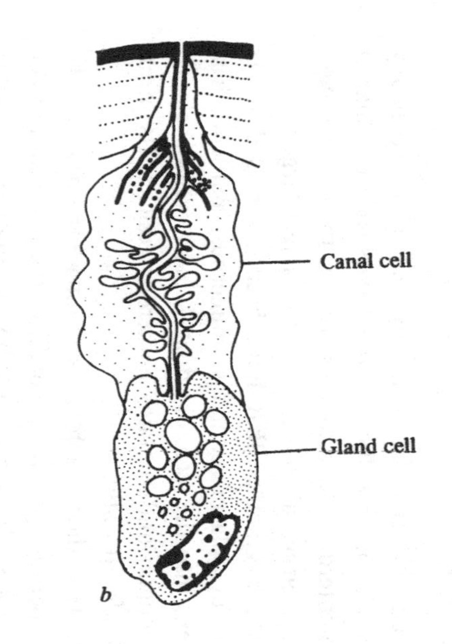

Fig. 205. Transverse sections of the epidermal glands of *Lithobius forficatus* as seen with the electron microscope. *a*, telopodal glands; *b*, a small epidermal gland (after Keil, 1977).

base of which leads a duct which is distally triangular, proximally round in section. The outer wall of this duct bears rod or Y-shaped processes which interdigitate with microvilli of the outer sheath cell. Gland cell 1 is large, sac-like and contains a very large vacuole filled with secretory granules that have been elaborated in its cytoplasm. The cytoplasm contains numerous mitochondria. Gland cell 2 opens into the secretory duct. It has a well-developed endoplasmic reticulum and contains numerous vesicles. Its membrane is much folded distally.

Keil suggested that there might be a closing apparatus for the secretory canal operated by the microvilli of the outer sheath cell which contain long filaments possibly composed of actin. Secretory granules are released through the membrane of the gland cell 1 which then reforms. These granules are mixed with the secretion of gland cell 2 which is probably an enzyme, producing the final secretion which is very hygroscopic.

The telopodal glands of the antennae and body surface must have other functions and in these, gland cell 2 is missing and gland cell 1 does not have secretory granules in its vacuole.

Fig. 206. Transverse section of the coxal glands of *Henicops* sp. from Java (after Herbst, 1891).

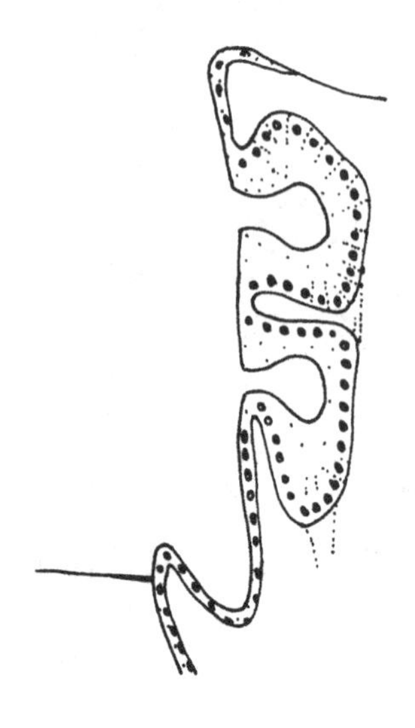

The second type of gland cell recognised by Rilling (Keil's 'small epidermal gland') consists of an elongated canal cell which is fused with one and exceptionally, two secretory cells (Fig. 205*b*). This complex does not always stain well and can be mistaken for a bipolar nerve cell, as for example by Duboscq (1898), inside a leg spine. The openings are very small and distributed all over the body both on the sclerites and the membranes. Most setae are flanked by a pair of these glands cells. Rilling (1968) suggested that they produced a secretion that reduced the permeability of the cuticle. They are presumably the same as the gland cells described by Blower (1951). These glands are also associated with the Tömösváry organ (Tichy, 1973).

No detailed study has been made of the coxal glands of lithobiomorphs. In *Bothropolys fasciatus* (Newport) and *Henicops* sp. (Fig. 206) the gland has a short wide duct leading from a glandular epithelium covered with a connective tissue sheath (Herbst, 1891). Willem (1897) considered the glands to be defensive and Latzel (1880), Verhoeff (1902–25) and Attems (1926) believed them to produce the sticky fibres now known to be secreted by the telopodal glands. It seems likely that they may be osmoregulatory as are the coxal glands of geophilomorphs.

Anal glands are present in the anamorphic stadia of *Lithobius* and a *Lamyctes* from Java (Herbst, 1891).

Scutigeromorpha

In *Scutigera* the tergites and sternites show at an equal distance from the mid-line and the lateral border two zones almost devoid of setae. In these regions the cell nuclei are irregular in shape and in some cases completely fragmented. Duboscq suggested that amitosis might be taking place. Similar nuclei are found in *Lithobius* but only on the dorsal surface.

In the antenna of *Scutigera coleoptrata* the epidermal cells are rounded with small nuclei in the regions where there are no gland cells but where they lie between gland cells they are elongated as are their nuclei. The gland cells have ampullae like those of *Scolopendra* and are mostly in groups of 10 to 12 (Fuhrmann, 1922).

Defence mechanisms

In addition to the use of the poison claws in defence, centipedes produce a number of defensive secretions and show defensive behaviour patterns.

Geophilomorpha

In a number of geophilomorphs the secretions of sternal glands are luminescent. The subject has been reviewed by Harvey (1952) and Minelli (1977).

Harvey stated that the earliest record of luminous chilopods is probably that of Oviedo, who found them on the island of St Domingo in 1520. The subject was mentioned by many early naturalists. Harvey gives a full bibliography of the nineteenth-century observations. In Europe the following species have been reported as luminescent: '*Geophilus simplex*' Gervais (Macé, 1886) a species of uncertain status, *Geophilus electricus*, *Necrophloeophagus longicornis*, *Haplophilus subterraneus* and *Strigamia crassipes* (Gazagnaire, 1890) and *Geophilus carpophagus* (Duboscq, 1898). Detailed studies were made by Brade-Birks & Brade-Birks (1920) on *Geophilus carpophagus* and by Koch (1927) on *Strigamia crassipes*. References to *G. electricus* should be treated with caution: the name seems to have been applied rather generally to luminous geophilomorphs.

Outside Europe luminous species have been reported from Algeria (Gazagnaire, 1888, 1890), Java and Krakatoa after the volcanic explosion (Harvey, 1952), Singapore where the species 'often to be found among ones papers' exuded blue luminous matter when hurt (Ridley, 1936) and on guano in the Deer Cave in Gunong Mulu National Park in Sarawak where a geophilomorph producing a bright green phosphorescence is common (Lewis, unpublished data). *Orphnaeus brevilabiatus* (Newport) is widespread: it is luminescent in Micronesia and the East Indies (Haneda, 1939) and Dahomey (West Africa) where the red secretion is strongly phosphorescent (Brölemann, 1926). *Orphnaeus* sp. from Eritrea secretes a pale yellow liquid which has a bright green phosphorescence (Lewis, 1969*a*). In America *Geophilus vittatus* emits a faint but distinct blue-green glow that lasts for some seconds (Jones *et al.*, 1976).

Dubois (1886) believed that the luminous seccretion of geophilomorphs was produced by epithelial cells in the gut and Macé (1887) was of the opinion that in *Geophilus simplex* and *Necrophloeophagus longicornis* it was produced from *glandes preanales* in the last two body segments.

Histological studies by Brade-Birks & Brade-Birks (1920) and Koch (1927) revealed the existence of two types of gland cells. White gland cells seen through the integument presumably contain protoluciferin and are full of large eosinophil granules; secondly there are non-staining fine granular mucous cells. Brade-Birks & Brade-Birks noted the almost complete disappearance of the white glands in *Geophilus carpophagus* after a particularly extensive display of luminescence. No 'white glands' are present in *Geophilus insculptus*, a non-luminous species.

Koch (1927) compared the structure of the glands in the luminous species *Strigamia acuminata* and the non-luminous *Clinopodes linearis*. In the latter species the gland cells are also whitish and visible through the integument and on stimulation the white material is secreted through pore fields. Both eosinophil and mucous glands are found in *C. linearis* but since the former differ from those of *S. acuminata* and are of a fine granular type staining weakly in eosin, Koch believed them to be the cells producing the luminous secretion.

The milky appearance of the secretion indicates that is full of globules. These were termed *vacuolides* by Dubois (1886) since they contain a small vacuole in the centre. He pointed out that they were not fatty but proteinaceous in nature. He described the formation of crystals from the vacuole.

Gazagnaire (1888) noticed the *vacuolides* in *Orya barbarica* and Koch (1927) not only observed crystal formation in *S. crassipes* but also in *Himantarium gabrielis* and *Geophilus insculptus* which are non-luminous. Brade-Birks & Brade-Birks (1920) also noticed crystals in the secretion of both luminous and non-luminous forms.

Harvey (1952) and Thomas (1896) described a wave of darkness sweeping over the illuminated geophilomorph from tail to head 'then a second or so later glowing all over again' the 'manoeuvre was repeated several times'. Both Dubois and Koch observed that the weak stimulation resulted in luminescence without the expulsion of the luminous secretion which led Koch to suppose that the luminescence was at first intracellular. Brade-Birks & Brade-Birks did not observe intracellular luminescence in *G. carpophagus*. On electrical stimulation a few seconds elapse before the secretion appears. It is very bright at first but quickly falls to a lower intensity and the light then slowly disappears over a period of thirty seconds. In other forms the light of the secretion lasts longer.

Brade-Birks & Brade-Birks described the viscous secretion of *G. carpophagus* as practically colourless with a fruity smell and a strongly acidic reaction, Gazagnaire (1890) the secretion of *Orya barbarica* as yellowish with a bluish-green luminescence, while Koch (1927) described the milky secretion of *S. crassipes* as weakly fluorescent, bluish in reflected light, a weak yellow in transmitted light. He found a few bacteria in the secretion but was unable to culture any luminous forms and concluded that the luminescence was not bacterial in origin.

Harvey states that the light of geophilomorphs is undoubtedly a chemiluminescence not a crystalloluminescence. He was unable to demonstrate the luciferin–luciferase reaction with one specimen of a 'geophilid' at Buitenzorg, Java but 'additional tests should be made. In fact the chemistry of myriapod luminescence is quite untouched and offers a very promising field of investigation'.

The function of the secretion of the sternal glands. Gazagnaire (1890) reviewed the work on European geophilomorphs: the ten reported observations of luminescence took place between mid-September and late November. He noted that records were often of more than one individual and although chilopods were usually solitary and males were attacked by females there appeared to be a season of 'truce' and this was associated with reproduction. Fabre (1855) observed the deposition of spermatophores by '*Geophilus convolvens*' between 25 September and 15 November. Later observations have made the seasonal appearance of luminescence rather doubtful. Duboscq (1898) reported that in spring *Geophilus carpophagus* emitted a pale blue luminescence, Brade-Birks & Brade-Birks reported the species to luminesce in December, January, February and April and captive specimens from April to September. Koch (1927) tabulated 23 observations of luminescence in England, France and Germany. He concluded that luminescence can take place throughout the year. That luminous species are often to be found to be non-luminous can be explained by the fact that once the secretion is exhausted it takes at least three to four weeks for it to be reformed.

Harvey (1952) regards the absence of eyes in geophilomorphs as making it unlikely that the luminescence is for recognition or sexual attraction. He regarded it as being probably used for protection, quoting Thomas (1902) who observed a luminous centipede scattering its secretion over red ants which had attacked it and Ridley (1936) who observed the beetle *Harpalus ruficornis* to attack a '*Geophilus electricus*'. The 'mouth, face and legs' of the former 'were covered with patches of luminous matter which seemed to annoy it'. He quoted from a letter published in Kirby and Spence's *Entomology* of a *Carabus* running round a *Geophilus electricus* 'as if wishing but afraid to attack it'.

When *Strigamia crassipes* was confined with a beetle the centipede on being touched by the beetle's antennae lighted momentarily and then turned the underside of the body against the beetle and covered it with slime. Brade-Birks & Brade-Birks (1920) produced luminescence by handling, immersion in water, an electric current, attack by ants and contact with another individual.

Harvey observed that a number of workers have noted the peculiar smell of the secretion and also that it is acid, suggesting that it would be an excellent repellent.

Plowman (1896) noted that streaks of light are left for a few seconds in the trail of a '*Scolopendra electrica*' as it crawls so that it is often difficult to say exactly where the creature is. Thomas (1902), who examined a *Geophilus* being attacked by ants and which was reacting by discharging a luminous secretion, noted that when she placed her hand over a glass into which she had placed the animal a strange prickly sensation such as is caused 'by slight contact with electricity' was experienced.

Recently the secretion of some geophilomorphs has been found to contain noxious chemicals. Schildknecht, Maschwitz & Krauss (1968) found the secretion of *Pachymerium ferrugineum* to be proteinaceous and cyanogenetic and speculated that the hydrogen cyanide might be generated by the dissociation of a cyanohydrin such as mandelonitrile, but they identified no such cyanogenetic compound. Jones *et al.* (1976) investigated *Geophilus vittatus, G. cayugae* Chamberlin, *Strigamia bothriopa, S. chionophila* and *S. icterica* from Ithaca, New York, and *Orphnaeus brasilianus* from Florida. Most of the work was done on *Geophilus vittatus.* This species ejects a clear viscous secretion in response to manipulation or pinching. The fluid visibly contaminates the body of the animal as well as the substrate on which it walks. The pH of the fluid ranged from 6.0 to 6.5. Emission of hydrogen cyanide was readily demonstrated by the blue colour that developed on strips of filter paper impregnated with copper acetate-benzidine acetate reagent held close to their bodies.

Gas-chromatographic analysis showed the presence of benzaldehyde. Thin-layer chromatography revealed the presence of mandelonitrile and benzoic acid. The other species studied all produced sticky cyanogenetic secretions: the presence of protein and benzoyl cyanide was verified in a specimen of *Strigamia*, probably *S. bothriopa.*

The two cyanogenetic components in the secretion of *G. vittatus*, mandelonitrile and benzyol cyanide, are the obvious sources of the hydrogen cyanide as well as of the other two aromatic compounds in the mixture, benzaldehyde and benzoic acid. Mandelonitrile dissociates directly to benzaldehyde and hydrogen cyanide while benzoyl cyanide gives benzoic acid and hydrogen cyanide on hydrolysis. The most interesting chemical feature of the secretion is the presence of benzoyl cyanide, an active acylating agent as well as a cyanide precursor which has not, hitherto, been isolated from any natural source.

Brade-Birks & Brade-Birks (1920), Weil (1958), Schildknecht *et al.* (1968) and Jones *et al.* (1976) noted the deterrent effect that the secretion of geophilomorphs had on ants. The latter authors noted that *Strigamis bothriopa* was immediately attacked by the ant *Formica exsectoides* when introduced into the ants' foraging area. Within seconds after having clamped themselves to the body of the centipede, the ants were seen to relinquish their hold. Visibly contaminated by the sticky secretion, the ants performed intensive cleaning activities, vigorously at first and then more awkwardly as the secretion hardened on exposure to air. Ants frequently became stuck together and virtually immobilised: some became stuck to the substratum. Those more massively wetted died. The centipedes sustained no injuries as long as their secretory supply lasted. Once depleted of the secretion they were vulnerable: a single *S. bothriopa* introduced into the arena after having been milked exhaustively was overrun by a swarm of ants, doused in the ants' acid spray and eventually dismembered and killed. Weil (1958) noted that *Myrmica rubida* became entangled in secretion when attacking *Necrophloeophagus longicornis.*

Jones *et al.* noted that an *Orphnaeus brasilianus* from Florida curled round her eggs with sternites facing outwards. Prodding with forceps caused discharge of the defensive secretion. The secretion appears to affect the ants primarily mechanically by virtue of its viscocity derived from its proteins, but it could also have acted in chemical ways. Some of its components might have acted as topical irritants, perhaps inducing the prompt cleansing reactions elicited on contact, while others such as hydrogen cyanide could have acted as toxins where the ants were heavily contaminated by the secretion.

Scolopendromorpha

Not only do scolopendromorphs bite other species of animal, there is also a danger of biting taking place when two individuals of the same species meet. Ritualised meeting reactions avoid this problem. When two *Scolopendra cingulata* meet, each attempts to grasp the posterior region of the trunk of the other with its terminal legs (Klingel, 1960*a*). Pairs may lie locked together for an hour before separating. If specimens meet head on in a confined space, they retreat, turn round and move back hind end first with terminal legs outspread. Only very occasionally do animals bite each other, in five out of nine cases this resulted in death. Brunhuber (1969) has described similar behaviour in the South African species *Cormocephalus anceps anceps.* In this species head to head encounters involving the antennae resulted in violent avoiding reactions, the animals rushing away in opposite directions. Encounters made head to tail resulted in the hindmost animal immediately recoiling and moving rapidly in the opposite direction. Alternatively, after head to head contact, both centipedes rapidly swung round the posterior part of their bodies and attempted to grip each other with their terminal, and then, by the more posterior legs. In encounters between two males, both centipedes gripped each other dorsally behind each other's head; this position was maintained for periods sometimes lasting several hours. In the case of male meeting female, it was usually the male that attempted to grip the female; the females gripped less tenaciously. In female to female encounters the avoiding reaction rather than the defence posture was usual.

Babu (1964) demonstrated the presence of giant fibre systems in the ventral nerve cord of *Otocryptops* sp., and *Scolopendra* spp. 'startle times' compare favourably with those of other arthropods (Chapter 5).

A number of observations suggest that some scolopendromorphs have poison glands in their walking legs. Norman (1897) wrote of '*Scolopendra morsitans*' in Texas: 'It is a popular belief that the tips of all the legs are poisonous and inflict severe wounds merely by the animal crawling over the naked skin, especially if the claws are pressed into the skin: there is, of course, no evidence for this belief'. In southern California *Scolopendra heros* produces a reddish streak after it has crawled upon the body. 'It makes tiny incisions with its numerous feet ... when alarmed the centipede drops into each incision some kind of venom that causes intense irritation so that the affected part becomes inflamed and the two rows of punctures show white against the flesh' (Cloudsley-Thompson, 1958*b*). Dr J. Anderson (personal communication) informs me that a colleague of his suffered in this way when a large Nigerian species crawled along his arm.

There is also evidence that scolopendromorphs produce noxious chemicals from other structures: *Otostigmus aculeatus* Haase from Tonkin (Vietnam) produces a phosphorescent secretion when touched which has an acrid phosphorous smell and produces a redness of the skin followed by blisters, scabs and a loss of skin which 'goes bad' (Houdemer, 1926). The secretion was released through fine openings situated symmetrically on the sides of the ventral surface. The author was not sure whether these were on the sternite or the intersegmental membrane. The South African *Cormocephalus nitidus* Porat gives all a fetid odour when irritated. Nothing is known of the identity of the gland producing the secretion nor of its chemical nature (Lawrence, 1968).

The terminal legs of scolopendrids are not involved in locomotion and show considerable variation in structure. In some cases this is related to their functioning in distracting predators in addition to their possible role as posterior antennae (Jangi, 1961).

Cloudsley-Thompson (1961) described how the long slender terminal legs of the West African species *Rhysida nuda togoensis* slowly bend and straighten when detached to the accompaniment of a faint creaking sound. Stridulation sounds were also produced by a terminal leg of a *Scolopendra* sp. when the prefemur and femur or femur and tibia were moved.

Lewis (1971, unpublished data) was unable to repeat Cloudsley-Thompson's observations of stridulation in *Scolopendra morsitans* from Nigeria.

Alipes crotalus (Gerst.) of South and East Africa has large leaf-like terminal legs which it vibrates rapidly to produce a rustling or fluttering sound when disturbed (Lawrence, 1953). The East African

Fig. 207. The distal segments of a terminal leg of *Alipes grandidieri grandidieri* (after Enghoff & Enghoff, 1976).

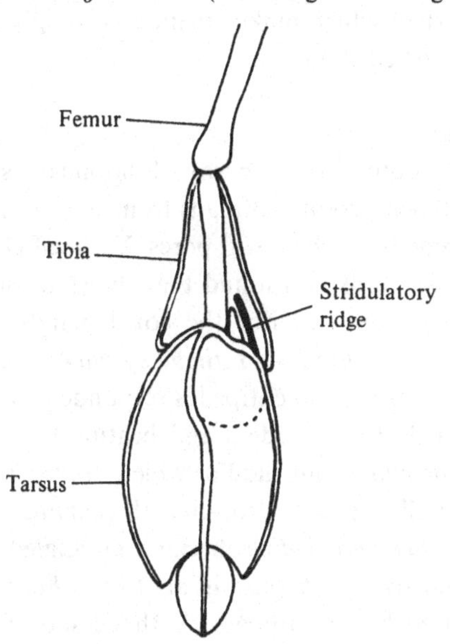

Alipes grandidieri grandidieri (Lucas) runs with its terminal legs directed backwards and held above the ground. When irritated the animal swings the anal legs from side to side and stridulates. The legs are sometimes autotomised: when this happens they continue to stridulate. The tibia and tarsus are expanded into membranous plates on which a smaller area is particularly thin (Fig. 207). The thickened edge of the tibia is densely covered with fine transverse furrows which rub against longitudinal furrows on the thickened edge of the tarsus when the two are rubbed together by sagittal bends of the tibia-tarsus joint (I. B. Enghoff & H. Enghoff, 1976, unpublished report). In another African species, *Asanada sokotrana* Pocock the terminal legs are readily autotomised and when detached perform wriggling movements (Lewis, unpublished data).

Many tropical scolopendromorphs are brightly coloured, *Ethmostigmus trigonopodus* in Nigeria is black with orange legs, *Asanada sokotrana* red. It is possible that these bright colours may indicate that the species are poisonous or distasteful.

The most conspicuous defence mechanism is seen in *Rhysida intermedia* Attems, it is a very vigorous wriggling of the body (Enghoff, personal communication). *Rhysida nuda togoensis* in Africa and *Otostigmus* sp. in Sarawak make similar rapid writhing movements when disturbed which makes them exceedingly difficult to catch (Lewis, unpublished data).

Lithobiomorpha

Latzel (1880) noted that many lithobiids especially *Lithobius grossipes* C. Koch produce threads from the end legs and that these appear to come from the coxal pores. Verhoeff (1902–25) observed *Bothropolys asperatus* and found that the slimy secretion was produced from the telopodal not the coxal glands. He described a number of observations on *Lithobius forficatus* confined with lycosid spiders and ants. The centipedes responded by turning the posterior end towards the intruders and beating the terminal legs up and down, throwing out liquid which forms a strand bearing numerous, equally spaced drop-like thickenings. In one experiment a *Camponotus herculeanus* became entangled in two elastic threads produced by a *Lithobius*, in another a *Formica rufa* was partially immobilised by two threads. Verhoeff showed that a

Lithobius which has lost its terminal legs uses its fourteenth pair in the defensive reaction. He thought that the coxal glands may also be used to produce the defensive secretion when the terminal legs are missing.

The colour markings of *Lithobius variegatus* make it inconspicuous against its natural background of leaves and stones and it tends to remain motionless when disturbed (Eason, 1964).

Two small British species *L. curtipes* C. L. Koch and *L. duboscqui* Brol. have the habit of curling up rather than running away when disturbed. Eason (1964) suggested that this is a valuable protective reflex since it usually results in the animal rolling off a leaf or twig into some crevice in the soil or litter where it is difficult to see and relatively safe from possible predators.

The small North American species *Nampabius michagenensis* Chamberlin and *Tidabius tivius* (Chamberlin) exhibit death-feigning behaviour. When roughly disturbed they contort their slender bodies into a rigid arc, remaining in this position for two or three minutes, after which they unbend and run beneath a protective covering (Johnson, 1952).

Scutigeromorpha

In the cavernicolous Malayan species *Scutigera decipiens* (Verhoeff) autotomised legs produce a surprisingly loud creaking sound due to a stridulatory organ consisting of a minute transverse slit, the distal margin of which bears a row of outwardly directed hooks which must rub against some other part of the leg (Annandale, Brown & Gravely, 1913).

Potential predators find it difficult to reach the body of *Scutigera* because the legs are so long. When a predator grasps a leg it is immediately autotomised (Herbst, 1891). The autotomised leg continues to move for some time distracting the attacker whilst the centipede flees (Haase, 1884*a*).

Predators

Centipedes reported as prey items are seldom identified accurately: many references refer simply to 'centipedes' or 'chilopods' but such references do give some indication of the wide range of centipede predators. Amongst the invertebrates that have been reported is the carabid beetle *Pterostichus madidus*, a specimen of which was found to contain a chilopod 'maxilliped' (Luff, 1974). Chilopod remains are occasionally present in the gut contents of *Lithobius forficatus* and *L. variegatus* (Lewis, 1965).

Centipedes have been recorded in the gut contents of a number of American salamanders: *Plethodon jordani metcalfi* and *P. jordani shermani* (Whitaker & Rubin, 1971), *P. glutinosa* and *P. jordani* (Powders & Tietven, 1974). *P. longicornis*, *P. yonahlosse* and *P. glutinosus* (Rubin, 1969) and *P. wehrlei* (Hall, 1976). These authors reported percentage frequencies ranging from 9 to 32.

Guyetant (1967) found that centipedes were rare or absent in the stomach contents of French frogs and toads (*Rana temporaria*, *R. esculenta* and *Bufo bufo*).

Easterla (1975) stated that snakes fairly commonly preyed on centipedes. Pianka (1968) found a centipede in one of a sample of 67 *Varanus eremius* in Australia. J. White found centipedes in the pellets of the burrowing owl *Speyotyto cunnicularia* (Cloudsley-Thompson & Crawford, 1970). There have been a number of centipedes identified in the diet of mammals: centipedes appear in the guts of the pigmy shrew (*Sorex minutus*) in Ireland (Grainger & Fairley, 1978) and form between 1.2 and 11.4 per cent by volume of the food of four species of shrews in Indiana (Whitaker & Russel, 1972). In Uganda the Link rat (*Deomys ferrugineus*) a mainly insectivorous, nocturnal forest dweller includes centipedes in its diet (Delany, 1975) and the African false vampire bat (*Cardioderma cor*) which in East Africa perches on low vegetation at night listening for terrestrial prey, regularly takes centipedes and has been observed eating specimens 60–80 mm long (Vaughan, 1976). These large centipedes were presumably scolopendrids. Haltenorth & Diller (1977) reported that three mongooses: the grey meercat (*Suricata suricatta*) of South Africa, the banded mongoose (*Mungos mungos*) of savanna south of the Sahara and the dark or long-nosed mongoose (*Mungos obscurus*) of West African rain forests, as well as the Wild cat (*Felis silvestris*) of Europe, Africa and parts of Asia eat centipedes.

There are a number of references on predators in which the centipede prey is identified to order or further. These cases are dealt with below.

Geophilomorpha

Verhoeff (1902–25) suggested that mice and voles might be possible predators of geophilomorphs: direct evidence of predation by mammals was provided by Rudge (1968) who showed that the common shrew (*Sorex araneus*) feeds on geophilomorphs. None were found in 66 guts examined from shrews from an open hillside planted with conifers in Stirlingshire but they formed an average one per cent of the diet (eight per cent occurrence) in rough grass in Berkshire and an average two per cent (seven per cent occurrence) in hedgerows and copses in Devonshire. Three specimens of *Geophilus carpophagus* were found in the stomach contents of a sample of 41 *Salamandra atra* in Austria (Fachbach, Kolossa & Ortner, 1975).

In a three year study of *Strigamia maritima* Lewis (1961) saw no predation of the centipede in the field. Laboratory experiments using the more common predatory animals in the habitat were carried out: the carabid beetle *Pogonus chalceus* readily accepted adolescens I *Strigamia* grasping the centipede's posterior end and eating it from behind, gradually working forwards. The adolescens III stadium was never eaten, although beetles were observed to make futile grabs at these larger specimens. The staphylinid beetle *Cafius xantholoma* only successfully attacked adolescens I *Strigamia* and the carabid *Dichirotrichus obsoletus* refused the centipede even after a week's starvation. Neither *Lycosa arenicola*, nor immature linyphiid spiders accepted the centipede as food. Plateau (1878) found that in captivity *Cryptops* ate *Himantarium.* Latzel (1880) found a scorpion feeding on *Dicellophilus carnioleusis.*

The primitive ant *Amblypone pluto* Gotwald and Levieux (subfamily Ponerinae) which occurs in the savanna of the central Ivory Coast in West Africa and whose colonies consist of about 30 individuals with a dozen or so larvae and several cocoons, feed exclusively on geophilomorphs (Levieux, 1972). The workers, which are 6 mm long, hunt singly, attacking *Schendylurus paucidens* Silv. and *Pleuroschendyla* spp. which are about 30 mm long. The centipedes are pricked by the *aguillon* which causes a progressive locomotor paralysis in the prey. Once immobilised the prey is transported to the brood cavity and eaten alive. The workers pierce the body through an intersegmental membrane with their mandibles and feed on the haemolymph and soft parts. The workers feed first, the females second and the males third. The larvae feed last. They are deposited on or near the centipede and carried away again when they have finished feeding. The workers do not seem to regurgitate food for the larvae as they do in other Ponerinae.

Scolopendromorpha

With the exception of Cloudsley-Thompson's (1949) observation of a *Cryptops hortensis* being eaten by an immature spider of the genus *Theridion*, all references to scolopendromorph predators are to vertebrates.

The South African Cape black-headed snake or centipede eater (*Aparallactus capensis*), a small slender snake found in cryptozoic habitats in open bush country, feeds largely on centipedes which may be up to 130 mm long and twice the diameter of the snake's body. After a considerable battle, the prey is seized firmly at or just behind the head and swallowed head first. The bite of the centipede appears to have little or no effect on the snake (FitzSimmons, 1962). The author gave no information on the food of the Angolan centipede eater (*Aparallactus capensis bocagii*). Clark (1967) found a scolopendrid centipede 140 mm long broken just behind the head in a juvenile *Vipera ammodytes meridionalis* of snout/vent length 195 mm from Euboia Islands, Greece. Khanna (1977) quoted observations of Sharma & Vazirani of a *Scolopendra morsitans* in the gut contents of a Saw-scaled viper *Echis carinatus* and of the yellow-bellied house gecko *Hemidactylus flaviviridis* feeding on *S. morsitans*.

The monitor lizards *Varanus exanthematicus* and *V. niloticus* occasionally feed on centipedes in Senegal: three in 28 specimens and four in 32 specimens respectively (Cisse, 1972). Millipedes are the major constituent of the diet of *V. exanthematicus*.

In Russia, rooks (*Corvus frugiligus*) occasionally take *Scolopendra cingulata* (Voinstvenskii, Petrusenko & Boyarchuk, 1977).

Ryan & Croft (1974) found large green scolopendrids to be a major item in the diet of foxes (*Vulpes vulpes*) in summer and autumn in New South Wales, Australia, being present in 28 of the

56 stomachs examined. Schubart, quoted by Cloudsley-Thompson (1949) found two *Scolopendra* 20 cm long in the stomach of an armadillo.

Lithobiomorpha

Both vertebrates and invertebrates prey on lithobiomorphs. One *Bothropolys multidentatus* (Newport) was found in a sample of the stomach contents of the American salamander *Ambystoma jeffersonianum* (Auerbach, 1952). In Austria, two *Lithobius forficatus* were found in the stomach of *Salamandra salamandra salamandra* and two in 41 *Salamandra atra* (Fachbach, Kolossa & Ortner, 1975). *Lithobius* is readily accepted by the green lizard *Lacerta viridis* (Cloudsley-Thompson, 1949) and by domestic poultry (Eason, 1964).

The common shrew feeds on *Lithobius*. No centipedes were found in the gut contents of 66 shrews from a Scottish hillside but *Lithobius* formed on average two per cent of the diet in Berkshire and four per cent in Devonshire. The shrew eats specimens of *L. variegatus* in two minutes. In most cases the prey are located by nasal contact but occasionally the shrews orientate to the slight scuffling sounds made by the centipedes (Rudge, 1968). *Bothropolys multidentatus* is numerous in mole tunnels in Michigan and is undoubtedly captured and eaten by moles: fragments of the centipedes are found in the tunnels (Johnson, 1952).

Bristowe (1941) wrote: 'So far as my field and laboratory experience goes centipedes are not attacked by any of the hunting spiders'. Cloudsley-Thompson (1949) confirmed this. Bristowe saw *Lithobius* overcome in the laboratory by *Ciniflo similis, Theridion tepidariorum, T. denticulatum, Tegeneria atrica, Aranea diadema* and *Zygiella literata.* 'Not infrequently I have noticed that the meal is left uncompleted as though the flavour is not particularly agreeable.' On five occasions Roberts (1956) found corpses of *L. variegatus* in webs of cryptozoic spider *Amaurobius terrestris* in the field. The spider also took *L. variegatus* in the laboratory.

Verhoeff (1902–25) observed a scorpion (*Euscorpius* sp.) in the South Tyrol carrying a *Lithobius.*

Roberts (1956) examined the gut contents of seven carabid and two staphylinid species and found remains of *L. variegatus* in three of the carabids. *Nebria gyllenhali* contained only larval stadia (eight in 100 beetles) as did *Nebria brevicollis* (two in 17). *Abax parallelopipedus* contained mostly larvae but also some epimorphic and maturus remains (22 in 294).

In laboratory experiments, *Carabus violaceus* and *Abax parallelopipedus* took all stages of *L. forficatus*. *Nebria* spp. and *Loricera pilicornis* took only larvae and epimorphic stadia. Roberts suggested that carabids were of some importance as predators of *Lithobius*.

Murakami (1958*a*) observed the scutigeromorph *Thereuouema hilgendorfi* to feed on *Bothropolys* 'in nature'.

Scutigeromorpha

There appear to be no records of predation on scutigeromorphs; Verhoeff (1902–25) suggested that geckoes were likely predators.

20

Parasites

The literature on the parasites of centipedes was reviewed by Cloudsley-Thompson (1949) and additional data were provided by Remy (1950).

Ectoparasites

Acari

Like other arthropods, centipedes are frequently found to have mites attached to them but these have received little attention. The six-legged larvae of trombidiids occasionally attach themselves to the legs of centipedes. It seems that they feed on the host as their soft-skinned abdomens increase in size the longer they remain attached. Gamasid mites, which are normally free-living humus-dwellers, are sometimes specific parasites of myriapods: none are found on centipedes in South Africa although *Antennophorus* and related genera occur on tropical Scolopendras and ants in other parts of the world (Lawrence, 1953).

The resting stage (deutonymph or hypopus) of several species of Tyroglyphidae is found on the appendages of almost all orders of ground-living arthropods. They are minute and do not harm their hosts as they have no mouth parts for feeding; they attach themselves by means of suckers at the posterior end of the body (Fig. 208). Although they are not, strictly speaking, parasites it is convenient to deal with them here.

Lewis (1960) reported that the littoral geophilomorph *Strigamia maritima* frequently carried hypopi of the tyroglyphid *Histiostoma* sp. Specimens normally carried up to ten hypopi, the largest number on one specimen was 43, but the number often becomes much greater in laboratory cultures. Adolescens I *Strigamia* are far less heavily infested than other stages, possibly because they offer less suitable attachment sites due to their small size. Generally between 70 and 100 per cent of the post adolescens I stadia carry hypopi but in March, April and early May very few hypopi are found on the centipede (Lewis, unpublished data).

Hypopi similar to those found on *S. maritima* are found on *Strigamia acuminata*, *Haplophilus subterraneus*, *Necrophloeophagus longicornis*, *Pachymerium ferrugineum* and *Chaetechelyne vesuviana* but none were found on a sample of twenty *Hydroschendyla submarina* collected in Devon in June 1959, presumably because conditions were unsuitable for the mite in the littoral rock crevices inhabited by this species (Lewis, unpublished data).

L. variegatus in Hampshire carries hypopi of *Rhizoglyphus echinopus* (Roberts, 1956).

Endoparasites

A variety of endoparasites comprising Protozoa, Nematoda, Nematomorpha, Diptera, Hymenoptera and Fungi have been recorded from centipedes.

Fig. 208. *a*, dorsal and *b*, ventral view of the deutonymph of the tyroglyphid mite *Histiostoma* sp. (after Lawrence, 1953).

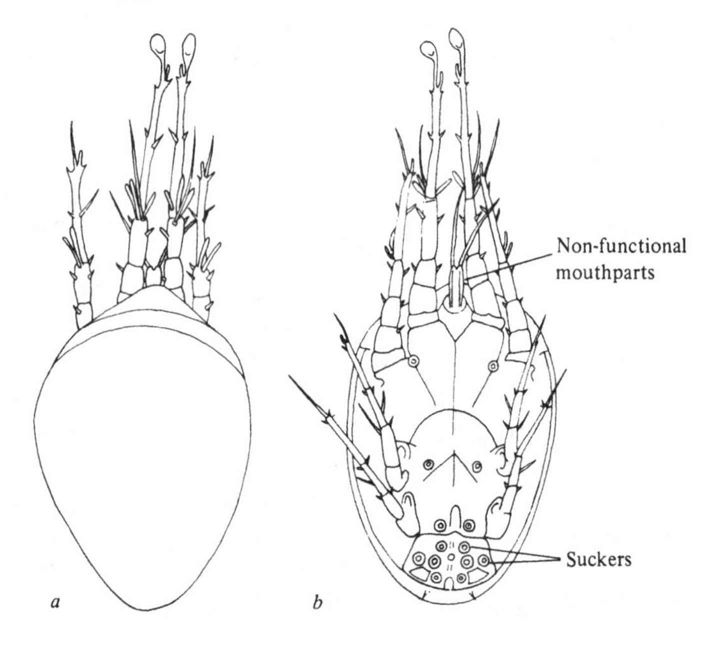

The most common parasites appear to be members of the protozoan class Sporozoa and belong to the groups Gregarinida, Coccidiida and Adeleida. A comprehensive account of these parasites was given by Grassé (1953). Gregarine infection takes place by the ingestion of oocysts, each of which contains eight sporozoites. These are released into the lumen of the gut when the oocysts are ingested, enter gut cells and are then termed trophozoites. The trophozoites protrude from the gut cells (Fig. 210*a*) and eventually come to lie free in the gut lumen where they form gametocytes which associate in pairs to form association cysts (gametocysts) (Fig. 209*h*). Inside the cyst the gametocytes undergo multiple fission to produce gametes. A considerable amount of parent cytoplasm remains unused in gamete formation and this forms the residual body which often assists in the rupture of the cysts either by

Fig. 209. Protozoan parasites of geophilomorphs.
Actinocephalus stella: *a*, *b* and *c*, young trophozoites; *d*, *e* and *f*, older trophozoites; *g*, trophozoite free in gut lumen; *h*, association cyst; *j*, sporocyst (after Ormières, 1966). *Rhopalonia geophili*: *k*, trophozoite; *l*, *m* and *n*, dehiscence of gametocyst; *o*, *p*, *q* and *r*, dehiscence of sporocyst (after Léger).

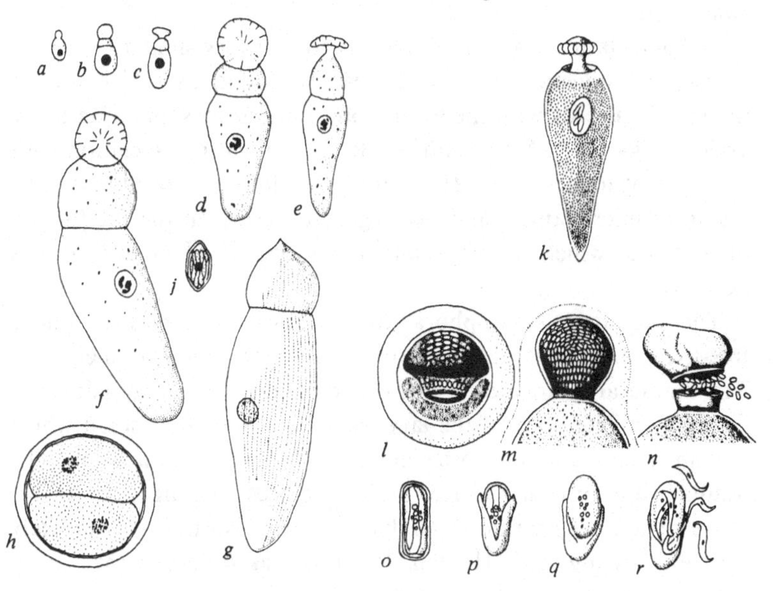

swelling or by forming tubes. The association cysts pass out of the centipede with the faeces. Inside the association cysts the gametes fuse to form sporocysts or oocysts (Fig. 209*j*) by means of which re-infection takes place (Fig. 209*o–r*).

The gregarines of centipedes are of complex structure, the cell consisting of an attachment region, the epimerite, which is borne on a protomerite. This is itself separated by a septum from the main part of the cell, the deutomerite, which contains the single nucleus (Fig. 209*f*, *k*).

Infection by Coccidiida (e.g., *Eimeria*) takes place by ingestion of sporocysts containing sporozoites. The sporozoites enter intestinal cells where they undergo several cycles of asexual reproduction (schizogony) not seen in the gregarine parasites of centipedes. During schizogony the trophozoites divide to form numerous merozoites which infect further intestinal cells. The sexual phase is heralded by the formation from some trophozoites of large (female) macrogametes and from others of a large number of microgametes. A microgamete fuses with a macrogamete to form a zygote which leaves the intestinal cell forming an oocyst and passes out of the animal. Inside the oocyst four sporocysts develop, each containing eight sporozoites. Infection takes place when the oocyst is swallowed.

In Adeleida (e.g., *Adelea*) the life-history is very similar to that of Coccidiida. Infection by sporozoites is followed by schizogony in the intestinal cells with the formation of merozoites and after a few cycles, of gametes. Microgametes develop in close association with the macrogametes (Fig. 210*a*) and only four are produced. The fusion of microgamete and macrogamete results in the production of a zygote which develops into an oocyst in which sporozoites develop.

The phylum Nematophora, the gordian or hair-worms, have free-living adults that are long, up to one metre, and extremely thin. The larvae are parasites of arthropods and occasionally parasitise centipedes. The body of the larva is divided into a presoma which has an invaginable proboscis armed with spines and a trunk. The adults live in or near water or in damp places. The digestive tract is more or less degenerate in adults and larvae so that it is unlikely that food is ingested (Hyman, 1951). It was supposed that carni-

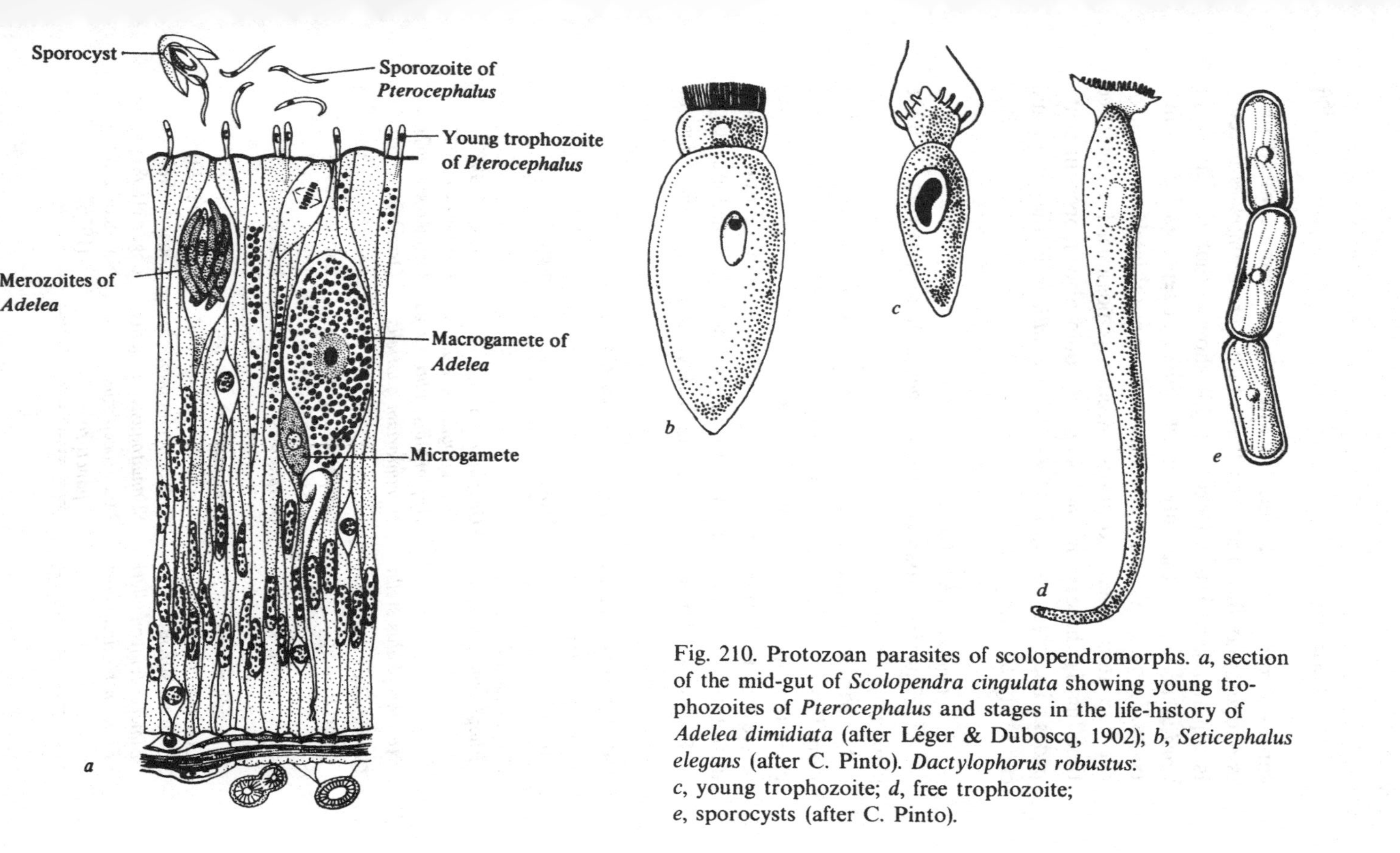

Fig. 210. Protozoan parasites of scolopendromorphs. *a*, section of the mid-gut of *Scolopendra cingulata* showing young trophozoites of *Pterocephalus* and stages in the life-history of *Adelea dimidiata* (after Léger & Duboscq, 1902); *b*, *Seticephalus elegans* (after C. Pinto). *Dactylophorus robustus*: *c*, young trophozoite; *d*, free trophozoite; *e*, sporocysts (after C. Pinto).

vorous arthropods were infected when they consumed insects such as midges, may flies and alder flies which were themselves infected as aquatic larvae but Dorier (1925) showed that the larvae of *Gordius aquaticus*, the only nematomorphan reported from centipedes, forms cysts which can survive for several weeks in water or a damp atmosphere, so direct infection is possible. He suggested that this is the case in millipedes: it could also happen in centipedes. The normal host of *Gordius aquaticus* is the phryganid caddis fly *Stenophylax*.

Geophilomorpha

The protozoan parasites of geophilomorphs (Fig. 209) are listed in Table 14.

No parasites were found in 750 specimens of *Strigamia maritima* from Sussex, England, nor were any seen in twenty squashes of the guts nor in four sets of serial sections (Lewis, 1961). It was suggested that this absence of parasites may have due to the unsuitability of the littoral habitat for their intermediate stages.

The oxyurid nematode *Cephalobellus* occurs in *Clinopodes lin-*

Table 14. *The protozoan parasites of Geophilomorpha*

Parasite	Host	Reference
Gregarinida		
Rhopalonia geophili	*Geophilus* sp.	Labbé (1899)
	Stigmatogaster gracilis	Ormières (1967)
	Haplophilus subterraneus	Ormières (1967)
Actinocephalus stella	*Himantarium gabrielis*	Ormières (1967)
Coccidiida		
Coccidium sp.	*Geophilus* sp. and *Stigmatogaster gracilis*	Labbé (1899)
Cyclopspora sp.	*Geophilus* sp. and *Stigmatogaster gracilis*	Labbé (1899)
Coccidium sp.	*Himantarium gabrielis*	Labbé (1899)
Eimeria pfeifferi	'*Geophilus ferrugineus*'	Labbé (1899)
Eimeria simoni	*Himantarium gabrielis*	Léger (1898)
Eimeria hagenmulleri	*Stigmatogaster gracilis*	Léger (1898)
Eimeria mecistophori	*Mecistocephalus punctifrons*	Narasimhamurti (1976)

earis (Weil, 1958). Stadium 8 is particularly strongly infected. The worms occur from segment 15 to segment 28. Young still under the care of the mother are parasitised: in these young animals the attack proves lethal in a few days. In older animals the attacks cause no apparent damage.

Lawrence (1953) listed mermithid nematodes as parasites of geophilomorpha but gave no details.

The tachinid *Exoristoides harrisi* has been reared from a '*Geophilus* sp.?' from Iowa, USA (Reinard, 1935).

Scolopendromorpha

The protozoan parasites reported from scolopendromorphs (Fig. 210) are listed in Table 15.

Lewis (1966) found that only four specimens of a sample of 37 *Scolopendra morsitans* from Khartoum, Sudan, contained gregarines, probably *Grebnickiella*. The number of parasites varied from one to three. He contrasted this with the high degree of infection by gregarines found in *Lithobius* spp. in England (see below). No metazoan parasites were found in this sample.

Cornwall (1916) observed that gregarines were often seen in the mid-gut of *Ethmostigmus platycephalus spinosus* in India, sometimes

Table 15. *The protozoan parasites of Scolopendromorpha*

Parasite	Host	Reference
Gregarinida		
Dactylophorus robustus	*Cryptops hortensis*	Labbé (1899)
Gregarina actinotus[a]	*Scolopocryptos sex-spinosus*	Labbé (1899)
Pterocephalus (= *Grebnickiella*) *nobilis*	*Scolopendra cingulata*	Labbé (1899)
Echinomera magalhaesi	*Scolopendra* sp.	Ormières (1976)
Grebnickiella pixellae	*Scolopendra morsitans*	Misra (1942)
Grebnickiella gracilis	*Scolopendra* sp.	Ormières (1970)
Seticephalus elegans	*Scolopendra* sp.	Grassé (1953)
Adeleida		
Klossia bigemina	*Cryptops savignyi*	Labbé (1899)
Adelea dimidiata	*Scolopendra cingulata*	Labbé (1899)

[a]an uncertain species.

in enormous numbers and the thin–walled 'crop' often harbours numerous nematode worms.

Dorier (1929) referred to a record of the nematomorphan *Gordius aquaticus* from *Scolopendra* sp. but did not give a full reference. He also recorded the species from *Scolopendra cingulata* from Morocco.

The most unusual case of parasitism in centipedes is the reputed occurrence of leeches of the genus '*Herpodella*' in the gut of *Scolopendra morsitans* and *S. cingulata* in South India (Rajulu, 1965). *S. cingulata* is a circum-Mediterranean species so this may be a mis-identification. Twenty-one of the 25 centipedes examined contained leeches. The centipedes were found 'in wet mud below stones and logs'. One specimen contained 34 leeches. The largest specimen of the parasite measured a little less than 5 mm in length. A photomicrograph shows a specimen staining darkly and uniformly. There is no indication of internal structure. Photographs of sections of *Gregarina garnhami* from the desert locust (Harry, 1970) have a similar appearance.

No parasitic leeches are found in the centipedes in the rainy season but at this time the leeches have been found free-living. Centipedes collected in January, the beginning of the dry season, showed small-sized specimens attached to the spiracles. This phenomenon merits further attention.

No metazoan parasites were found in more than 1000 specimens of *S. amazonica* from northern Nigeria (Lewis, unpublished data).

Plateau (1878) found large quantities of a fungal mycelium, possibly *Peziza* in the oesophagus of a *Cryptops* sp. but it seems most likely that it was food rather than a parasite. Balbiani (1889) described three species of fungi from the oesophagus of *Cryptops*: *Omphalocystis plateaui* (in *C. savignyi*) and *Mononema moniliforme* and *Rhabdomyces lobjoyi* in *C. hortensis*. An unidentified yeast may invade the blood of *Scolopendra* which may succumb to the attack (Duboscq, quoted by Remy, 1950).

Lithobiomorpha

The gregarines of lithobiomorphs (Fig. 211) are relatively well known. They are listed, together with the other known protozoan parasites of lithobiomorphs in Table 16.

Fig. 211. Protozoan parasites of lithobiomorphs. *a* and *b*, *Echinomera hispida*; *c*, *E. horrida*; *d*, *E. lithobii*; *e* and *f*, *E. caudata*; *g* and *h*, *Acutispora pulchra* (from Ormières, 1966); *j* and *k*, dehiscence of a spore of *Urobarrouxia caudata* (after Léger).

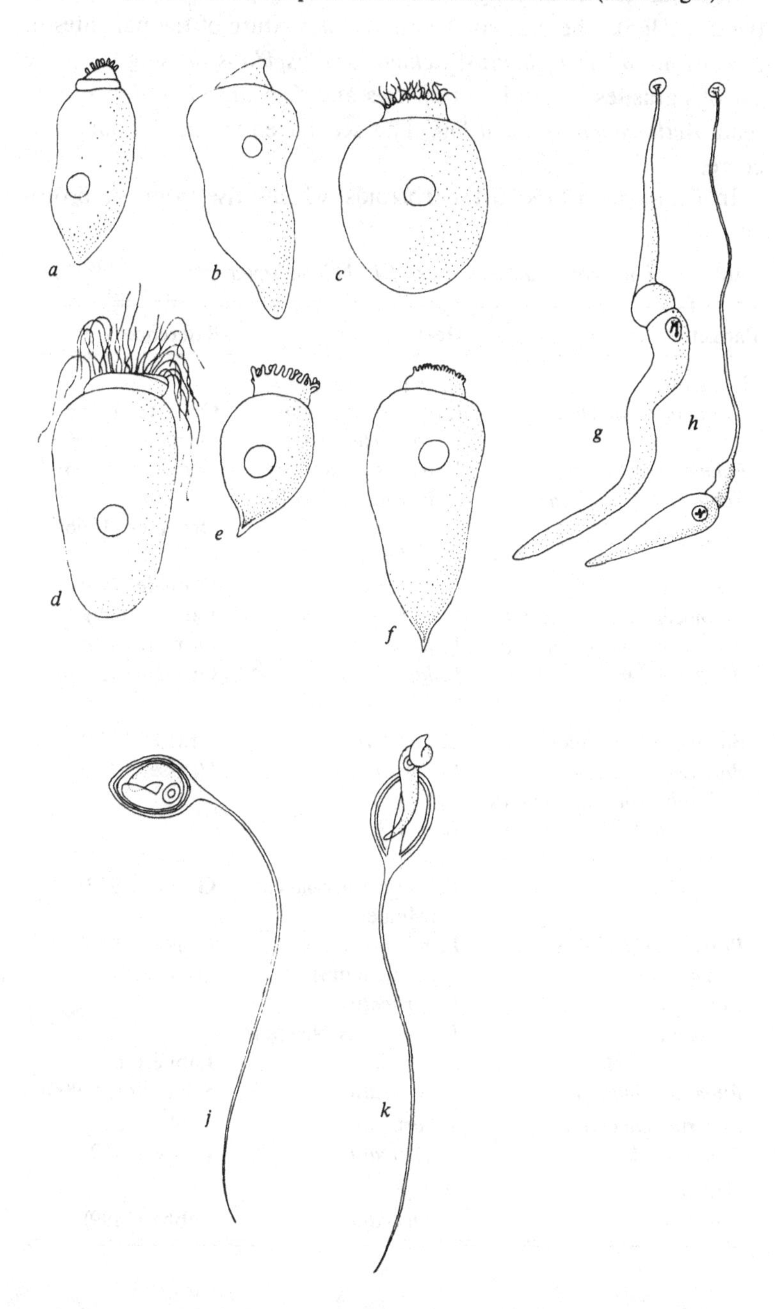

There are errors in the list of Sporozoa in Verhoeff (1902–25), which was taken directly from Labbé (1899): *L. martini* appears twice (p. 365). The first entry contains a mixture of the parasites of *L. martini* and *L. forficatus*, *Echinomera hispida* is missing from the list of parasites of the latter species and *Cephalus dujardini* should read *Actinicephalus dujardini*. The second entry for *L. martini* is correct.

In *Echinomera lithobii* the 'rhizoids' which arise from the proto-

Table 16. *The protozoan parasites of Lithobiomorpha*

Parasite	Host	Reference
Gregarinida		
Echinomera hispida	*Lithobius forficatus*	Ormières (1966)
	L. coloradensis Cook	Ormières (1966)
Echinomera lithobii	*L. piceus* L. Koch	Ormières (1966)
Echinomera caudata	*L. inermis* L. Koch	
	L. lapidicola	Ormières (1966)
	L. melanops	
Echinomera horrida	*L. calcaratus*	Ormières (1966)
Actinocephalus dujardini	*L. forficatus*	Labbé (1899)
Acutispora macrocephala	*L. forficatus*	Ormières (1966)
Acutispora pulchra	*L. lapidicola*	Ormières (1966)
Coccidia		
Barrouxia schneideri	*L. forficatus*	Grassé (1953)
Barrouxia (Urobarrouxia) caudata	*L. martini*	Grassé (1953)
Barrouxia (Echinospora) labbei	*L. mutabilis*	
	L. inermis pyrenaicus (Meinert)	Grassé (1953)
Barrouxia (Echinospora) ventricosa	*L. pilicornis hexodus* (Brölemann)	Grassé (1953)
Coccidium sp. and *Eimeria* sp.	*L. forficatus*	
	L. castaneus Newport	
	L. martini	Labbé (1899)
Eimeria schubergi	*L. forficatus*	Schaudinn (1900)
Eimeria schneideri	*L. forficatus*	Labbé (1899)
Bananella lacazei	*L. forficatus*	Labbé (1899)
Adeleida		
Adelea ovata	*L. forficatus*	Labbé (1899)

merite and penetrate the host cells appear progressively during the development of the gregarine (Ormières & Marquès, 1976). About a hundred arise from the flat protomerite. Each enters a single host cell but remains surrounded by invaginated cell membrane: all the cells in the area are penetrated. The epimerite forms a *gouttière circumcellulaire* around each cell boundary (Fig. 212). The rhizoids are neither chitinous nor intercellular as earlier authors believed.

In *E. hispida*, *E. horrida* and *E. caudatum* (family Dactylophoridae) the mature gametocysts contain a dark hemisphere containing the oocysts which is produced into a cone surmounted by a dark ring. This protrudes into the clear hemispherical *pseudocyst* (Fig. 213*a–d*) which is presumably the residual body and which brings about dehiscence of the gametocyst (Ormières, Marquès & Puissegur, 1977). The gametocyst of *Acutispora* has a gelatinous covering and dehiscence is by simple rupture (Ormières, 1966).

Roberts (1956) compared the numbers of dactylophorid gregarines in *L. variegatus* from a Hampshire woodland in October 1952 and February 1953 and from January to July 1955. He found the degree of infection ranged from 50 to 75 per cent and was similar in males and females. The intensity of infection was low, the mean

Fig. 212. Diagram showing the relationship of the rhizoids of *Echinomera lithobii* and the cells of the host gut (after Ormières & Marquès, 1976).

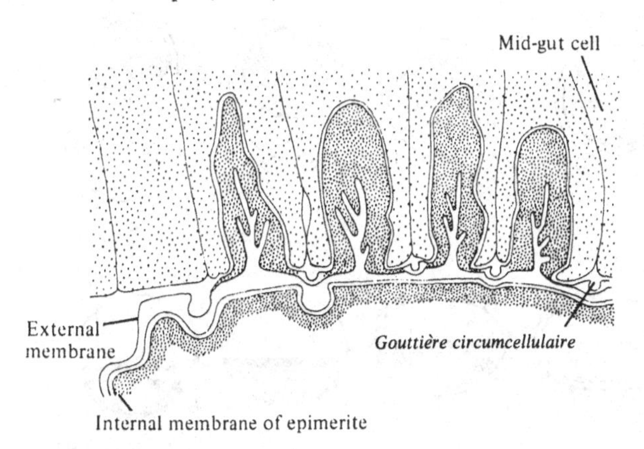

being about 10 parasites per animal but in a small number of females the intensity was very high (above 80 trophozoites per animal). Roberts found no parasites in the larval stages of *L. variegatus*. Sporozoites were very abundant in March, April and May almost to the exclusion of adult trophozoites. There was no evidence that the parasite was harmful.

Lewis (1967) investigated the parasites of both *L. forficatus* and *L. variegatus* in a Yorkshire woodland for a period of 16 months. *Echinomera hispida* virtually disappeared from *L. forficatus* in the winter months rising to between 50 and 80 per cent infected in spring and autumn but falling in the months of May, June and July. By contrast the slightly larger *Echinomera* sp. is present in relatively high numbers in *L. variegatus* throughout the year. *E. hispida* produces association cysts from April to October but *Echinomera* sp. produces them throughout the year, suggesting that it is better adapted to cold conditions. In both centipedes the intensity of infection was higher in females than in males. Lewis suggested that the low number of gregarines in the summer months might be due to the detrimental effect of dry conditions on the spores of the gregarines. It was suggested that the apparent lack of success of *L. forficatus* in damper woodlands noted by Roberts

Fig. 213. Diagrams showing the mode of dehiscence of the gametocysts of *a–d* the Dactylophoridae and *e–h Trichorhynchus* (after Ormières, Marquès & Puissegur, 1977).

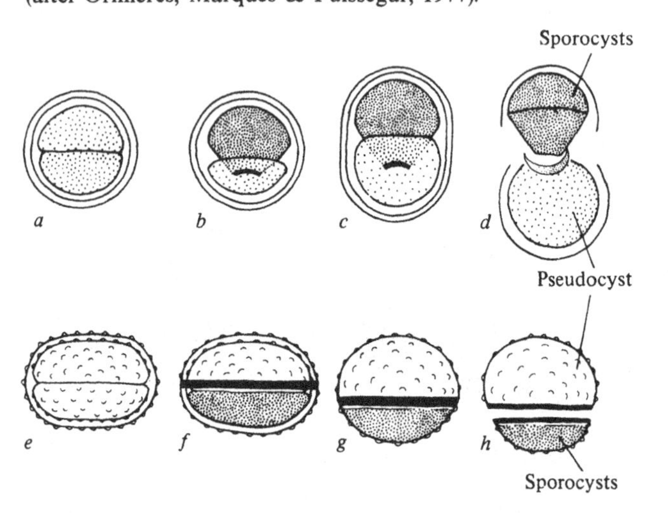

(1956) and again by Lewis could be ascribed to *E. hispida*: it was found that the incidence of infection was higher in damper woodlands. Harry (1970) regarded this hypothesis as untenable as the centipedes contained relatively low numbers of gregarines. He reports that experimental infection of locusts by more than 100 000 gregarines did not appear to affect moulting and feeding. He pointed out that because the survival of the parasite depends on the survival of the cell to which it is attached, it is unlikely to destroy its essential anchor. Although the attached gregarines may impair the functioning of the epithelial cells directly, it is more likely that they exert their effect on the physiology of digestion and absorption between the intestinal cells and the food in the lumen.

L. variegatus in Yorkshire, England is heavily infected by Coccidia as judged from the presence of oocysts in the mid-gut: about 50 per cent of the population are parasitised (Lewis, 1967).

Dorier (1929) quotes, incompletely, a record of the nematomorph *Gordius aquaticus* from *L. forficatus*.

Both wasps and flies have been reported as parasites of lithobiomorphs. Newman (1867) found 21 pupae of the proctotrupid wasp *Proctotrupes calcar* firmly attached by their posterior ends to the arthrodial membranes of the ventral side of a dead *L. forficatus* which they had presumably killed. E. H. Rübsaumen, quoted by Verhoeff (1902–25), saw larvae of *Cryptoserphus ater* issuing from a *Lithobius forficatus* and according to O. Schmiedeknecht, also quoted by Verhoeff, the genera *Disogus* and '*Codrus*' also parasitise this species.

Thompson (1939) found, on one occasion, a planidium in *L. forficatus* which closely resembled *Perilampus hyalinus* and was in all probability a chalcidoid wasp of the family Perilampidae. He considered that it was probably a secondary parasite of one of the centipede's tachinid parasites which had entered an unparasitised specimen: this is a fairly common occurrence with chalcids.

The first dipterous parasite of *Lithobius* was discovered by Giard (1893) and described as *Thryptocera lithobii*. It was, in all probability *Helocera delecta* (Wood & Wheeler, 1972).

Thompson (1915) described the three instars of a second tachinid parasite of *Lithobius forficatus*, *Loewia foeda* in England. Wood & Wheeler (1972) have recorded *L. foeda* from *L. forficatus* in New York State. The larval cuticle is thin and although the tracheal system is filled with gas, the spiracles are extremely minute in the first instar and absent in the second. Only the third instar is amphipneustic with large, globular posterior spiracles. The larvae remain free in the body cavity of the host during their entire development. This is in contrast with the usual condition in tachinids where the larvae breathe through an opening in the integument of the host or of its tracheal wall and become partially enveloped in a sleeve of host tissue the 'respiratory funnel' (Wood & Wheeler, 1972). These authors suggested that the thinning of the larval cuticle and loss of spiracles may have arisen because a secure connection between the parasite and the host did not evolve, possibly because the tissues of the centipede could not form a respiratory funnel.

Most tachinids scatter their active, newly hatched larvae in suitable habitats to locate the appropriate host unaided. Larvae searching in this way undoubtedly come into contact with a wide range of potentially new, untried hosts which the adult fly could never reach. Wood & Wheeler believe that an association between coleopterous or lepidoterous larvae is the primitive one and that the host relationship between *L. foeda* and a centipede is a derived condition.

Thompson (1939) dissected 300 specimens of *L. forficatus* from various English localities. They contained two species of tachinids, *Loewia foeda* and a second unidentified species with a very heavily armoured larva resembling the American *Gymnochaeta alcedo.* Many gravid female tachinids were also dissected but Thompson never found these larvae and concluded that this species may lay undeveloped eggs in the vicinity of the host. He found no other parasites in the sample, nor were any eggs of the parasites found on the bodies of the *Lithobius* examined. About 7.5 per cent of the population was parasitised. Almost half the specimens parasitised contained more than one larva but only in one case were both larvae alive at the time of dissection. The two species appear to be specific parasites of *Lithobius.*

No metazoan parasites were recorded from a sample of 385 *L. forficatus* and only one tachinid larva was found in 371 *L. variegatus* from Yorkshire, England (Lewis, 1965).

A parasite of doubtful affinites, *Chytridiopsis schneideri* occurs in the intestinal epithelium of *Lithobius mutabilis* (Léger & Duboscq, 1909). The life cycle of the parasite was described in detail for *Chytridiopsis socius* from the beetle *Blaps mucronata.* Small ovoid corpuscles 1.5–2 μm in diameter occur in the intestinal epithelial cells, usually between the nucleus and the base of the cell. This grows to a size of about 20 × 7 μm to form a multinucleate schizont which forms amoeboid schizozoites which escape into the intestine and infect other cells. Cysts develop later. In *C. schneideri* these are 20 μm in diameter, some contain macrospores, others microspores. Grassé (1953) stated that *Chytridiopsis* had affinities with the fungi.

Some inclusions of the blood cells of *Lithobius* resemble intracellular parasites (Duboscq. 1898).

Fig. 214. *Trichorhyncus pulcher. a*, trophozoite; *b*, sporocysts (after A. Schneider).

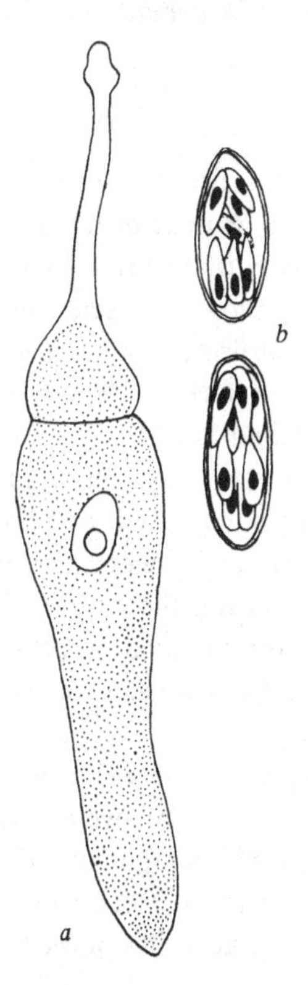

Scutigeromorpha

The sporozoan *Gregarina megalocephala*, an uncertain species, and *Trichorhynchus pulcher* have been recorded from *Scutigera* (Labbé, 1899). *Trichorhynchus* has no fixing rhizoids as do dactylophorids such as *Echinomera* but a well–developed neck and lobate epimerite (Fig. 214*a*). Cyst formation and dehiscence are characteristic (Fig. 213*e–h*) and Ormières, Marquès & Puissegux (1977) created a new family, Trichorhynchidae, for this genus.

Lawrence (1953) stated that 'Among the Myriopoda the carnivorous Chilopoda afford a sharp contrast to the vegetarian Diplopoda since no parasitic nematodes proper have been recorded from any of them, although they are occasionally parasitised by Mermithidae. In general it seems a fair assumption that phytophagous orders of cryptozoa, especially those which consume a certain amount of soil with their food, are more likely to become infected and harbour far larger numbers of nematodes than those orders which are predatory'.

Pseudoparasitism

A number of cases have been recorded in which centipedes have been voided from the mouth, anus or nose of human beings. The most detailed review of this subject is that of Blanchard (1898) who cited case histories in detail. The following species have been involved, *Haplophilus subterraneus gervaisi*, *Chaetechelyne vesuviana*, *Geophilus carpophagus*, *G. electricus*, '*G. cephalicus*', *Lithobius forficatus* and *Scutigera coleoptrata*. Of these *G. carpophagus* was the most common. Cases may last for many years and a large number of centipedes may be involved. Shipley (1914) reported a case of a woman of sixty-eight who passed some hundreds of *Haplophilus subterraneus* during the course of twelve or eighteen months. The centipedes were vomited or passed from the anus, some emerged from the nose. As many as seven or eight a day left the alimentary canal.

Blanchard suggested that the centipedes were either introduced into the alimentary canal with vegetables, or were swallowed intentionally. Cloudsley-Thompson (1958*b*) thought that the explanation of pseudoparasitism probably lies in the fact that centipedes seem to exert a morbid fascination on the appetites of the hysterical and insane.

21

Physiology and ecology

A number of topics of an ecological nature namely behaviour, food and feeding, respiration, the reproductive system and reproductive behaviour, life-histories, predators, and defence mechanisms and parasites have been the subject of previous chapters.

Many of the remaining ecological data are fragmentary and widely scattered and any account must needs reflect the interests of the particular author and his reading and cannot be comprehensive.

Water relations of terrestrial centipedes

Water loss experiments

A number of investigators have shown that centipedes lose water rapidly at low humidities. Auerbach (1951) investigated the 'desiccation death time' of various centipedes from Michigan, USA (Table 17): Roberts (1956) and Vaitilingham (1960) measured 'survival times' of British woodland centipedes at six different humidities; their results for 55 per cent relative humidity are shown in Table 18. Lewis' (1963) results, for six British species are shown in Table 19. Palmen & Rantala (1954) found the geophilomorph *Pachymerium ferrugineum* in Sweden to survive for between 38 and 109 hours at 34 per cent relative humidity at 20°C.

It is clear from these results that, by and large, geophilomorphs are more resistant to desiccation than lithobiomorphs or scolopendromorphs. Unfortunately there are no data for scutigeromorphs.

There are some discrepancies in the results of different authors: Lewis (1963) recorded the desiccation death time of *Brachygeophilus truncorum* as 15–17 h at 35 per cent relative humidity, Vaitilingham (1960) found its mean survival time to be 5 h at 55 per cent and whereas Roberts (1956) showed that *Lithobius variegatus* had a lower survival time than *L. forficatus*, Curry (1974)

Table 17. *The desiccation death times of centipedes from Michigan, USA, at 35 per cent relative humidity and 22.5–25°C (data from Auerbach,* 1951)

Species	Average length (mm)	Average weight[a] (g)	Mean death time (h)
Geophilomorpha			
Arenophilus bipuncticeps	37.2	—	7.3 ± 0.36
Geophilus rubens	43.0	0.05	27.8 ± 1.7
Lithobiomorpha			
Sonibius politus	9.4	0.008	2.7 ± 0.13
Pokabius bilabiatus	11.2	0.015	2.4 ± 0.11
Nadabius iowensis	12.8	0.018	3.4 ± 0.29
Bothropolys multidentatus	19.0	0.06	5.4 ± 0.44
Neolithobius voracior	21.8	—	5.5 ± 0.18
Lithobius forficatus	21.3	—	6.6 ± 0.40
Scolopendromorpha			
Scolopocryptops sexspinosus	38.0	0.10	11.2 ± 0.85

[a] Weights estimated from Auerbach's data.

Table 18. *Mean survival times of centipedes from woods in Hampshire, England at 55 per cent relative humidity and* 19 ± 1°*C. Data from Roberts* (1956) (R) *and Vaitilingham* (1960) (V)

Species	Length (mm)	Survival time (hours)
Geophilomorpha		
Brachygeophilus truncorum (V)	12–14[a]	5
Geophilus carpophagus (V)	about 40[a]	48
Lithobiomorpha		
Lithobius variegatus (R)	21–24	25 ± 5.1
Lithobius forficatus (R)	21–24	51 ± 10.6
Lithobius muticus (R)	13–16	8.6 ± 2.7
Lithobius lapidicola (R)	9–12	8.6 ± 1.8
Lithobius duboscqui (R)	9–12	3.2 ± 0.5
Lithobius curtipes (V)	8–11[a]	10
Lithobius crassipes (V)	9.5–13.5[a]	5

[a] Data from Eason (1964).

showed that at 0 per cent relative humidity *L. variegatus* lost water more slowly than *L. forficatus.*

Large specimens of a particular species lose water more slowly than small ones (Auerbach, 1951; Palmen & Rantala, 1954; Lewis, 1963; Curry, 1974). At 97 per cent relative humidity adult *Pachymerium ferrugineum* survive for 312–960 h at room temperature, young individuals 35–56 h.

Lithobius forficatus and *L. variegatus* attempt to run up the sides of the container at low humidities but go into a period of quiescence when 20 to 30 per cent of their body weight has been lost (Curry, 1974). When about 40 per cent of the body weight has been lost the animals no longer move when prodded with a blunt seeker but this is not the point of death: at least 50 per cent of the animals in this condition recover if placed dorsal surface down on damp filter paper.

When first placed in a desiccator *Geophilus insculptus* walks round the bottom of the tube. When about 50 to 60 per cent of the body weight has been lost stimulated animals walk backwards dragging the anterior end. This species loses water relatively rapidly at first and then more slowly (Curry, 1974), possibly because water is lost initially through the exposed arthrodial membranes between the sclerites. As water is lost the sclerites are pulled

Table 19. *Desiccation death times of centipedes from the south of England at* 35 *per cent relative humidity and* 19 ± 2°C. *Data from Lewis* (1963). *Lengths of adults from Eason* (1964)

Species	Length (mm)	Weight (mg)	Death time (hours)	Percentage weight loss
Geophilomorpha				
Haplophilus subterraneus	50–60	1.4–60	5–16	42.3 ± 5.8
Necrophloeophagus longicornis	about 30	10–55	11.5–21.5	41.1 ± 3.3
Brachygeophilus truncorum	12–14	2.0–4.1	15–17	—
Strigamia maritima	up to 40	1.2–55	14–32	42.6 ± 3.8
Scolopendromorpha				
Cryptops hortensis	about 20	2.0–13.5	4–9.5	41.1 ± 4.8
Lithobiomorpha				
Lithobius variegatus	16–24	20–95	7.5–15	39.8 ± 1.7

together cutting down water loss. A second phase of accelerated water loss could be due to a breakdown in some internal water retaining mechanism, perhaps in the epidermis, and a second fall perhaps due to a decrease in the amount lost through the mouth and anterior spiracles as the anterior end dries out.

Mead-Briggs (1956) showed that *Lithobius* sp. had considerably higher transpiration rates into dry air or air with a constant saturation deficit than either insects (*Blatta*, *Periplaneta*, *Calliphora*, *Glossina* and larvae of *Tenebrio*) or of the woodlouse *Oniscus*. In this investigation surface area was calculated as S = KW 2/3 where W was the weight and K a constant found to be 12 in the case of *Lithobius*.

The scolopendromorph *Rhysida nuda togoensis* from Ghana has a similar, perhaps slightly higher rate of water loss than *Lithobius*. The rates for *Ethmostigmus trigonopodus* from Malawi and *Scolopendra clavipes* from Tunisia (calculated from Cloudsley-Thompson, 1956) are lower than those for *Lithobius* (Cloudsley-Thompson, 1959). The lowest rate of water loss was recorded from *Scolopendra* spp. from New Mexico (Cloudsley-Thompson & Crawford, 1970). In none of these species was there any indication of a critical temperature such as that observed in insects when the cuticular layer of wax or fat becomes disorientated at high temperatures giving rise to a sudden increase in permeability. There is no evidence of the uptake of moisture from unsaturated air in *Scolopendra* (90, 95 and 98 per cent relative humidity) nor of the absorption of water through the cuticle, as happens in *Lithobius*, but they will drink free water.

The site of water loss

Chilopods may lose water by defaecation, by evaporation through the anus, through the spiracles, the mouth and the integument.

The anterior end of *Haplophilus subterraneus* dries out far more rapidly than the rest of the body suggesting that the mouth may be important as a site of water loss (Lewis, 1963). It is possible that the shape of the mouthparts may in some species reflect the need to close the pre-oral cavity efficiently. The importance of atrial trichomes in reducing evaporation from the spiracles is discussed in Chapter 11.

Currently there is no precise information on the importance of spiracles, integument, mouth and anus in water loss.

Damaged *Lithobius* with some legs missing do not seem to differ in their desiccation rates from intact animals (Curry, 1974). This may either indicate that the cuticle may be as important as the spiracles as a site of water loss, or alternatively that when a leg is lost the wound is very effectively sealed.

Lithobius duboscqui and *L. curtipes* curl up when disturbed, and *L. muticus* and *L. piceus* exhibit 'partial curling'. This, Vaitilingham (1960) suggested, may cut down water loss but there seems to be no evidence to suggest that this is so.

The significance of differing rates of water loss

Different mean death times of similar size groups of different species reflect differences in the habitats occupied by the species. *Lithobius forficatus* which is found in more xeric habitats has a higher mean death time than either *Bothropolys multidentatus* or *Neolithobius voracior* (Chamberlin). The mean death times of the small lithobiomorph *Sonibius politus* (McNeill) and the geophilomorph *Geophilus rubens*, which are woodland species commonly living under bark, are greater than the similar prairie forms *Pokabius bilabiatus* (Wood) and *Arenophilus bipuncticeps* (Wood). This may reflect the greater climatic stress to which the former are subject: the prairie forms avoid dry conditions in midsummer by burrowing deep in the soil (Auerbach, 1951).

Similarly *L. forficatus* in England which is common in drier woods tolerates longer exposures than *L. muticus* and *L. lapidicola* which require a saturated atmosphere for survival and are confined to the lower litter horizon. *L. duboscqui*, the species most susceptible to desiccation (it is also the smallest) is found only in the humus layer where the relative humidity never falls below 100 per cent (Roberts, 1956).

The geophilomorph *Haplophilus subterraneus* found in soil and humus has a low survival time whereas *Necrophloeophagus longicornis* and *Brachygeophilus truncorum* which are often found in more superficial habitats have higher survival times as does *Strigamia maritima* which is a littoral species occurring in shingle and rock crevices (Lewis, 1963).

There seems little doubt that centipedes living in more superficial habitats have lower rates of water loss than those living deeper in humus or soil. In some cases the latter are smaller and therefore expected to be less resistant to desiccation as they have a larger surface area to volume ratio.

Because of their elongated shape, geophilomorphs have a larger surface area than lithobiomorphs of the same volume and yet many species show a very high resistance to desiccation when compared to lithobiomorphs. This may, in part, be due to their possession of an extensive pleural covering of sclerotised pleurites.

High permeabilities in the sluggish soil-dwelling form *Haplophilus subterraneus* might be an adaptation to allow cutaneous respiration when the animal is submersed by heavy rains when tracheal respiration will be less effective. The same could be true of the lack of pleural sclerotisation in other orders, which being more fleet can escape dry conditions more rapidly than geophilomorphs and can thus afford a relatively higher permeability.

In humidity choice chambers, centipedes invariably choose the higher of the two humidities offered (Chapter 6). Even *Pachymerium ferrugineum*, which is particularly resistant to desiccation, shows a preference for the higher humidity when offered a choice between 97 and 100 per cent relative humidity, 48 out of 50 individuals aggregating in the moister half after 48 min (Palmen & Rantala, 1954).

There is, however, some evidence to suggest that a saturated atmosphere may be harmful. Duboscq (1899) observed that centipedes kept in a saturated atmosphere became oedematous and Brunhuber (1970*b*) reported that young *Cormocephalus nitidus* needed to be kept under drier conditions than adults for their first year. If the culture boxes were kept at a relative humidity preferred by the adults, many died of fungal attack. Larvae increased in volume while still within the egg membrane and this could only be accounted for by an increase in body water but the source of this was unknown, it was possibly absorbed from the atmosphere.

Lamyctes fulvicornis has a reputation for favouring the banks of streams but Eason (1964) considered that this was probably undeserved. Chamberlin (1912) however wrote '*L. fulvicornis* prefers the immediate vicinity of water and was found in Illinois and Wisconsin among the stones and sticks at edges of streams. At times it may be seen to go beneath the water if the rock upon which it is running becomes partially submerged'. Latzel (1880) reported that this species occurred on the bank of the Vienna river and that specimens are occasionally covered by the water.

Uptake of water (endosmosis)

The longevity of centipedes when immersed in fresh water has been investigated by Plateau (1890), Hennings (1903), Verhoeff (1902–25), Schubart (1929) and Vaitilingham (1960). Their results are summarised in Table 20. Approximate LD 50s in hours estimated from Vaitilingham's data are *Geophilus carpophagus* 84, *Brachygeophilus truncorum* 58, *Lithobius duboscqui* 36, *L. curtipes* 32, *L. forficatus* and *L. variegatus* 7. It is not clear from his account whether the animals were allowed to float or forcibly immersed.

Blower (1955*a*) showed that geophilomorphs were able to survive immersion for much longer than 24 h (exact times were not given) the time depending on the species and the amount of air in solution. Lithobiids were only able to survive for a few hours. Blower pointed out that immersion experiments allow an assessment of the permeability to be made as the effect of the tracheal

Table 20. *Survival times of centipedes immersed in fresh water*

Species	Survival time (hours)	Authority
Geophilomorpha		
Geophilus carpophagus	67	Plateau (1890)
Necrophloeophagus longicornis	144–355	Plateau (1890)
	208	Verhoeff (1925)
Pachymerium ferrugineum	168–1536(16–18 °C)	Schubart (1929)
Strigamia acuminata	12–15	Verhoeff (1925)
Strigamia maritima	70–80	Hennings (1903)
Scolopendromorpha		
Cryptops savignyi	6	Plateau (1890)

system, possibly important in water loss, is eliminated. He concluded that the resistance to water uptake is correlated with the amount of lipoid secreted onto the surface of the cuticle: the lipoid at the surface and in the epidermal glands is more readily demonstrable in geophilomorphs than in lithobiomorphs.

Behaviour when immersed in fresh water

When immersed in water *Lithobius forficatus* at first struggles violently, sometimes swimming (like *Nereis*) by a series of rapid flexions, but its motions soon become sluggish (Cloudsley-Thompson, 1945). According to Vaitilingham (1960) individuals of this species move about actively for a few minutes and either float on the surface or come to rest on the bottom of the container. *Brachygeophilus truncorum* and *Geophilus truncorum* before coming to rest on the water surface sometimes move about violently, curling and uncurling their bodies.

The South-East Asian scolopendromorph *Scolopendra subspinipes* swims when immersed. Only the top of the head and the tergites, with the exception of the last three are above water level. With the exception of the first three or four pairs, the legs are held against the sides of the body. The posterior end of the body beats to left and right ten to twelve times in five seconds. The Bornean species *Arrhabdotus octosulcatus* is unable to swim and when immersed walks slowly along the bottom of the tank with only the head out of the water (Lewis, unpublished data). Both species occur in tropical rain forest but *A. octosulcatus* may be exclusively arboreal. *S. subspinipes* is more likely to be immersed during floods after heavy rain.

When put in water *Scutigera coleoptrata* bends the trunk and makes weak rowing movements with the legs. The tergites remain unwetted. After half an hour it sinks and dies of suffocation (Verhoeff, 1938).

Cause of death of animals immersed in fresh water

The most obvious possible causes of death of immersed animals are asphyxiation and osmotic effects. A bubble collects at the posterior end of the body and several small bubbles collect on the legs and trunk of immersed *Brachygeophilus truncorum* and

Geophilus carpophagus and in lithobiids small bubbles of air are present in the position of the spiracles (Vaitilingham, 1960).

Blower (1955*a*) showed that the survival of immersed geophilomorphs depended on the amount of air in solution and experiments on littoral geophilomorphs suggest that cutaneous respiration is important (see below). Nevertheless *Haplophilus subterraneus* increases in weight rapidly in 10 per cent sea water in which specimens survive for 30–46 h. Specimens survive 12–21 h in 60 per cent sea water and only 4–8 h in 100 per cent sea water at 17 °C suggesting that osmotic factors are also important. After 14 h in sea water the sodium concentration of the body fluids rises from a mean of 205 ± 18 to 343 ± 27 mM/l: the sodium concentration of normal sea water is 453 mM/l (Binyon & Lewis, 1963).

The above data suggest that the causes of death may be complex: a shortage of oxygen may cause a breakdown in the animal's osmoregulatory system.

Chalande (1886) conducted a series of experiments on the effect of immersion in fresh water on the rate of heartbeat in *Geophilus carpophagus*, *Himantarium gabrielis*, *Scolopendra cingulata* and *Cryptops hortensis*. In a specimen of *G. carpophagus* the normal rate of heartbeat was 22 contractions per min. This fell to nine after 55 min immersion: contraction stopped after 10 h. The animal was removed and dried and the heart resumed beating at 21 contractions per min.

Vaitilingham's results for survival times of lithobiomorphs upon desiccation and upon immersion are of considerable interest. The fact that the larger *Lithobius* spp. which show greater resistance to desiccation have a low survival time when immersed suggests that surface area/volume ratio is important in these phenomena. The larger animals may owe their resistance to desiccation in part at least to their low surface area/volume ratio and if cutaneous respiration is important on immersion this would also be responsible for their rapid death when under water.

Vaitilingham concluded that the longer survival time of small lithobiomorphs when submersed was probably due to a greater area of the body being covered by a lipid layer than in large species of *Lithobius*. If this were so, however, the small species would have a greater resistance to desiccation and this is not the case.

Function of coxal glands

It has been suggested that the coxal glands of the last pair of legs of geophilomorphs act as osmoregulatory organs, excreting water under wet conditions and absorbing it in dry environments (Rosenberg & Seifert, 1977). The ultrastructure of the cells suggests that they function in transport and preliminary experiments concerning the behaviour of the animals and the changes in the cells were interpreted as signs of reversal in the direction of transport.

Coxal glands are also found in lithobiomorphs and scolopendromorphs but not scutigeromorphs. They are lacking in the geophilomorph *Mesocanthus albus* Meinert which is found in relatively arid environments in Africa and in the scolopendrid *Asanada sokotrana* which is found in deserted *Trinervetermes* mounds in African savannas.

A decrease in the number and diameter of coxal pores has been observed in the lithobiid genus *Hessebius* which inhabits mostly dry mountain habitats of middle Asia and the Caucasus and with a number of species of the genus *Lithobius* living under similar conditions. A remarkably large diameter of coxal pores is seen in most of the lithobiid species inhabiting areas of high humidity – Pacific Islands, Primorye, Kolkhida Depression (Zaleskaja, 1975).

Water relations and ecology of littoral centipedes

A number of species of geophilomorph are found only on the sea shore between tidemarks, the following list is probably far from comprehensive:

Geophilus algarum Bröl.; England and Atlantic coast of France.

G. fucorum Bröl.; Mediterranean coasts.

G. fucorum seurati Bröl.; England and Algeria.

These three forms may represent one polytypic species (Lewis, 1962).

G. becki Chamberlin; Cabrillo Beach near San Pedro, California.

Brachygeophilus admarinus Chamberlin; under stones near low tide mark, SE Alaska.

Clinopodes poseidonis (Verhoeff); under stones and seagrass, Mediterranean.

C. poseidonis sudanensis Lewis; Red Sea coast of Sudan.
Nesogeophilus littoralis; Takakuwa, Japan.
Mixophilus indicus Silvestri; coiled in mud, Cooum River, Madras, India.
Strigamia maritima (Leach); Atlantic coasts of Europe.
S. maritima japonica (Verhoeff); Japan.
Hydroschendyla submarina (Grube); Atlantic coast of Europe, Mediterranean, Bermuda.
Lionyx hedgpethi Chamberlin; California under stones between tide marks.
Mecistocephalus manazurensis Shinohara, Japan.

Unidentified specimens have been recorded from the crust of *Melobesia* and *Vermetus* which covers all rocks at low tide level in bays in the Cape Verde Islands and from large empty barnacle shells at low tide level in the Galapagos Islands (Crossland, 1929).

In addition to these exclusively littoral species, there are geophilomorphs that are frequently found on the sea-shore but which also occur in terrestrial habitats; such are *Henia bicarinata* (Meinert) which occur in the *trottoir* of coralline algae in the Mediterranean and *Necrophloeophagus longicornis* not infrequently found on the sea-shore in England. Coastal areas and particularly sea-shores seem to be the 'preferred' habitat of *Pachymerium ferrugineum* (Palmen & Rantala, 1954) at least in Northern Europe.

Survival on immersion in sea water

Necrophloeophagus longicornis and *Strigamia maritima* survive longer when immersed in fresh water than they do in sea water (Table 21) but *Pachymerium* appears to survive longer in diluted sea water than in fresh water (the data for this species are for animals from different localities and by different authors). When immersed in 10 per cent, 50 per cent and 100 per cent sea water (Table 22), the terrestial *Haplophilus subterraneus* shows the shortest survival time, the upper littoral *Strigamia maritima* an intermediate and the mid-littoral *Hydroschendyla submarina* the greatest survival time. All three species survive longest in 10 per cent sea water and shortest in 100 per cent sea water (Binyon & Lewis, 1963).

Respiration under water

Suomalainen (1939) showed that *Pachymerium ferrugineum* floated on sea water for 2–8 days at 19–27 °C and for 16–31 days at 6–12 °C before sinking to the bottom of the experimental container. At first, animals that had sunk had many air bubbles on the body surface. Lewis (1960) noted the presence of bubbles of air on the cuticle of newly immersed *Strigamia maritima* but considered that they would be unlikely to be an important source of oxygen to the animals, because they soon disappear, probably being dislodged when the animals start moving. Bonnell (1929) reported that the Indian littoral centipede *Mixophilus indicus* trapped a bubble of air in a loop of the posterior end of the body but Rajulu (1972) was unable to confirm this.

Suomalainen considered that the spiracles of *Pachymerium* acted as physical gills. At 10 °C *Strigamia maritima* survived 84 h immersion in sea water but four of a group of five specimens in nitrogen

Table 21. *Survival times of centipedes immersed in sea water*

Species	Survival time (hours)	Authority
Necrophloeophagus longicornis	6–19 (July)	Plateau (1890)
	27–72 (Sept., Oct.)	
Strigamia maritima	30–40	Hennings (1903)
Pachymerium ferrugineum	24–95 days (19–27 °C)[a]	Suomalainen (1939)
	68–178 days (6–12 °C)[a]	Suomalainen (1939)
Mixophilus indicus	24–52	Rajulu (1972)
Scolopendromorpha		
Cryptops savignyi	up to 8	Plateau (1890)

[a] 0.5–0.6 salinity.

Table 22. *Longevity of centipedes (in hours) in normal and diluted sea water at 17 °C (from Binyon & Lewis, 1963)*

	10% sea water	60% sea water	100% sea water
Haplophilus subterraneus	30–46	12–21	4–8
Strigamia maritima	36–72	12–36	12–24
Hydroschendyla submarina	48–84	36–48	12–36

saturated water were dead after 24 h immersion, suggesting that the animals respire when immersed (Lewis, 1960). Lewis thought it possible that the spiracles acted as physical gills.

Two specimens of *Mixophilus indicus* 48 and 61 mm long survived 35 and 16 h respectively in deoxygenated water and animals of similar size survived 52 and 24 h in oxygenated water (Rajulu, 1972). The author considered that this suggested that cutaneous respiration was taking place. Wax sections of animals injected with 0.01 per cent reduced indigo carmine solution, which turns blue in the presence of oxygen, showed intense blue colour around the tracheae and a less intense but distinct blueing of the epidermis under regions of untanned cuticle.

After a quiescent period, immersed specimens become active and rise to the surface and attempt to escape. This behaviour pattern occurs earlier in large specimens (Fig. 215*a*). Specimens 48 mm long remove oxygen from water more rapidly than specimens 61 mm long. After removal from water the rate of oxygen uptake increases in both size categories but rises and falls back to normal more rapidly in the smaller specimens (Fig. 215*b*). The pH of the blood in immersed specimens falls steadily from 7.2 to 6.7 in 41 mm

Fig. 215. Immersion experiments using *Mixophilus indicus*. *a*, time taken for immersed specimens to exhibit escape behaviour (see text); *b*, rates of oxygen consumption after removal from sea water. ●-----●, specimens 4.8 cm long; ○----○, specimens 6.1 cm long (after Rajulu, 1972).

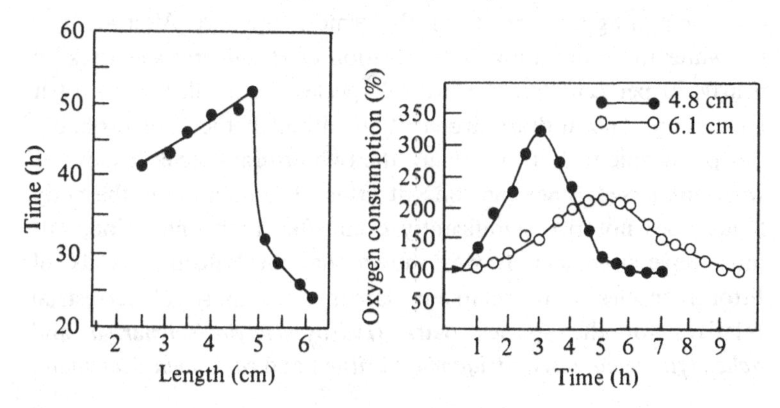

specimens and to pH 5.8 in 61 mm specimens in 5 h. Rajulu did not consider the possible function of the spiracles as physical gills.

Rajulu noted that only the head capsule and the leg bases of *Mixophilus* 28–48 mm long are tanned, but in specimens over 50 mm long the tergites are also tanned. These observations suggest that *Mixophilus* is a remarkable animal. One would expect the sclerites at least to be lightly tanned. The late tanning of the sclerites is likewise unusual. The possibility cannot be ruled out that the author was dealing with two species: this would account for the differences in rates of oxygen consumption on the return of specimens to air and the contrast between small specimens whose resistance to immersion increases with size and large specimens in which it decreases (Fig. 215*a*).

After removal of the animal from water the main longitudinal tracheal trunks show rhythmic contractions. Tracings obtained with a kymograph showed that the maximum number of movements is 22–25/s. The terrestrial species *Himantariums samuelraji* did not show such 'tracheal pulsation'. The mid-region of the tracheal trunks are much swollen in *Mixophilus* but not in *Himantarium* (Rajulu, 1970*b*). The author did not describe the mechanism responsible for the pulsation.

Osmoregulation

Total osmotic pressures of the body fluids of the terrestrial geophilomorph *Haplophilus subterraneus* and the littoral species *Strigamia maritima* and *Hydroschendyla submarina* vary from 44 to 50 per cent of the sea water value (Binyon & Lewis, 1963), sodium ions accounting for almost half the osmotic pressure. After 4.5 h in sea water the sodium ion concentration of *H. subterraneus* rises by nearly 70 per cent but the littoral species are unaffected by 14 h immersion. The authors argued that although they are probably less permeable than *Haplophilus*, the two littoral species lose water when immersed. Since the concentration of sodium ion in the body fluids does not rise significantly even after 14 h immersion, salt must have been lost. It was shown that the 'salivary glands' of littoral species were relatively larger than those of terrestrial relatives for the species pairs *Hydroschendyla submarina* and *Schendyla nemorensis*, *Strigamia maritima* and *Strigamia acuminata*, and *Haplophilus subterraneus* and *Henia bicarinata* and it was suggested that the salivary glands could well be the site of salt secretion. The recent work of Rosenberg & Seifert (1977) suggests that the coxal glands may also be important in osmoregulation.

Behaviour of littoral centipedes

When immersed in sea water in a narrow chamber filled with ballotini 3 mm in diameter *Strigamia maritima* became active after an initial period of immobility of variable duration but never longer than two hours at 16–19°C. Active animals make exploratory movements with their antennae and then crawl out of the water. In contrast, *Hydroschendyla* remain more or less stationary when immersed, even when left overnight (Lewis, 1960).

At Cuckmere Haven, Sussex, *Strigamia maritima* occurs in a shingle bank. At times small animals are very common at a given point whereas large animals are relatively rare. In April 1959 adolescens I *Strigamia* were far more numerous beneath *Obione* drift on the landward side of the shingle bank than they were at the edge of the salt marsh inland from the shingle. Lewis (1961) suggested that this was due to the fact that small worms and arthropods which would form a suitable food for the adolescens I were common in the *Obione* drift but absent from the salt marsh.

On 28 January 1959 *Strigamia* was particularly common on the seaward, south facing side of the shingle bank, 5 cm below the surface at the boundary of wet and dry shingle. The temperature here was 12°C at a depth of 30 cm in the shingle.

Clearly *Strigamia* is a mobile species concentrating in areas which are climatically favourable and have a good food supply. It is found only at the top of the shore, around the High Water Spring drift line on the seaward side of the shingle bank, but this may well be due to the unstable nature of this part of the beach, and the fact that the fauna is usually sparse here: Blower (1957) has found the species in coarse shale down to mid-tide level in the Isle of Man. The animal is found far lower down on the landward side of the shingle bank where it feeds on the isopod crustacean *Sphaeroma* and is liable to spasmodic immersion at high spring tides and during bad weather, but it is difficult to equate conditions here with those prevailing at a particular tidal level. Immersed animals migrate up the beach suggesting that even mature animals are not well adapted to resist prolonged immersion in sea water.

Mature females migrate into moist sandy areas behind the shingle bank to lay their eggs, which are very permeable and to brood their young which are rapidly desiccated in unsaturated air. Eggs collected in early June 1958 shrank rapidly in sea water, but three collected on 1 July 1959 showed only a slight initial shrinkage which was followed by recovery. It would appear that a regulatory mechanism develops in older eggs. The time of egg laying (May and June) appears to be geared to correspond with the season when the spring tides have their smallest amplitude and with the least stormy part of the year and is far better defined than in terrestrial species. Males also migrate up the beach to deposit their spermatophores and all stadia migrate into the top of the shingle bank to moult. Such migrations suggest that the spermatophores, eggs, larval stadia and moulting animals are unable to withstand immersion in sea water.

Similar observations have been made on populations at Plymouth, Devon. At Jennycliff Bay the population was found to be aggregated in two regions, at the storm line and at Mean High Water Springs some 1.5 m vertically lower down the shore. Premoult animals migrate out of the summer tidal zone into the storm line. Gravid females also migrate upshore, followed by mature males (Lewis, 1962).

Strigamia is also found in rock crevices in the *Xanthoria* and *Verrucaria* zones and the top of the *Chthamalus* zone at Plymouth where it is replaced by *Hydroschendyla submarina*. D'Arcy Thompson (1889) reported finding *S. maritima* 'close to the water mark of very low spring tides where it could not be exposed for more than two days a fortnight'. This seems improbable.

At Plymouth, *Hydroschendyla* is common in rock crevices where there is a heavy deposit of silt between the *Verrucaria* and *Fucus serratus* zones. This species lays its eggs and moults in the littoral zone. The eggs are impermeable showing to change in volume after 12 h in sea water or 12 h in unsaturated air (Lewis, 1962).

The success of littoral centipedes

Field observations suggest that populations of *S. maritima* and *H. submarina* at Plymouth are far more dense than are terrestrial centipede populations. This may in part be due to the absence of large predators on the upper shore and the absence of parasites, perhaps due to the unsuitability of the external environment for the development of their intermediate stages (Lewis, 1961). Further factors may be the abundance of food and the ameliorating effect of the sea on the climate of the littoral zone (Lewis, 1962).

Genetic isolation of littoral populations

The varied nature of coasts tends to break up littoral centipede species into a number of isolated populations such as might be expected to favour genetic divergence. Such differences are probably reflected in differences seen between populations of *Geophilus algarum* and *G. fucorum* which may represent a single polytypic species, and in the variation in leg numbers in different populations of *Strigamia maritima* (Lewis, 1962) and different populations of *Nesogeophilus littoralis* in Japan (Shinohara, 1961).

Resistance to high and low temperatures

The results of temperature preference experiments are discussed in Chapter 6: the effects of temperature extremes are discussed here.

The temperature range of activity of the geophilomorph *Pachymerium ferrugineum* in Finland extends from about 5 to 35°C. Below 5°C feeding activity ceases and cold stupor occurs at about 0°C. In saturated air adults survive for about 6 h at 37°C but less than 15 min at 45°C. When the temperature is lowered rapidly, about −15°C proved a critical level but in natural habitats no mortality could be observed even in places where the minimum temperature during winter was −22 to −26°C (Palmen & Rantala, 1954).

In contrast to *Pachymerium*, large specimens of *Scolopendra* collected in summer from New Mexico showed at best only moderate recovery (ability to walk within 24 h but without its previous speed and co-ordination) when supercooled to 3.1 ± 0.48 °C. A single supercooling was injurious or fatal suggesting that they were not able to withstand the rigours of winter surface temperatures in central New Mexico and probably burrow deep into the ground in the winter (Cloudsley-Thompson & Crawford, 1970).

A temperature of $-7°C$ for 12 h kills at least half the large *Scolopendra polymorpha* Wood exposed (Crawford & Riddle, 1974). The centipedes, from the Penoncillo Mountains in New Mexico, are found in shallow burrows and beneath rocks were they are practically immobilised by temperatures at or near 0°C so that it seems unlikely that they could go further underground in colder weather. The authors estimated that there was an average of about one day a year when half or more of these specimens might die of cold at a cold site on a north facing slope. They cautiously estimated a yearly mortality due to cold of 10 per cent of each breeding population.

Crawford, Riddle & Pugach (1975) reported good recovery of *S. polymorpha* following a single supercooling in the range -4.0 to $-9.5°C$ for specimens collected in February and acclimatised to 5°C. Measurements of seasonal changes in supercooling points, haemolymph melting points and polyhydric alcohol concentrations did not indicate any appreciable development of cold hardiness in this species. However, rates of oxygen consumption measured during the winter were lower than expected on the basis of measurements taken at other times of the year (January: 20.28 ± 2.15 $\mu l\ g^{-1}\ h^{-1}$, August: 96.62 ± 8.79 $\mu l\ g^{-1}\ h^{-1}$) suggesting that it spends part of the winter in a 'quiescent' state.

The effect of low temperatures on lithobiomorphs has been investigated by Roberts (1956). Maintained at $2 \pm 1°C$ for 24 h, *L. forficatus*, *L. variegatus* and *L. duboscqui* did not feed. *L. forficatus* is the most resistant to low temperatures, tolerating exposure to -4 to $-6°C$ for 24 h with 100 per cent survival. *L. duboscqui* could not tolerate more than 2 h at $-1°C$.

The larger specimens of *L. forficatus* and *L. variegatus* showed a greater tolerance of low temperature. At $-1°C$ *L. forficatus*, 22–24 mm long, froze after 168 h exposure, *L. variegatus*, 17–20 mm long after 48 h.

During the mild winter of 1952–53 the humus temperature never fell below 2–3°C but litter temperatures were between 0° and 1°C, temperatures which would be lethal to *L. variegatus* after long exposure. In winter, however, maturus *L. variegatus* leave the litter for the cryptozoic habitat while immature stadia become restricted to the lowest litter layer.

The winter of 1955–56 was more severe than that of 1952–53 and in February 1956, of 23 immaturus and 96 maturus *L. variegatus* collected from dead logs, four immaturus and 52 maturus individuals were dead. The dead specimens were all under bark on the upper sides of logs where temperatures were as low as −3 °C. Living specimens were either in or beneath the logs (Roberts, 1956).

Respiratory metabolism

Wignarajah (1968) measured rates of respiration of *Lithobius crassipes* and *L. forficatus* in northern England at 5, 10 and 15 °C. He found that the ratio live weight/respiration per unit weight gave a typical L-shaped curve. Respiratory rates were temperature dependent.

The small New Mexico lithobiomorph *Nadabius coloradensis* shows a trend of rapidly increasing oxygen consumption through June and July and a rapid decline during September to a low in October as measured at 15 °C. At high test temperatures (15 °C and above) starvation depressed the rate of oxygen consumption.

RQs of 0.44 to 0.6, regardless of nutritional state, suggest the predominant utilisation of lipid in respiration (Riddle, 1976).

Activity rhythms

Centipedes are rarely seen above ground during daylight. An exception is the North American geophilomorph *Strigamia chionophila* Wood which is common in litter of *Acer* forests and dense stands of *Betula* and *Populus* in Michigan. Johnson (1952) commonly observed this bright red species coiled on the topmost layer of dry leaves of the litter 'basking in subdued light' and Dowdy (1944) recorded a density of 0.85/m^2 for *Geophilus strigosus* McNiell collected from a field in Ohio using a sweep net as compared to a density of 1.92/m^2 in the soil.

No investigations appear to have been carried out on the diurnal rhythms of geophilomorphs. Luminous species have frequently been reported on the soil surface at night (see Chapter 19) and the littoral species *Strigamia maritima* feeds on rock surfaces at night (Turk & Turk, 1958; Blower, 1957) but Lewis (1961) observed it to be feeding in the day-time on a number of occasions: it seems possible that this species is active throughout the day and night and this could well be true of other species especially the more subterranean ones.

The North American *Scolopendra heros* prefers to remain underground on warm days and surfaces on cloudy, rainy days (Campbell, 1932). Manton (1965) concluded that *Scolopendra* is not adapted to moist forests and if this is so specimens might be expected to be driven onto the soil surface by a rising water table. *Scolopendra moristans* occasionally appears on the soil surface after heavy rain in Northern Nigerian savanna and on one occasion *Asanada sokotrana* was found on the surface in daylight (Lewis, unpublished data). The Bornean scolopendromorph *Arrhabdotus octosulcatus* may be an arboreal species and hence would normally be exposed to daylight: it is heavily pigmented (Lewis, unpublished data).

Actograph experiments have shown that *Scolopendra clavipes* from Tunisia shows an endogenous rhythm of nocturnal activity persisting for several days in darkness at constant temperature and humidity (Cloudsley-Thompson, 1956) as does the African species *Rhysida nuda togoensis* from Ghana (Cloudsley-Thompson, 1959). *Scolopendra* spp. from New Mexico show a marked diurnal rhythm when illuminated from 0.600–21.99 h in moist air at 27 °C. The rhythm persists in constant darkness but is almost lost in continuous light (Cloudsley-Thompson & Crawford, 1970). Mead (1968) showed that in continuous darkness and constant relative humidity the European *S. cingulata* keeps up a daily circadian rhythm for 80 days: forced oscillations can be obtained with periods much shorter or longer than 24 h. The experiments were carried out with a humid substrate and an atmospheric relative humidity of 60 to 80 per cent. The activity ceases when the humidity is raised to 100 per cent. The period of activity lasts for about one hour in each twenty-four: occasionally specimens show a second phase of activity three or four hours after the first. Mead suggested that this was due to hunger. Behaviour in the actograph was affected by the background pattern of activity within the building showing a peak of activity on Thursday and a trough on Sunday (Mead, 1970).

When kept in a terrarium *Scolopendra cingulata* is active on average for 1–2h each eighth night (Klingel, 1960*a*).

Lithobius forficatus was found to be active on tree trunks after snow-melt in Canada (Monteith, 1976*b*) and it is occasionally found on paths during the day in England (Lewis, unpublished data). Nevertheless the species shows a diurnal activity rhythm: 30 individuals living together in continuous illumination (230 lux) and constant temperature show four times greater activity during the night than during the day (Amouriq, 1967).

In Malta *Scutigera* comes out from beneath stones and rubbish and makes 'a dart at some small insect ... seemingly not minding a blazing sun at all' (Sinclair, 1895) but in North America *S. coleoptrata* is clearly nocturnal (see Chapter 10) and in nine years in the Sudan and Nigeria no scutigeromorphs were observed to be active during the day (Lewis, unpublished data).

No experiments appear to have been carried out on activity rhythms in scutigeromorphs.

Cavernicolous centipedes

Centipedes are commonly found in caves. Some undoubtedly are accidental inhabitants, others (troglophiles) appear to show a preference for caves whilst yet other species, (troglobionts) are confined to caves and often show morphological adaptations to a cavernicolous mode of life.

Geophilomorphs found in caves are almost invariably common terrestrial species but *Thalthybius carvernicolus* Matic, Negrea & Fundora, taken from stones embedded in guano and on clay in five caves in Cuba may well be a troglobiont (Negrea, 1977).

The scolopendromorphs appear to have produced few truly cavernicolous forms. In Cuba, however the cryptopid *Otocryptops rubiginosus* (L. Koch) has been reported from thirteen caves where it sometimes forms rich populations: it is undoubtedly a troglophile. Two further Cuban cryptopids, *Cryptops* (*Trigonocryptops*) *troglobius* Matic, Negrea & Fundora and *Newportia leptotarsis* Matic, Negrea & Fundora have only been recorded from caves and show morphological characteristics of true cavernicoles: very long antennae and terminal legs, and a reduction of pigmentation.

A large number of cavernicolous Lithobiomorpha lack eyes: Matic (1960) published a key to 10 species of blind European *Lithobius*. Common morphological adaptions of cavernicolous lithobiomorphs in addition to the reduction or absence of ocelli are elongated appendages bearing elongated setae (Fig. 216), reduced pigmentation and enlarged Tömösváry organs. Troglobiont lithobiomorphs are common in Spain, Italy and the Balkans.

An extreme example, *Lithobius* (*Troglolithobius*) *sbordonii* Matic, has very long antennae composed of more than 100 segments, lacks eyes, and has a very large Tömösváry organ. The forcipular coxosternite is narrow, the forcipular teeth are atrophied, the forcipules are very long and feebly curved (Fig. 217). The tergites are very long and narrow, without prolongations and the legs,

Fig. 216. The cavernicolous lithobiomorph *Lithobius drescoi* (from Demange, 1958).

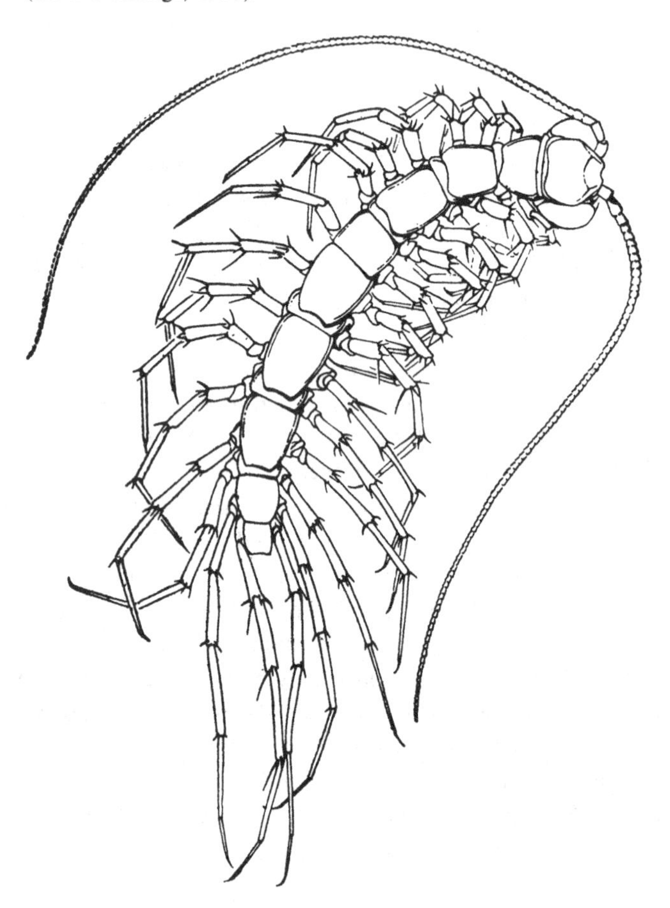

which have a divided tarsus, are very long so that they have a scutigeriform appearance (Matic, 1967).

Other lithobiid genera have produced cavernicolous forms, for example, *Eupolybothrus andreevi* Matic, an eyeless form from Bulgaria (Matic, 1964).

Matic (1973) regards *Scutigera coleoptrata* as a troglophile. *Thereuopoda decipiens cavernicola* a scutigeromorph from South-East Asia is cavernicolous and *Thereuonema higendorfi* has been reported from caves in Korea (Murakami & Paik, 1968).

General ecology

In this section scattered ecological data not discussed elsewhere have been assembled.

Fig. 217. *Lithobius* (*Troglolithobius*) *sbordonii*. *a*, dorsal view of head and first trunk segment; *b*, Tömösváry organ; *c*, claw of leg 13 (after Matic, 1967).

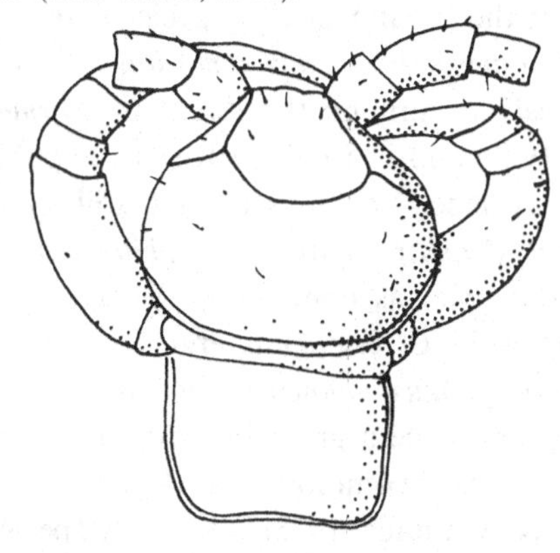

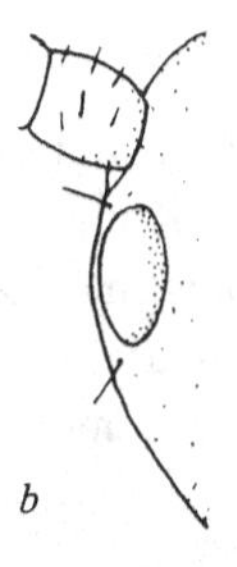

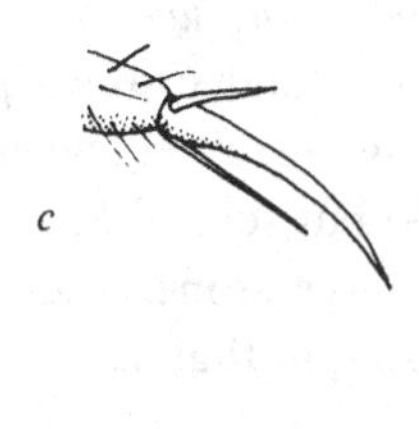

Geophilomorpha

Some geophilomorphs show a changing distribution with age. *Geophilus rubens* is an American forest species in which the majority of adults, especially females, are taken from logs whereas 90 per cent of the young are found in leaf litter (Johnson, 1952).

Geophilomorphs are usually solitary but group feeding occurs in *Strigamia maritima* and *S. acuminata* (see Chapter 10). *Strigamia maritima* may be present in large numbers in small crevices in chalk boulders and large numbers of *Mesocanthus albus* were found under the bark of an *Acacia* tree at Jebel Tozi in the Sudan (Lewis, unpublished data). It may be significant that both species have small poison claws.

A number of species have been reported from ants' nests. Holmquist (1928) listed *P. ferrugineum*, *Soniphilus embius* Chamberlin and *Arenophilus bipuncticeps* from the nests of the mound building ant *Formica ulkei.* Auerbach (1951) also recorded *P. ferrugineum* from the mound of this species and *P. ferrugineum* and *S. embius* from nests of *Lasius umbratus mixtus aphidicola. Schendyla nemorensis* occurs in the nests of *Formica rufra* in Finland (Palmen, 1948) and in England (Lewis, unpublished data).

There is some evidence to suggest that soil type influences distribution: Blower (1955*a*) regarded *Geophilus carpophagus* and *Brachygeophilus truncorum* as common in poorer mull soils and mor. In richer mull soils *G. insculptus* and *G. electricus* are most frequent. *Necrophloeophagus longicornis* is not often frequent in woodland and appears to be a grassland species. *Geophilus carpophagus* which is most common on heath and moorland (Fairhurst, Barber & Armitage, 1978) is the only species of geophilomorph commonly found in houses and outbuildings in the British Isles (Eason, 1964). It shows calcifuge tendencies (Fairhurst, Barber & Armitage, 1978).

Scolopendromorpha

Scolopendra and *Cormocephalus* are solitary and exhibit a ritualised fighting reaction when they meet (Klingel, 1960*a*; Brunhuber, 1969) but several *S. cingulata* may sometimes be found under the same large stone. *Asanada sokotrana* is a slow moving species and unusual in that several specimens are commonly found in close proximity: twelve specimens have been taken from one small termite mound (Lewis, 1973). The species, which has weak poison claws and unspined terminal legs, shows no ritualised fighting reactions.

S. cingulata burrows under stones using the forcipules and the first three or four pairs of legs to remove soil and small stones. In the laboratory specimens produced five to six chambers measuring 8–12 × 2–3 cm linked by passages 5–15 cm long. The burrows are not territories but temporary retreats used for a few days (Klingel, 1960*a*).

Lithobiomorpha

In the British Isles *L. forficatus* is widely distributed and is frequently found in suburban areas and owing to its habit of wandering in the open at night, finds its way into outbuildings and houses (Eason, 1964). These generalisations are also true of the species in America. In the Chicago area it is found in more xeric environments than other species of similar size (*Neolithobius voracior, Bothropolys multidentatus*) (Johnson, 1952). All occur in leaf mould in the spring, moving into logs or deep into the soil in summer and winter (Auerbach, 1951). There is no evidence of such massive seasonal movements in *L. forficatus* or *L. crassipes* in northern England (Wignarajah, 1968). Lloyd (1963) showed that *L. variegatus* moves out of beech branches into litter in severe weather in England.

In orchards in Ontario *L. forficatus* is present under, or adjacent to, tree trunks four days after the snow has melted. It moves from the trunks to board traps between the trees as the soil surface dries out. The species is not evident under board traps or in soil from mid-June to September. With the advent of rains and cooler temperatures they are again found under trees and are active until the winter freeze-up. Heavy rain, direct sunlight or hot dry days stimulate them to retreat to the shelter of tree trunks or into the adjacent soil (Monteith, 1976*b*). Both *L. variegatus* and *L. forficatus* ascend tree trunks at night (see Chapter 10).

Lithobius variegatus is the only centipede endemic to the British Isles. It is absent from the eastern counties, showing a strong preference for an oceanic climate. The single record from the

Channel Islands may be due to an introduction. It has been suggested that *L. validus* Meinert which is common in central Europe may be a subspecies of *L. variegatus*. *L. variegatus* is essentially a woodland species and is relatively uncommon in gardens, buildings and suburban areas. Its banded colour markings make it inconspicuous against its natural background (Eason, 1964). It is far more common on heath and moor than *L. forficatus* (Fairhurst, Barber & Armitage, 1978).

L. variegatus is generally distributed in woodland but some other species are patchily distributed. *Lithobius duboscqui* shows such a distribution in woodland, typically in the upper humus layer, although it has a tendency also to inhabit cultivated land and gardens. It is more common in calcareous areas (Fairhurst, Barber & Armitage, 1978). *L. muticus*, a nematode feeder, is very localised; in a plot 32 × 6 m, 29 specimens were confined to one sub-plot 8 × 3 m. In a second plot the 10 specimens found were likewise confined to one sub-plot. These two colonies were still in the same location after three years (Roberts, 1956). *L. lapidicola* was found in only one plot of the wood investigated. The favoured microhabitat of *L. melanops* is under the bark of logs.

Johnson (1952) attempted to determine the degree of compatibility between the sexes and the effects of enforced aggregation in *Lithobius forficatus*.

Males and females of the same size tolerated each other until the moulting period, when invariably the moulted animal was eaten by the other. Where the male or female was one-third to one-half the size of the other the larger specimen killed and ate the other within a few days.

Confined pairs, regardless of sex were observed in parallel juxtaposition as frequently as apart. When a pair were in close contact antennal explorations were frequently observed; the antennae of one gently playing over the tergites, antennae and posterior appendages of the other regardless of sex. Forcipular exploration of short duration occasionally was evinced, but this was limited to the tergites. If the forcipules approached the appendages, antennae or cephalic region of the other specimen, it turned and attacked, finally the couple settled down antennae quivering and the anal legs held high.

Scutigeromorpha

The scutigeromorphs are distributed largely in the tropics and subtropics. *Scutigera coleoptrata* is probably an indigenous species in the Mediterranean region but it has extended to much of Europe, Asia and America in houses. It is well established in the Channel Islands and may be indigenous there. It has occasionally been recorded from buildings in the British Isles (Eason, 1964). In North America *S. coleoptrata*, frequently referred to as *S. forceps*, is commonly known as the house centipede. It has spread from the southern states and reached Pennsylvania in 1849, New York in 1885 and Massachusetts in about 1890. It is now very common throughout New York and the New England states and it extends westward beyond the Mississippi to the Rocky Mountains and has, apparently, been introduced to the State of Washington (Johnson, 1952). In Michigan it is most frequent in damp basements.

In southern India the common house centipede *Scutigera longicornis* is frequently found in the burrow of the common Indian gerbil *Tatera indica.* The centipede was present in 17 of the 26 burrows excavated. There were one to three per burrow. The burrows have two openings in addition to several blind tunnels and there is a central bed chamber 60–125 cm below the surface. The centipede is never found in the blind tunnels. During the day it is found 'sleeping' in or near the bed chamber but away from the *Tatera.* At 06.00 h it is at the mouth of the burrow but moves further down the burrow as the morning progresses (Muruzesan & Azariah, 1972). The authors suggested that the behaviour is associated with temperature: the temperature of the burrow can be 19–20 °C below that outside.

There is a conspicuous lack of ecological data for scutigeromorphs. Their distribution in the tropics seems very patchy. They seem rare in Nigeria and the rain forests of Sarawak but common in the stony Red Sea Hills of the Sudan and in crevices in silt on the Nile bank at Khartoum (Lewis, unpublished observations). Since their long legs must restrict them to crevice habitats there are presumably many areas which do not offer suitable refuges.

Johnson (1952) noted differences in behaviour between a male and female *S. coleoptrata* in a terrarium in which a stratum leaf mould was piled to give surface plane of about sixty degrees. The female was always found higher than the male either on the leaf mould or clinging to the glass walls. The male generally selected a lower corner where the sticks and stems were loose. Much more pugnacious and venturesome, the alert female moved about relentlessly. Upon encountering the male, she explored him quickly with her antennae while he remained motionless except for reciprocal antennal explorations.

When not hunting or awaiting prey, both sexes were invariably busily occupied cleaning their bodies, passing the appendages one after the other through their mandibles removing adhering dust, water droplets, and mites. The entire cleaning process lasted about 20 min, following each short excursion around the terrarium. The cleaning started with the antennae and ended with the anal appendages; the female gonopods were also cleaned. After 35 days the male moulted and was eaten by the female.

Population density

Population density and biomass

Table 23 shows the available data on the density of specifically identified centipedes expressed as number per square metre together with biomass wet weight where this is available. A number of other studies have been carried out in which the centipedes were not identified (Auerbach, 1951; Bornebusch, 1930; Huhta & Koskenniemi, 1975; Salt, Hollick, Raw & Brian, 1948; Starling, 1944; Vaitilingham, 1960) producing densities varying from 1.8 to 648 animals/m^2. Bornebusch (quoted from Auerbach) gave figures varying from 7.0 in beech mull to 273 in beech humus: Salt, Hollick, Raw & Brian (1948) gave numbers of 327 and 648 for pasture in southern England in November and May respectively. The samples from pasture were extracted by flotation and consisted largely of small geophilomorphs.

The data are not strictly comparable as various extraction techniques were used and sample sizes varied considerably. Hand sorting is probably a very poor method especially for small species where gross underestimates may be obtained. Nevertheless a few generalisations are possible, namely that very high densities may be reached by small geophilomorphs and that the numbers of the larger *Lithobius* spp. do not appear to exceed 20/m^2. The density of small species of *Lithobius* can be much higher (250 in *L. duboscqui*) but since small species are often patchily distributed overall densities over a large area may be much lower.

In three samples of 0.25 m^2 from larger beds of the lichen *Cladonia* on shore rocks in Finland 12, 18 and 23 *Pachymerium ferrugineum* were collected and in similar samples from beds of *Rachomitrium* the numbers were 31, 38 and 52. Larger beds have relatively more individuals than small ones (Palmen & Rantala, 1954). These gives figures of 48–204 individuals/m^2, but the geophilomorph probably aggregates in the lichen beds.

The biomass of centipedes seldom appears to rise above 500 mg wet weight/m^2 (Table 23). Huhta & Koskenniemi (1975) gave figures of 252 and 96 mg/m^2 for total centipedes in pine woods in southern Finland and 33 in northern Finland. Albert (1976) demonstrated that there is a correlation between the biomass of *Lithobius curtipes* and the abundance of spiders. Where the biomass of *L. curtipes* is high, spider abundance is generally low and vice versa. The mean biomass of *L. mutabilis*, 113 mg/m^2 dry weight and *L. curtipes*, 38 mg/m^2 (total, 151 mg/m^2) is very similar to the biomass of spiders (165 mg/m^2) and predaceous beetles (100 mg/m^2 for both staphylinids and carabids). Roberts (1956) suggested that lithobiids may well replace staphylinids and spiders in damper habitats.

Wignarajah (1968) studied the energetics of *L. forficatus* and *L. crassipes* in a northern English woodland. In both species the proportion of energy flow through the juveniles is low, adults making a major contribution to energy flow in total population. His results are summarised in Tables 24 and 25.

Factors controlling population density

Roberts (1956) concluded that weather and predators were two of the more important factors controlling the numbers of *L. variegatus* but suggested that the low numbers of *L. forficatus* in a damp woodland were due to some additional factor not directly related to microclimate. Lewis (1965) concluded that the similarity in the food of these two species would seem to eliminate this factor as being important in influencing their distribution.

Lewis (1966) suggested that parasites might be responsible for

Table 23. *Population density and biomass of centipedes*

	Density/ m^2	Biomass wet weight (mg/m^2)	Habitat	Locality	Extraction technique	Authority
Geophilomorpha						
Necrophloeophagus longicornis	47.6	—	Arable soil	S. England	Hand-sorting	Morris (1922)
	0	—				
	26.9	—	Arable soil	S. England	Hand-sorting	Morris (1927)
	42.3	—				
	4.8	—	Pasture soil	N. Wales	Hand-sorting	Edwards (1929)
	67.3	—	Arable soil			
	25.6	—	Pasture soil	N. Wales	Hand-sorting	Thompson (1924)
Brachygeophilus truncorum	675	—	Woodland soil	S. England	Tullgren funnel	Roberts (1956)
	908	—				
	399	—				
Geophilus carpophagus	34.6	—	Woodland	S. England	Tullgren funnel	Roberts (1956)
Geophilus rubens	10–12	—	*Acer* and *Quercus* forest	Michigan, USA	—	Johnson (1952)
Geophilus strigosus McNeill	1.9–3.3	—	Old fields and woodland soils	Ohio, USA	Washing	Dowdy (1944)
Strigamia acuminata and *S. crassipes*	2–8	—	Woodland litter	S. England	Tullgren funnel	Vaitilingham (1960)
Schendyla nemorensis and *B. truncorum*	172	—	Beech litter	Holland	Tullgren funnel	Drift (1951)

Scolopendromorpha						
Scolopendra morsitans	0.16	140	Cotton soil	Lake Chad, Nigeria	Flooding	Lewis (1972*b*)
Lithobiomorpha						
Lithobius variegatus	1.6–9.6	20–145	Beechwood soil and litter	S. England	Tullgren funnel	Roberts (1956)
	3.25		Beechwood litter	S. England	Hand-sorting	Roberts (1956)
Lithobius forficatus	2–18	—	Beech litter	S. England	Tullgren funnel	Vaitilingham (1960)
	0.1	—	Beech litter and soil	S. England	Tullgren funnel	Roberts (1956)
	1.4	—	Sweet chestnut litter and soil			
	1.65	—	Oak litter	S. England	Tullgren funnel	Vaitilingham (1960)
	8.2–19.8	157–669	Birch-alder litter and logs	N. England	Tullgren funnel	Wignarajah (1968)
Lithobius crassipes	23–52	96–279	Birch-alder litter and logs	N. England	Tullgren funnel	Wignarajah (1968)
Lithobius curtipes	2–42	—	Litter	S. England	Tullgren funnel	Vaitilingham (1960)
	32	113	Beech litter	N. Germany	Kempson apparatus	Albert (1976)
Lithobius mutabilis	41	339	Beech litter	N. Germany	Kempson apparatus	Albert (1976)
Lithobius duboscqui	92–250	—	Soil and beech litter	S. England	Tullgren funnel	Roberts (1956)
	155	253				
	14	20				
Lithobius bilabiatus Wood	0.87	—	Woodland soil	Ohio, USA	Washing	Dowdy (1944)

the lack of success of *L. forficatus* in damp habitats but Harry (1970) regarded this hypothesis as untenable (see Chapter 20).

Nothing is definitely known of the factors controlling the number of centipedes. In *L. variegatus* 70 per cent of the mortality takes place in the larval stadia in the summer. There is also a high mortality in the maturus stadium after oviposition in May and June reducing its population by 75–80 per cent (Roberts, 1956).

Table 24. *Population density and biomass of* L. forficatus *and* L. crassipes *in log and litter in birch-alder woodland in northern England* (*from Wignarajah*, 1968)

	Mean number /m²	Mean live weight (mg/m²)	cal/m², ash free
L. forficatus			
Anamorphic stadia	3.1–5.5	8.2–14.4	10.8–19.2
Epimorphic stadia	4.8–11.9	124.7–308.6	162.1–419
Adults	0.27–3.9	24.4–350.7	36.8–545.2
L. crassipes			
Anamorphic stadia	3.99–14.3	2.9–10.3	4.0–14.5
Epimorphic stadia	9.6–17.5	33.4–61.2	49.1–87.9
Adults	3.88–19.7	40.9–207.1	63.0–328.4

Table 25. *Energy budget of* L. crassipes *and* L. forficatus *from Wignarajah* (1968), *in cal/m²/annum*

	Mean standing crop per annum	Respiration of standing crop	Production of standing crop	Assimilation of standing crop
L. crassipes	322	1268	74	1341
L. forficatus	535	1998	85	2083

22

Taxonomy

Centipede taxonomy is beset with problems. There is a paucity of adequate monographs so that whereas some areas of the world are well catered for, for example the British Isles (Eason, 1964), France (Brölemann, 1930), South Africa (Lawrence, 1955) and Madagascar (Lawrence, 1960), others are almost totally neglected. Attems' monographs on the Geophilomorpha (1929) and Scolopendromorpha (1930*a*), the former brought up to date by Attems in 1947, provide invaluable starting points for workers in these orders. Lists of new species are given in the *Zoological Record* and the Centre International de Myriapodalogie in Paris publishes annually a *Liste des travaux parus et sous presse* for the previous year but this gives titles only. There are no monographs for lithobiomorphs or scutigeromorphs and as a result these orders may well have been neglected by some taxonomists.

Many taxonomic descriptions are inadequate because too few characters have been taken into consideration: it is impossible to come to any objective conclusion about the relationship of some African species of *Lamyctes* (Lithobiomorpha) because the characters used to diagnose one species by one author are not used at all in the description of a second species by a second author. Difficulties arise also when the species is described on the basis of a few specimens or when the species is very widely distributed and known from a few widely separated localities. As further material of such species is examined, their status has to be re-evaluated.

Not all centipede taxonomists have appreciated that since subspecies are not reproductively isolated they interbreed when they meet and are therefore unlikely to occur as distinct populations in one locality. 'Subspecies' occurring in one locality are more likely to be good species or individual variants of one species.

As there is no metamorphosis in centipedes, immature stadia resemble adults and are sometimes described as separate species. Other causes of taxonomic difficulty are wear, damage and sexual dimorphism. The problems for each order are dealt with below.

Geophilomorpha

The more important diagnostic characters for geophilomorphs are the shape of the head capsule, the characteristics of the labrum, mouthparts and forcipules, the nature and distribution of the sternital pore fields and fossae, and the distribution of coxal pores and other characteristics of the last pair of legs. Leg number is also important. It is frequently necessary to dissect out the mouthparts and integumental characters can only be seen clearly after the specimen has been cleared in 60 per cent lactic acid.

A number of the above-mentioned characters change with the age of the specimen: immature animals have fewer setae, fewer coxal pores and fewer sternital fossae than mature animals and are sometimes described as separate species. Bagnall (1935) described two small British species: *Geophilus pusillimus* and *G. anglicanus* both with a single coxal pore. Blower (1955*b*) suggested that the former was an immature *Strigamia acuminata*. The latter was probably an immature *Brachygeophilus truncorum*. Any small geophilomorph with one coxal pore is likely to be a young adolescens stadium and should be treated with great caution by the taxonomist.

A few species show marked sexual dimorphism, the most extreme case being that of the Himalayan genus *Tygarrup*, males of which were placed in a separate genus *Brahamaputrus* by Verhoeff (Crabill, 1968).

Some widely distributed geophilomorphs are highly variable. Northern European populations of *Pachymerium ferrugineum* have 41 to 47 pairs of legs, specimens from Palestine 67 or 69 pairs. The designation of subspecies on the basis of leg numbers alone is a procedure fraught with danger.

Less problems arise from wear and damage in geophilomorphs than in scolopendromorphs (see below) but in species with teeth along the inner edge of the poison claw, erosion may take place.

Lewis (1962) found that six of the ten specimens of *Hydroschendyla submarina* (Grube) the inner edge of the poison claw was smooth: in taxonomic works it is described as being crenulated.

As yet unexplained is the occasional appearance of very large specimens in a particular species. *Geophilus carpophagus* usually reaches a maximum length of 40 mm but may be as long as 60 mm (Eason, 1964).

Few suggestions have been made about new taxonomic characters in the Geophilomorpha which may indicate that their taxonomy is in a relatively satisfactory state. Recently, however, Misioch (1978) has shown that common European centipedes, for example *Necrophloeophagus longicornis, Geophilus carpophagus* and *Brachygeophilus truncorum* are far more variable than the standard works suggest. Demange (1961) used the arrangement of the tracheal system as a major distinguishing character of his new species *Orya panousei* and referred to an unpublished thesis by Rubatat in which the importance of the tracheal system in taxonomy is demonstrated. Crabill (1969) considered 'tracheotaxy' as a generic criterion in the Himantariidae. A character which may be of taxonomic value is the distribution of sensory pegs on the antennae. These vary considerably from species to species.

There are many areas of the world where it seems unlikely that our knowledge will ever be sufficient to make an objective assessment of the true status of some closely related 'species'. Such may be the case with three species of *Pleuroschendyla* from central Nigeria: '*P. hausa*' from Kaduna, '*P. robertsi*' from Minna, a town 70 miles south of Kaduna and '*P. guiniensis*' from Bida, 110 miles south of Kaduna (Lewis, in manuscript). The three populations show consistent but small differences. It is possible that they are part of a cline (a series of adjacent populations in which a gradual change of characters occurs) but very extensive collection would be required to resolve this point.

Scolopendromorpha

The mouthparts of scolopendromorphs are little used in specific discrimination although the characters of the forcipular segment are important. A large number of the characters used in scolopendromorph taxonomy change with the age of the animal.

Lewis (1968) showed that in *Scolopendra morsitans* the total number of antennal segments, the number of glabrous (hairless) basal antennal segments, of tergites with complete paramedian sutures, of marginate tergites and of forcipular coxopleural teeth change with age, as does the coloration of the living animal. A further cause of variation is the loss and subsequent regeneration of appendages: regenerated antennae may have abnormally high or abnormally low numbers of segments and when the last pair of legs is regenerated the spine pattern is frequently atypical (Fig. 27). Kraus (1957) suggested that abnormal spinulation of the last pair of legs in a population of *Digitipes katangensis* Kraus from Zaire was caused by the radioactivity of the rocks in the area but it could well have been due to regeneration. In several species the forcipular coxopleural teeth may present an eroded appearance which is presumably due to wear.

Sexual dimorphism is another cause of variation. In *Scolopendra morsitans* the dorsal side of the prefemur and femur of the last pair of legs is flattened in adult males and has a marginal ridge (Fig. 160) and males tend to have a higher number of marginate tergites than females (Lewis, 1969*b*). Such differences may not always be ascribed to sexual dimorphism as the external genitalia of males are often concealed within the anal segment so that the sex may not be determined. Slight pressure will, however, often cause them to evert, thus allowing the sex to be determined.

Lewis (1978) pointed out that decisions as to whether a paramedian suture is complete or incomplete, whether a tergite does or does not show signs of margination or whether an antennal segment is or is not glabrous depends on the opinion of the observer. The appearance of the character may vary depending on whether the specimen is examined in spirit or is dry and may, as is the case with the sulcus on the femur of the last pair of legs in *Asanada* spp., vary in appearance with the angle at which they are illuminated (Lewis, 1973). An important character in separating species of *Otostigmus* is the presence or absence of 'spicules' on the posterior tergites; these are very difficult to see in specimens viewed under artificial light but are clearly visible in natural light (Lewis, unpublished data). Scolopendrids are best examined from several angles under both natural and artificial light after the spirit in which they have been preserved has been blotted off.

Perhaps the most problematical scolopendromorph is *S. amazonica* (Bücherl). This was first described as a subspecies of *Scolopendra morsitans* L. from Brazil and was elevated to specific status by Jangi (1959) referring to the smaller of two sympatric forms of *S. morsitans* from India. The two species were separated by a number of characters, the most important of which was the presence of a tarsal spine on the twentieth pair of legs in *S. morsitans*. African populations lack this spine but in other respects may resemble *S. morsitans* (Lewis, 1969*b*): it is clear that the two species do not occur in Africa. A number of workers regard *S. morsitans* as being highly polymorphic and *S. amazonica* invalid, yet if Jangi is to be believed, two species of *S. morsitans*-like scolopendras occur in India. Arguments about the correct nomenclature for these populations are unlikely to be particularly fruitful at this stage of our knowledge of their geographical variation.

Würmli (1975, 1978) concluded that the characters of '*morsitans*' and '*amazonica*' are not correlated with each other but combine randomly: *S. morsitans* being a polymorphic species.

It has been suggested that the characteristics of the male reproductive system could be used in the taxonomy of scolopendromorphs (Demange & Richard, 1969).

Lithobiomorpha

The number and disposition of the spines or spurs on the legs is of major importance in the taxonomy of the Lithobiomorpha. There can be a maximum of three dorsal and three ventral spines at the end of the coxa, trochanter, prefemur, femur and tibia (Fig. 93). Formulae are used to describe the distribution (spinulation or plectrotaxy) of these spines. Crabill (1962) discussed the significance of spinulation: Ribaut (1921) showed that once a spine appeared on an anterior leg its serial homologues would be present without interruption on all succeeding legs until the posterior limit of the series is reached. He noted that the posterior limit of a particular spine was more diagnostic of a species than the anterior limit, this is because the anterior limit of the spine series varies with age, the older the specimen the farther forward the spur series extends. Crabill pointed out that a praematurus form of one species can be confused with a maturus specimen of another closely related species. He had 'no doubt that many species owe their existence to this very error ... an immature *Nadabius aristeus* Chamberlin could be confused with a mature *N. pullus* (Bollman), if other non-plectrotaxic criteria were discounted ... there are many lithobiid species, so called, distinguishable from other species solely on the basis of plectrotaxy'.

Having considered three *Nadabius* species, Crabill concluded that the posterior limits of spine series are more indicative of categories above the species level (species group, subgenus, genus) than solely of species, as Ribaut contended, and used with care the anterior limit of a particular spine series was a good specific criterion.

Variation in spinulation occurs within species, thus in *Lithobius duboscqui* Brölemann in addition to minor individual differences there is a variety *L. duboscqui caernensis* Eason without spines on any of the legs: it occurs in north and mid-Wales, in some parts of which it appears to be the predominant if not the only form of the species (Eason, 1964). Crabill (1962) stated that 'many forms that are today viewed as discrete species purely on the basis of plectrotaxy will eventually be recognised as intraspecific variants'. There are 'dangerous possibilities inherent in designating a new form solely on the basis of a single specimen's ventral plectrotaxic correlative criteria. Nonetheless, this very practice has plagued lithobiid systematists in the United States and abroad in the past and it still enjoys great favor today'.

Eason (1970*a*) has investigated certain non-plectrotaxic characters in *Lithobius*. After examining 50 adult female *Lithobius forficatus* from Iceland, he suggested that the variation observed in the form of the claws and spurs of the gonopods (Fig. 218*a–e*), might be due to wear or damage. Forms such as *L. forficatus sorrentinus* Verhoeff and *L. forficatus cheruscus* Verhoeff characterised by such gonopodal variation are thus simply individual variants. Examination of a series of *Eupolybothrus grossipes* (C. L. Koch) from the southern Tyrol of Austria and northern Italy showed that immature post-larval stadia have narrow posterior projections on the short tergites and deeply emarginate posterior borders on the large tergites so that the posterior angles appear to project (Fig. 218*f–j*). As development proceeds the projections tend to get broader and the emargination shallower. Verhoeff described three forms of *E. fasciatus* (= *grossipes*) depending on this and other immature characters. They are obviously invalid (Eason, 1970*a*).

In addition to problems of specific and subspecific discrimination, the generic classification of the Lithobiidae is in disarray owing to a lack of co-ordination between the systems used in America and Europe and a failure on the part of Europeans to accept one another's views. The lithobiids are so difficult to group that on the one hand Attem's genera *Alokobius*, *Monotarsobius* and *Pokabius* are so large that their distribution gives little more information than does that of the family as a whole, whereas many of Chamberlin's genera are too small to give any more information than that given by the distribution of a single species (Eason, 1974). The secondary sexual modifications in the posterior legs of males (see Chapter 17) must be used with caution for they are not all homologous and even very similar structures on the same segment

Fig. 218. *Lithobius forficatus* female gonopods. *a–c*, Iceland; *d*, Scotland; *e*, normal *Eupolybothrus grossipes*; *f–j*, successive stages in the development of the thirteenth tergite (after Eason, 1970).

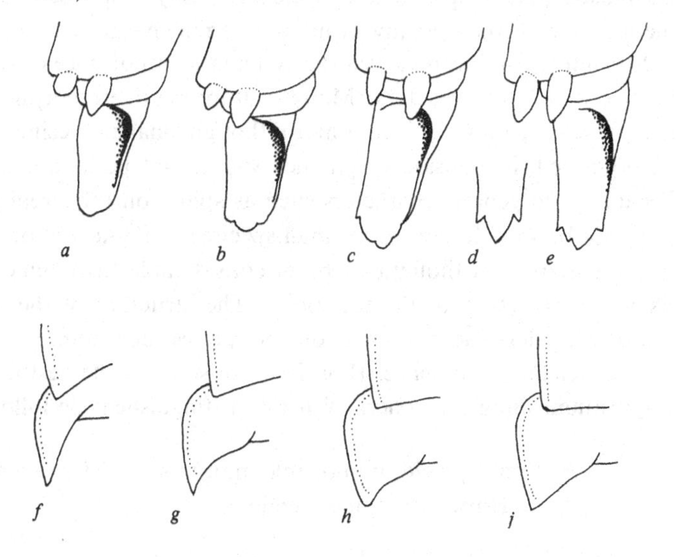

of the same leg, Eason considered, may have been acquired independently.

There are signs of a revival of interest in lithobiomorph taxonomy: in a recent study of the *Lithobius* species from Mt Canigou in the Pyrenees, Tobias (1974) found the pigment pattern of the head capsule to be of considerable taxonomic value. Unfortunately most taxonomy is still carried out on museum specimens so that such characters are likely to be of limited value for some time to come.

It seems likely that the number of European species now recognised will be reduced: Borek (1967) has shown that *Lithobius dadayi* Tömösvary and *L. rupivagus* Verhoeff which were differentiated from *L. lucifugus* L. Koch on the basis of antennal segment number are merely varieties or possibly subspecies of *L. lucifugus.*

Scutigeromorpha

The taxonomy of the scutigeromorphs is less advanced than that of the lithobiomorphs due to an extreme shortage of material. This is due to the agility of the animals and possibly to patchy distribution. The animals are very delicate so that captured specimens are always more or less damaged. The legs are readily automised and a specimen collected clumsily can lose many of them and with them, many of its taxonomic characters.

Würmli (1974) has reviewed the systematic importance of various scutigeromorph characters. Many mature scutigeromorphs measure at least 20 mm in length and although smaller specimens can be described, he considers that they should not be named as new because even generic characters such as spines on the tergal plates may not be well-developed in small specimens. Pigmentation of the living specimen is thought to be of considerable taxonomic value as is the structure of the forcipules. The structure of the female gonopods offers the best criterion for species separation.

Exoskeletal prominences (Fig. 219) are some of the best features for distinguishing the genera. Würmli distinguished the following:

A. Simple, rigid, immovable, non-innervated prominences
   1. Hairs: short and slender.

2. Spinules, spinulae (*Haardornchen*): short sturdy hairs (Fig. 219*a*).
3. Spiculae (*Haarspitzen*): spike-shaped long hairs (Fig. 219*b*).
4. Short spiculae.
5. Spines, spinae: very often associated with bristles (Fig. 219*c*).

Fig. 219. Spines and setae of scutigeromorphs. *a*, spinulae from the tergal plate of *Psellioides* sp.; *b*, spicule from the tergal plate of *Thereuonema* sp.; *c*, spine and bristle of the tergal plate of *Thereuonema*; *d*, spine-bristle from the second maxilla of *Allothereua* sp.; *e*, spines and a bristle from the tergal plate of *Allothereua* sp., and *f*, of *Thereuonema* sp.; *g*, *tarsale sinuatum* of second tarsus, eighth leg of gen. n., sp. n. (after Würmli, 1974).

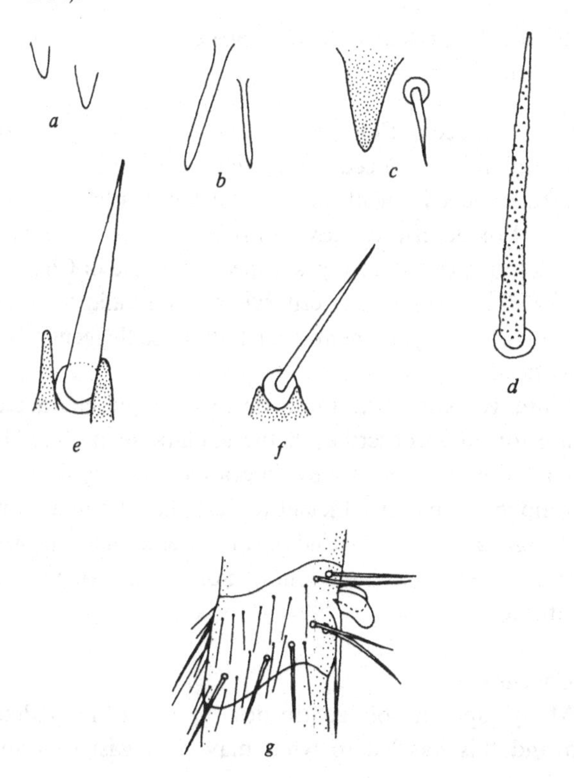

B. Moveable, innervated prominences
  1. Bristles, setae (Fig. 219*c*).
  2. Spine-bristles, spinosetae: enlarged, thickened bristles. The microsculpture on the spine-bristle (especially of the second maxilla) (Fig. 219 *d*) can be a criterion for the distinction of species.

The exoskeletal prominences on the antennae and legs (Fig. 219) may also prove to be of value: Würmli distinguished:

1. Hairs (moulting hairs, *Hautungshaare*).
2. Bristles.
3. 'Resilient sole hairs' (*federnde Sohlenhaare, Crines appressi subpedales*): hairs thickened at the base, pressed against the under surface of the article and extending to the middle of the under surface of the next one (Fig. 219*g*).
4. Tarsal papillae (*Tarsal zapfen*): often very specific in form.

Würmli concluded: 'For promoting the taxonomy of the Scutigeromorpha we need investigations into variability, post-embryonic development and allometry, ecology, behaviour and genetics. For descriptive taxonomy it is urgently necessary to find and redescribe the older types (especially those of Chamberlin). The search for new systematic criteria that are valid for all stadia must continue (possibly pigmentation?). Perhaps the generic classification may need revision'.

'We do not yet know the importance of single characters and thus we are forced to describe all the somatic features... It does not seem advisable to erect new species on the basis of female gonopods alone .... it is insufficient to describe a limited number of segmental organs .... it is not advisable to erect new species from isolated males (even mature ones) nor from younger female pseudomaturae'.

Conclusion

Many species of centipedes have been inadequately described and this has led to what may be a vast synonomy. A number of recent studies have, however, led to an increase in our knowledge of the causes of variation in centipedes and allow of a re-evaluation of taxonomic characters. Now seems to be the time for revision and consolidation rather than continued description of new species, a process all too easy in some families. When the anarchy left, perhaps unavoidably, by previous authors has been eliminated the time will again come for the description of new material.

23

Relationships of the chilopod orders

Verhoeff (1902–25) gave a detailed review of the development of ideas on the relationships of the chilopod orders from Aristotle and Pliny onwards. The term Chilopoda was introduced by Latreille (1817) and Newport (1844) erected the families Scutigeridae (Cermatiidae), Lithobiidae, Scolopendridae and Geophilidae. Brandt (1841) separated the scutigerids as a family Schizotarsia from the remaining centipedes (Holotarsia).

Haase proposed an alternative division of the centipedes into the Anamorpha and Epimorpha. The Anamorpha (Lithobiomorpha and Scutigeromorpha) may be briefly characterised as having 15 heteronomous leg-bearing segments, long or very long legs, female genitalia forceps-like and larvae which hatch with four or seven pairs of legs: Tömösváry organs are present. The Epimorpha (Scolopendromorpha and Geophilomorpha) have an elongated body with at least 21 leg-bearing segments, legs short to long, female genitalia palp-like or absent and the young hatching with a full complement of legs: Tömösváry organ absent. Misioch (1978) suggested that there was some post-embryonic increase in leg number in *Necrophloeophagus longicornis* and *Geophilus electricus* but this does not materially alter the concept of anamorphs and epimorphs.

Pocock (1902) proposed a subclass Pleurostigma for centipedes with lateral spiracles and a subclass Notostigma for the scutigeromorphs with their dorsally opening tracheal system. Verhoeff (1902–25) used the terms Pleurostigmophora and Notostigmophora. He classified the subclass Pleurostigmophora thus:

Superorder Anamorpha
- Order Heterostigmata
  - Suborder 1. Craterostigmomorpha
  - Suborder 2. Lithobiomorpha

Superorder Epimorpha
Order Scolopendromorpha
Order Geophilomorpha

Verhoeff regarded the Scutigeromorpha and Geophilomorpha as being distantly related but he recognised similarities between the Scutigeromorpha and Lithobiomorpha. He saw the homonomous segmentation, well-developed intercalary segments, low antennal segment number, poorly developed sense organs, small head and the structure of the legs in geophilomorphs as primitive characters: the structure of the mouthparts, poison claws and tracheal system as advanced ones. He regarded the scolopendromorphs and lithobiomorphs as intermediate in position: with the heteronomy, the well-developed coxae, numerous antennal segments and ocelli of the lithobiomorphs as advanced characters.

The scutigeromorphs were, for Verhoeff, the culmination of centipede evolution, showing similarities to the insects with respect to their legs and compound eyes. He listed many morphological features characteristic of the group. He regarded the second maxillae and forcipules as primitive.

Verhoeff visualised the primitive centipede as having 21–23 segments, with spiracles on all segments except the last. The head was anamorph-like, the second maxillae and forcipules as in *Scutigera*, the legs and intercalary segments as in geophilomorphs, genital appendages in both sexes and anamorphic development.

Attems (1930*b*) believed the division of the Chilopoda into Anamorpha and Epimorpha to be phylogenetically more ancient than the division into Notostigma and Pleurostigma. He included *Craterostigmus* as a suborder of the Lithobiomorpha.

Brölemann (1930) believed that the evolution of myriapods was dominated by the process of tachygenesis or shortening of the body which was correlated with early maturation. He considered that the division of the centipedes into Anamorpha and Epimorpha was unjustified, regarding the differences as being due to the time of hatching, an arbitrary criterion on which to divide the class in his opinion. He preferred to regard anamorphic development as a further step in the process of the contraction of development with consequent shortening of the body.

Brölemann regarded the Geophilomorpha with their variable

number of homonomous segments, small head and presence of intercalary segments as the most primitive centipedes. The Scutigeromorpha, he considered to be highly advanced. The separated coxae of the maxillae and forcipules were, he thought, due to the reappearance of larval characters. He considered *Craterostigmus* to be a scolopendromorph undergoing contraction, which had not reached that degree characteristic of the Lithiobiomorpha.

Fahlander (1938) discussed the relationships of the chilopod orders in considerable detail: he omitted the Craterostigmomorpha considering that there was insufficient data available for their relationships to be established with certainty. He was of the opinion that the compound eyes of the Scutigeromorpha were not fundamentally different from those of the Crustacea and Insecta: the Scutigeromorpha could be the most primitive centipedes and he considered that Verhoeff's separation of this order under the Notostigmophora was valid. He pointed out that the swollen head capsule, the anteriorly situated mouth and dorsally sited antennae need not, as Verhoeff had suggested, be advanced characters. He doubted that author's assertion that the Notostigmophora had more highly developed sense organs than the Pleurostigmophora, for, with the exception of the eyes and the shaft organ the sensilla are poorly developed. He saw the reduction of the eyes, flattened head and the development of antennal sense organs in the Pleurostigmophora as the result of the adoption of a cryptozoic mode of life. The reduction in the number of head glands and the maxillary nephridia and tentorium would be correlated with this. He agreed that the maxillae and forcipules were primitive in *Scutigera* and held that the Tömösváry organs were also primitive. Tömösváry organs are also present in Diplopoda, Symphyla and primitive insects. He also believed the gonopods of scutigeromorphs and lithobiomorphs were primitive.

He saw the fusion of tergites, the structure of the antennae, legs, blood system, respiratory system and absence of coxal glands in *Scutigera* as advanced characters, as was the fusion of the coxa of the gonopods of the female with the sternites.

Fahlander drew attention to the fact that both a large and small segment number in centipedes had been regarded as a primitive character. He suggested that brooding might be a secondary development and that the reduction of the gonopods and associated glands in epimorphs might be associated with this. He believed that primitive centipedes showed a weak heteronomy and spiracles on each segment as in *Plutonium.*

The Lithobiomorpha, he regarded as the most primitive Pleurostigmophora, sharing with the scutigeromorphs the primitive characters of maxillary kidneys, Tömösváry organ, well-developed female gonopod, 15 pairs of legs and hemianamorphic development. In the Epimorpha, Fahlander saw the large number of testes and tergite heteronomy in the scolopendromorphs as primitive characters. The segmentally arranged spiracles and two-segmented gonopods of geophilomorphs were primitive, the presence of only one pair of testes and the absence of eyes, secondary.

Manton's work on the morphology of centipedes led her to the conclusion that a large number of segments and epimorphic development represent a primitive condition in the group. An animal roughly resembling a geophilomorph but with a small number of segments (about 30–35) and feeble powers of telescoping the sclerites and stretching the pleural region would, she considered, have the potentiality of developing into either a geophilomorph or a scolopendromorph. Manton considered that it was possible that the evolution of tergite heteronomy, the reduction in the number of spiracles and the evolution of 15 pairs of legs may have arisen independently in scolopendromorphs and lithobiomorphs (Manton, 1952*b*).

Manton (1965) believed that the affinities of *Craterostigmus* lie with the Epimorpha rather than the Anamorpha for the following reasons: the presence of intercalary tergites and sternites, the general form of the limbs, a number of characteristics of the muscle arrangement and the epimorphic gait. Further epimorph characters are the basic form of the forcipular segment and its musculature, the lack of gonopods in the female and the brooding habit of the female over the egg mass. She believed that *Craterostigmus* shows particular resemblance to the Scolopendromorpha. It probably evolved from a fleet ancestor with pronounced tergite and muscle heteronomy: the presence of 15 pairs of legs is convergent with the Anamorpha.

Crabill, whose views are quoted very briefly in a footnote in Kaestner (1967), does not accept Manton's views. He believes *Craterostigmus* to belong to the Anamorpha close to the Lithobiomorpha. The evidence includes the anamorphic development, construction of the forcipules and the previously overlooked organs of Tömösváry which are always present in the Anamorpha but never in the Epimorpha. Crabill (in press) demonstrates that *Craterostigmus* is, as Attems (1926) believed, clearly related to the Lithobiomorpha.

Demange (1967) regarded the Geophilomorpha as the most primitive chilopods and the Scutigeromorpha as the most advanced.

Prunesco (1965*b*) regarded the scolopendromorphs, with their multiple pairs of testes, as more primitive than either the geophilomorphs or the lithobiomorphs. He believed the homonomy of the Geophilomorpha to be secondary, pointing out that a certain degree of heteronomy occurs in other arthropod groups. The appearance of spiracles on segments 2–21 in *Plutonium* was secondary and *Craterostigmus* was clearly intermediate between anamor-

Fig. 220. *a*, relationships of the orders of chilopods (after Prunesco, 1970*b*); *b*, relationships of the orders of chilopods based on the most recent evidence.

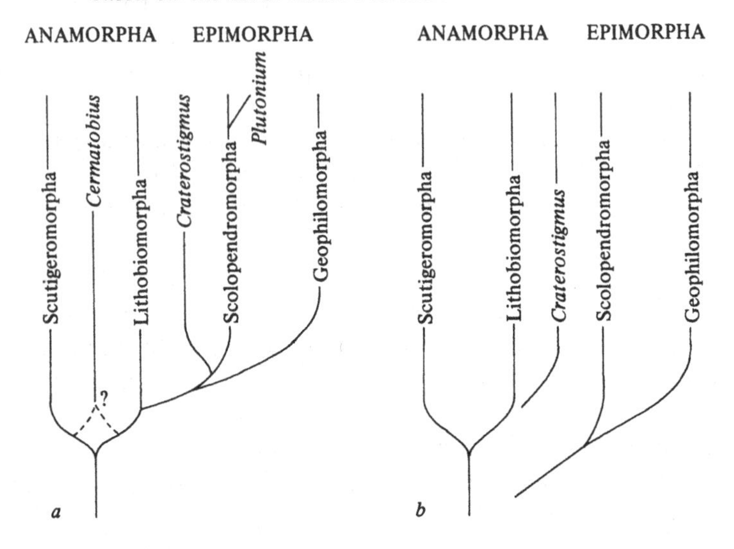

phic and epimorphic centipedes. He regarded it as a possibility that the small and large tergites of *Craterostigmus* might be equivalent to the diplosegments of millipedes. He considered the lithobiomorphs and scutigeromorphs to be the most primitive orders of centipedes.

It has recently been shown that *Cermatobius*, the sole occupant of the lithobiomorph family Cermatobiidae and considered by Prunesco (1970*b*) to be an early offshoot of either the Scutigeromorpha or the Lithobiomorpha, is an *Esastigmatobius* (Henicopidae) (Würmli, 1977).

Prunesco's (1970*b*) phylogenetic tree is shown in Fig. 220*a*. Essentially the same relationship of orders was suggested by Shinohara (1970) after a consideration of the appendages, tracheal systems, segmentation, genital system, karyotypes and post-embryonic development of centipedes. Shinohara regarded the primitive chilopod as heteronomous.

Conclusions

Although the views of the authors reviewed above are very divergent, some measure of agreement between them can be seen. Clearly the scutigeromorphs show a number of unique characters suggesting an early separation from other centipedes but there are striking similarities with the lithobiomorphs which I feel may indicate a common ancestry. Scutigeromorphs undoubtedly retain primitive characters, in particular the mouthparts, head glands, nephridia and tentorium but show specialisations in the legs and respiratory system. It has been argued both that the compound eye of *Scutigera* is primitive and that it is advanced. Paulus (1979) regards the eye of *Scutigera* as a secondary, having been derived from the eye of *Lithobius* and thus not homologous with the faceted eyes of insects and Crustacea. The homology of the *Strinocellen* of *Scutigera*, described by Knoll (1974), is not clear.

It is usually assumed that scutigeromorphs have evolved as fast runners but there is little evidence that they chase prey. The conflicting accounts of feeding suggest that the species pounces on its prey rather than chases it. The legs may have evolved as sensory structures rather than rapid locomotors: long legs are also seen in cavernicolous lithobiids. Scutigeromorphs appear to be found mainly under stones, in crevices of rock or mud: *S. coleoptrata's* preferred habitat is caves.

The scutigeromorphs may then have evolved as wide crevice dwellers possibly retaining or developing the compound eye for use in dull environments.

Craterostigmus is related to the lithobiomorphs.

There seems to be general agreement that the epimorphic orders are related and that the geophilomorphs are highly specialised forms and not primitive. It seems difficult to avoid concluding that the geophilomorphs must have evolved from heteronomous ancestors. The relatively localised distribution of many geophilomorph genera and species suggests that they are evolving relatively rapidly.

As to the early stages in the divergence of the chilopod orders there is not, in my opinion, sufficient evidence to decide which order is closest to the primitive archetypal centipede. The probable relationships of the chilopod orders are shown in Fig. 220*b*. No study of their relationships based on cladistic phylogenies has yet been made.

24

The classification of the Chilopoda

Opinions differ as to the correct classification of the Chilopoda. A simple classification is given here to show the family and order to which the genera mentioned in the text belong. All families are listed but not all subfamilies.

Subclass Epimorpha

Order Geophilomorpha (about 1000 species)

The classification of the Geophilomorpha given here is based on that of Attems (1929) with modifications based on Crabill (1960*c*, 1970). Attems distinguished ten families, namely the Himantariidae, Schendylidae, Oryidae, Mecistocephalidae, Geophilidae, Soniphilidae, Gonibregmatide, Sogonidae, Neogeophilidae and Azgethidae. Within the Geophilidae he distinguished the subfamilies Geophilinae, Dignathodontinae, Pachymerinae, Chilenophilinae and Aphilodontinae. Crabill (1960*c*) showed that *Azygethus atopus* Chamberlin for which the family Azygethidae was created is *Orphnaeus brevilabiatus* and thus the Azygethidae is a junior synonym of Oryidae. Crabill (1970) added a new family, the Eriphantidae and proposed several changes in the classification which are adopted here. The data on distribution are largely from Attems (1929).

Family Himantariidae

Himantarium (Mediterranean region, Madagascar), *Stigmatogaster*, *Haplophilus* (Europe, N Africa), *Meinertophilus* (Mediterranean region), *Mesocanthus* (Africa, India), *Nesoporogaster* (southern France, Balearic Islands), *Nothobius* (California).

Family Schendylidae

Schendyla (Europe, N Africa) (Crabill regards *S. nemorensis* as cosmopolitan), *Hydroschendyla* (Europe,

Bermuda), *Haploschendyla* (Mediterranean region), *Pectiniunguis* (N and S America), *Thalthybius* (Brazil, Seychelles, Mariana Islands, Cuba), *Lionyx* (N America).

Family Oryidae

Orya (Mediterranean region, Madagascar), *Orphnaeus* (pantropical), *Diphtherogaster* (S Africa).

Family Mecistocephalidae

Mecistocephalus (pantropical), *Dicellophilus* (Europe, N America, Japan).

Family Geophilidae (including Attems' Sogonidae and Soniphilidae)

Geophilus (Europe, N and S America, China, Australia), *Brachygeophilus* (Europe, N Africa, N America), *Necrophloeophagus* (Europe, Tunis), *Pleurogeophilus* (Europe, N America, Japan, N and E Africa), *Clinopodes* (Europe, Mediterranean region, Russia), *Arenophilus* (N America).

Family Chilenophilidae (including Attems' Pachymerinae) *Pachymerium* (Palaearctic, N and S America, S Africa, New Zealand), *Eurytion* (S Africa), *Arctogeophilus* (Europe, N America).

Family Eriphantidae

Eriphantes (Mexico).

Family Dignathodontidae

Strigamia (Europe, Caucasas, Japan, N America), *Henia* (Mediterranean region from the Canaries to the Caucasas), *Chaetechelyne* (France and Mediterranean region), *Mixophilus* (India).

Family Aphilodontidae

Philacroterium (S Africa)

Family Gonibregmatidae (Indo-Australian region)

Family Neogeophilidae (Mexico)

Order Scolopendromorpha (about 550 species)

The classification given here is based on Attems (1930*a*).

Family Scolopendridae

Subfamily Scolopendrinae

Scolopendra (all tropical and warm regions), *Trachycormocephalus* (Africa, India, Mesopotamia, Caucasas), *Cormocephalus* (all tropical and warm regions),

Campylostigmus (New Caledonia), *Asanada* (Middle East, Africa, India, Phillipines).

Subfamily Otostigminae

Otostigmus (tropics and subtropics), *Digitipes* (Congo), *Alipes* (Africa south of the Sahara), *Ethmostigmus* (Africa and Indo-Australian region), *Rhysida* (Africa, Indo-Australian region, S America), *Arrhabdotus* (Borneo).

Family Cryptopidae

Subfamily Cryptopinae

Cryptops (worldwide), *Paracryptops* (India, New Guinea, Guyana).

Subfamily Theatopsinae

Plutonium (Sicily, Sardinia, Campania), *Theatops* (N America, Mediterranean region, Sandwich Islands).

Subfamily Scolopocryptopinae

Scolopocryptops (N and S America, east Asia, Phillipines, Sunda Islands, New Guinea, New Zealand), *Newportia* (Central America, Antilles, S America).

Subclass Anamorpha

Order Lithobiomorpha (about 1100 species)

The classification of the Lithobiomorpha is in a state of total confusion: there appears to be no concensus of opinion as to the size of genera (Eason, 1974) or of families. The classification given here is based on notes kindly made available by Dr E. H. Eason.

Family Lithobiidae (North Temperate region). There are many genera, the North American ones not corresponding with the European ones.

American genera mentioned in the text: *Lithobius*, *Alokobius*, *Nampabius*, *Nadabius*, *Sonibius*, *Tidabius*, *Pokabius*, *Paitobius*, *Neolithobius*.

European and Asian genera mentioned in the text: *Lithobius*, *Harpolithobius*, *Dakrobius*, *Hessibius*, *Monotarsobius*.

Family Ethopolidae

Eupolybothrus (Europe), *Bothropolys* (N America, Asia).

Family Watobiidae (N America) (a family of doubtful validity)

Family Gosibiidae (N America and eastern Mediterranean) (a family of doubtful validity)

Family Pseudolithobiidae

Pseudolithobius (North America), *Osellaebius* (Turkey).

Family Pterygotergidae (a family of doubtful validity)

Pterygotergitum (Sinkiang).

Family Henicopidae (A well-defined family often given higher status. Essentially confined to the tropics and southern hemisphere. A few holarctic representative are cosmopolitan.)

Subfamily Henicopinae

Lamyctes (world-wide, often introduced), *Esastigmatobius* (Far East).

Subfamily Anopsobiinae

Anopsobius (S. America, New Zealand), *Catanopsobius* (Chile), *Dichelobius* (SW Australia, New Caledonia).

Cermatobius until recently the sole genus of the family Cermatobiidae and thought to show similarities to the Scutigeromorpha is in fact a synonym of *Esastigmatobius* (Henicopidae) (Würmli, 1977).

Order Craterostigmomorpha

Craterostigmus (Tasmania, New Zealand). The affinities of this strange genus lie with the Lithobiomorpha. There are at least two species (Crabill, personal communication).

Order Scutigeromorpha (about 130 species)

Family Scutigeridae

Scutigera (Mediterranean region, S Africa, N America, *S. coleoptrata* is often introduced), *Thereuopoda* (India, Ceylon, China, Japan, Borneo), *Thereuonema* (N Africa, Middle East, China, Japan), *Podothereua* (Bismarck Archipelago), *Allothereua* (Australia).

The numbers of species given above for each order are taken from Schubart (1960).

BIBLIOGRAPHY

Adensamer, T. (1894). Zur Kenntnis der Anatomie und Histologie von *Scutigera coleoptrata. Verh. zool.-bot. Ges. Wien*, **43**, 573–8.

Albert, A. M. (1976). Biomasse von Chilopoden in einem Buchenaltbestand des Solling. *Verh. ges. Okologie, Göttingen*, 93–101.

Amouriq, L. (1967). Rôle de l'éclairement dans le rythme nycthéméral d'activité motrice chez *Lithobius forficatus* (L.) (Lithobiidae, Chilopodes). *Revue Comp. Animal*, 57–62.

Andersson, G. (1976). Post-embryonic development of *Lithobius forficatus* (L.), (Chilopoda: Lithobiidae). *Ent. scand.*, **7**, 161–8.

Annandale, N., Brown, J. C. & Gravely, F. H. (1913). The limestone caves of Burma and the Malay Peninsula. *J. Asiat. Soc. Bengal* (NS). **9**, 391–424.

Applegarth, A. G. (1952). The anatomy of the cephalic region of a centipede *Pseudolithobius megaloporus* (Stuxberg) (Chilopoda). *Microentomology*, **17**, 127–71.

Aron, M. (1920). Sur l'existence d'une double spermatogènese chez le *Cryptops* (Myriapode). *C. r. Séanc. Soc. Biol.*, **83**, 241–2.

Attems, C. G. (1926). Chilopoda. In *Handbuch der Zoologie* (ed. W. Kukenthal & T. Krumbach), vol 4, 239–402, Berlin: Walter de Gruyter.

Attems, G. (1929). Myriapoda 1. Geophilomorpha. In *Des Tierreich*, vol. 52, Berlin: Walter de Gruyter.

Attems, G. (1930*a*). Myriapoda 2. Scolopendromorpha. In *Das Tierreich*, vol. 54, Berlin: Walter de Gruyter.

Attems, C. (1930*b*). The Myriapoda of South Africa. *Ann. S. Afr. Mus.*, **26**, 1–431.

Attems, C. 1947. Neue Geophilomorpha des Wiener Museums. *Annln. naturh. Mus. Wien*, **55**, 50–149.

Auerbach, S. I. (1951). The centipedes of the Chicago area with special reference to their ecology. *Ecol. Monogr.*, **21**, 97–124.

Auerbach, S. I. (1952). Centipedes in the diet of salamanders. *Nat. Hist. Miscellanea*, no. 103, 1–2.

Babu, K. S. (1964). Through-conducting systems in the ventral nerve cord in centipedes. *Z. vergl. Physiol.*, **49**, 114–19.

Baerg, W. J. (1924). The effect of the venom of some supposedly poisonous arthropods (centipedes and scorpions). *Ann. ent. Soc. Am.*, **17**, 343–52.

Bagnall, R. S. (1935). Notes on British chilopods (centipedes) 1. *Ann. Mag. nat. Hist.*, Ser. 10, **15**, 473–9.

Bähr, R. R. (1965). Ableitung lichtinduzierter Potentiale von den Augen von *Lithobius forficatus* L. *Naturwissenschaften*, **52**, 459.

Bähr, R. (1967). Electrophysiologische Untersuchungen an den Ocellen von *Lithobius forficatus* L. *Z. vergl. Physiol.*, **55**, 70–102.

Bähr, R. (1971). Die Ultrastruktur der Photoreceptoren von *Lithobius forficatus* L. (Chilopoda: Lithobiidae). *Z. Zellforsch. mikrosk. Anat.*, **116**, 79–93.
Bähr, R. (1974). Contribution to the morphology of chilopod eyes. *Symp. Zool. Soc., Lond.*, no. 32, 388–404.
Balbiani, E.-G. (1889). Sur trois entophytes nouveaux du tube digestif des Myriapodes. *J. Anat. Physiol., Paris*, **25**, 1–45.
Balbiani, E.-G. (1890). Étude anatomique et histologique sur le tube digestiv de *Cryptops*. *Arch. Zool. exp. gén.*, **8**, 1–82.
Bannister, W. H. & Needham, A. E. (1971). Connective tissue pigment of the centipede *Lithobius forficatus* L. *Naturwissenschaften*, **58**, 58–9.
Barth, R. (1967). Histologische Studien an der Giftdrüsen von *Scolopendra viridicornis* Newp. *An. Acad. Bras. Ci.*, **37**, 179–93.
Bartmeyer, A. & Schmalfuss, (1933). Tausenfussbiss-Vergiftung. *Samml. Vergiftungsf.*, **4**, 209–10.
Bauer, K. (1955). Sinnesökologische Untersuchungen an *Lithobius forficatus*. *Zool. Jb. (Physiol.).*, **65**, 267–300.
Bedford, J. & Leader, J. P. (1975). The composition of the hemolymph of the New Zealand centipede *Cormocephalus rubriceps* (Newport). *Comp. Biochem. Physiol.*, **50**, 561–4.
Bedini, C. (1968) The ultrastructure of the eye of a centipede, *Polybothrus fasciatus* (Newp.). *Monit. Zool. Ital.* (NS), **2**, 31–47.
Bedini, C. & Mirolli, M. (1967). Sensory cilia in the temporal organs of *Glomeris* (Myriapoda, Diplopoda). *Naturwissenschaften*, **14**, 373–4.
Bekker, E. G. (1926). Zur phylogenetischen Entwicklung des Skeletts und der Muskulature der Ateloceraten (Tracheaten). 1. Das Tergalskelett und die Dorsalmusculatur von Chilopoden. *Zool. Zh.*, **6**, 3–64 (in Russian, German summary).
Bekker, E. G. (1949). Contribution to the knowledge of the evolution of the external skeleton and musculature of Tracheata (Atelocerata). 2. Pleural and sternal skeleton and musculature of Chilopoda Epimorpha. *Zool. Zh.*, **28**, 39–58 (in Russian).
Bennett, D. & Manton, S. M. (1963). Arthropod segmental organs and Malpighian tubules with particular reference to their function in Chilopoda. *Ann. Mag. nat. Hist.*, Ser. 13, **5**, 545–56.
Bergsøe, V. & Meinert, F. (1866). Danmarks Geophiler. *Naturh. Tidsskr.*, **3**, 81–108.
Bertheau, Ph. (1971). Histologie comparée des tubes de Malpighi de quelques Chilopodes (Myriapodes). *C. r. hebd. Seanc. Acad. Sci., Paris*, **272**, 2913–15.
Biegel, J. H. (1922). Beiträge zur Morphologie und Entwicklungsgeschichte des Herzens bei *Lithobius forficatus* L. *Rev. Suisse Zool.*, **29**, 427–80.
Binyon, J. & Lewis, J. G. E. (1963). Physiological adaptations of two species of centipede (Chilopoda: Geophilomorpha) to life on the shore. *J. mar. biol. Ass. UK*, **45**, 49–55.
Blackman, M. W. (1905). The spermatogenesis of *Scolopendra heros*. *Bull. Mus. comp. Zool. Harv.*, **48**, 1–138.
Blanchard, R. (1898). Sur le pseudo-parasitisme des myriapodes chez l'homme. *Arch. Parasitol.*, **1**, 452–90.
Blower, G. (1950). Aromatic tanning in Myriapod cuticle. *Nature, Lond.* **165**, 569.
Blower, G. (1951). A comparative study of the chilopod and diplopod cuticle. *Quart. J. micr. Sci.*, **92**, 141–61.

Blower, G. (1952). Epidermal glands in centipedes. *Nature, Lond.*, **170**, 166–7.

Blower, G. (1955*a*). Millipedes and centipedes as soil animals. In *Soil Zoology* (ed. D. K. M. Kevan), pp. 138–51. London: Butterworth.

Blower, G. (1955*b*). Yorkshire centipedes, *Naturalist, Hull*, 137–46.

Blower, J. G. (1957). Feeding habits of a marine centipede. *Nature, Lond.*, **180**, 560.

Boisduval, J. B. A. D. de (1867) *Essai sur l'entomologie horticole, comprenant l'histoire des insectes nuisables a l'horticulture.* Paris, Librairie d'Horticulture de E. Donnaud.

Bonnell, B. (1929). Geophilid centipedes from the bed of the Cooum River (Madras). *J. Asiat. Soc. Bengal* (NS), **25**, 181–4.

Borek, V. (1967). Beitrag zur kenntnis der variabilität der Art *Lithobius lucifugus* L. Koch, 1862 (Chilopoda). *Věst. čsl. Spol. zool.*, **31**, 109–15.

Borek, V. (1969). Fund eines Gynandromorphs von *Monotarsobius austriacus* Verhoeff, 1937 (Chilopoda). *Pr. kraj. Mus. Hradci. Králové*, **10**, 33–4 (in Czech, German summary).

Bornebusch, C. H. (1930). The fauna of the forest soil. *Forst. ForsVaes. Danm*, **11**, 1–225.

Bouin, P. (1903). Sur l'existence d'une double spermatogenèse et de deux sortes de spermatzoides chez *Scolopendra morsitans. Archs Zool. exp. gén.*, **1**, Notes et Revues iii–vi.

Bouin, P. (1920). Sur la dimégalie des spermies dans certaines doubles spermatogénèses – sa signification. *C. r. Séanc. Soc. Biol.*, **83**, 432–6.

Bouin, P. (1934). Recherches sur l'evolution d'un chromosome special (hétérochromosome?) au cours de la double spermatogénèse chez *Scutigera coleoptrata* (Lin.). *Archs Zool. exp. gén.*, **75**, 595–613.

Brade-Birks, H. K. & Brade-Birks, S. G. (1920). Luminous Chilopoda. *Ann. Mag. nat. Hist., Ser.* 9, **5**, 1–30.

Brade-Birks, S. G. (1929). Notes on Myriapoda. 33: The economic status of Diplopoda and Chilopoda and their allies. *J. S.-E. agric. Coll., Wye*, **26**, 178–216.

Brandt, J. F. (1841). Ueber die in der Regentschaft Algier beobachteten Myriopoden. In *Reisen in Alger* III (ed. M. Wagner). Leipzig: Leopold Voss.

Brehm, A. E. (1877). *Tierleben* 9. Die Insekten, Tausendfüssler und Spinnen (ed. E. P. Taschenberg). Leipzig: Verlag des Bibliographischer Institut.

Breucker, H. (1970). Die Struktur des samenbleitenden Gangsystems bei *Geophilus linearis* Koch (Chilopoda). *Z. Zellforsch. mikrosk. Anat.*, **108**, 225–42.

Briot, A. (1904). Sur le venin des Scolopendres. *C. r. Séanc. Soc. Biol.*, **57**, 476–7.

Bristowe, W. S. (1941). *The Comity of Spiders*, vol. 2. London: Ray Society Publication no. 128.

Britten, H. (1920). Food hunting habits of *Lithobius forficatus* L. *Lancs. Chesh. Nat.*, **13**, 118.

Brocher, F. (1930). Observations biologiques sur la ponte et les premiers stades due *Lithobius forficatus* L. *Rev. suisse Zool.*, **37**, 375–83.

Brölemann, H. W. (1926). Myriapodes recueillis en Afrique Occidentale Française par M. l'Administrateur en chef L. Duboscq. *Archs Zool. exp. gén.*, **65**, 1–159.

Brölemann, H. W. (1930). Elements d'une Faune des Myriapodes de France. Chilopodes. *Faune de France*, Vol. 25. Paris: Librairie de la Faculté des Sciences.

Brunhuber, B. S. (1969). The mode of sperm transfer in the scolopendromorph centipede: *Cormocephalus anceps anceps* Porat. *Zool. J. Linn. Soc.*, **48**, 409–20.
Brunhuber, B. S. (1970*a*). The formation of the scolopendromorph spermatophore. *Bull. Mus. Hist. nat., Paris*, **41**, Suppl. No. 2, 1969 (1970), 24–8.
Brunhuber, B. S. (1970*b*). Egg laying, maternal care and development of young in the scolopendromorph centipede, *Cormocephalus anceps anceps* Porat. *Zool. J. Linn. Soc.*, **49**, 225–34.
Brunhuber, B. S. & Hall, E. (1970). A note on the accessory glands of the reproductive system of the scolopendromorph centipede, *Cormocephalus anceps anceps* Port. *Zool. J. Linn. Soc.*, **49**, 49–59.
Bücherl, W. (1940). Sobre a Musculatura de *Scolopendra viridicornis* Newp. *Mem. Inst. Butantan*, **14**, 65–95.
Bücherl, W. (1942). Contribuição ao estudo dos órgãos sexuais externos das espécies do gênero *Scolopendra* Linné mais frequentes no Brasil. *Mem. Inst. Butantan*, **16**, 37–58.
Bücherl, W. (1946). Ação do veneno dos escolopendromorfos do Brasil sôbre alguns animais de laboratório. *Mem. Inst. Butantan*, **19**, 181–98.
Bücherl, W. (1971). Venomous chilopods or centipedes. In *Venomous animals and their venoms*, vol. 3, pp. 169–96. New York: Academic Press.
Camatini, M. (1970). The structure of striated muscle fibres in some Chilopoda. *Bull. Mus. Hist. nat., Paris*, **41**, Suppl. no. 2, 1969 (1970), 31–34.
Camatini, M. & Castellani, C. L. (1974). Atypical myofilaments array of visceral muscle fibres of *Lithobius forficatus* testis. *J. Submicr. Cytol.*, **6**, 353–65.
Camatini, M. & Castellani, C. L. (1978). Myofilaments array of some visceral muscle fibres of *Lithobius forficatus* Linnaeus and *Pachyiulus enologus* B. *Abh. Ver. naturwiss. Ver. Hamburg*, **21/22**, 243–55.
Camatini, M., Franchi, E., Saita, A. & Bellone, L. (1977). Spermiogenesis in *Scutigera coleoptrata* (Myriapoda Chilopoda). *J. Submicr. Cytol.*, **9**, 373–87.
Camatini, M. & Saita, A. (1972). Comparative analysis of myofilaments arrangement in the phylum Anthropoda. *Boll. Zool.*, **39**, 173–85.
Camatini, M., Saita, A. & Cotelli, F. (1974). Spermiogenesis of *Lithobius forficatus* L. at ultrastructural level. *Symp. Zool. Soc., Lond.*, no. 32, 231–5.
Cameron, J. A. (1926). Regeneration in *Scutigera forceps*. *J. Exp. Zool.*, **46**, 169–79.
Campbell, M. I. (1932). Some observations on *Scolopendra heros* under laboratory conditions. *Trans. Kans. Acad. Sci.*, **35**, 80.
Chalande, J. (1885). Recherches anatomiques sur l'appareil respiratoire chez les chilopodes de France. *Bull. Soc. Hist. nat. Toulouse*, **19**, 39–66.
Chalande, J. (1886). Recherches sur le méchanisme de la respiration chez les myriapodes. *Bull. Soc. Hist. nat. Toulouse*. **20**, 137–62.
Chalande, J. (1905). Recherches histologiques et anatomiques sur les myriapodes du sud-ouest de la France. *Bull. Soc. Hist. nat. Toulouse*, **38**, 46–154.
Chamberlin, R. V. (1912). The Henicopidae of America north of Mexico. *Bull. Mus. Comp. Zool. Harv.*, **57**, 1–36.
Chamberlin, R. V. (1920). The myriopod fauna of the Bermuda Islands, with notes on variation in *Scutigera*. *Ann. ent. Soc. Amer.*, **13**, 271–303.
Chapman, R. F. (1971). *The insects structure and function* (2nd edn.) London: E.U.P.
Cisse, M. (1972). L'alimentation des varanides du Senegal. *Bull. Inst. Fr. Afr. noire* (*A*) **34**, 503–15.

Clark, R. J. (1967). Centipede in stomach of young *Vipera ammodytes meridionalis*. *Copeia*, 224.

Cloudsley-Thompson, J. L. (1945). Behaviour of the common centipede *Lithobius forficatus*. *Nature, Lond.*, **156**, 537–8.

Cloudsley-Thompson, J. L. (1949). The enemies of myriapods. *Naturalist, Hull*, 16–17.

Cloudsley-Thompson, J. L. (1950). Epicuticle of arthropods. *Nature, Lond.*, **165**, 692–3.

Cloudsley-Thompson, J. L. (1951). Supplementary notes on Myriapoda. *Naturalist, Hull*, 16–17.

Cloudsley-Thompson, J. L. (1953). A note on the littoral terrestrial arthropods of the Isle of Man. *Entomologist*, **86**, 11–12.

Cloudsley-Thompson, J. L. (1955). Some aspects of the biology of centipedes and scorpions. *Naturalist, Hull*, 147–53.

Cloudsley-Thompson, J. L. (1956). Studies in diurnal rhythms VI. Bioclimatic observations in Tunisia and their significance in relation to the physiology of the fauna, especially woodlice, centipedes, scorpions and beetles. *Ann. Mag. nat. Hist.*, Ser. 12, **9**, 305–29.

Cloudsley-Thompson, J. L. (1958*a*). The effect of wind upon the nocturnal emergence of woodlice and other terrestrial arthropods. I. *Ent. mon. Mag.*, **94**, 106–8.

Cloudsley-Thompson, J. L. (1958*b*). *Spiders, scorpions, centipedes and mites*. London: Pergamon.

Cloudsley-Thompson, J. L. (1959). Studies in diurnal rhythms IX. The water relations of some nocturnal tropical arthropods. *Ent. exp. appl.*, **2**, 249–56.

Cloudsley-Thompson, J. L. (1961). A new sound producing organ in centipedes. *Ent. mon. Mag.*, **96**, 110–13.

Cloudsley-Thompson, J. L. & Crawford, C. S. (1970). Water and temperature relations, and diurnal rhythms of scolopendromorph centipedes. *Ent. exp. appl.*, **13**, 187–93.

Cole, L. C. (1946). A study of the cryptozoa of an Illinois woodland. *Ecol. Monogr.*, **16**, 49–86.

Cornwall, J. W. (1916). Some centipedes and their venom. *Indian J. Med. Res.*, **3**, 541–57.

Cornwell, W. S. (1934). Notes on egg-laying and nesting habits of certain species of North Carolina myriapods and various phases of their life histories. *J. Elisha Mitchell Sci. Soc.*, **42**, 289–91.

Crabill, R. E. (1955). On the reappearance of a possible ancestral characteristic in a modern chilopod (Chilopoda: Scolopendromorpha: Cryptopidae). *Bull. Brooklyn Ent. Soc.*, **50**, 133–6.

Crabill, R. E. (1960*a*). Concerning the aberrant genus *Nothobius*, with a redescription of its type-species (Chliopoda: Geophilomorpha: Himantariidae). *Ent. News*, **61**, 87–99.

Crabill, R. E. (1960*b*). A remarkable form of sexual dimorphism in a centipede (Chilopoda: Lithobiomorpha: Lithobiidae). *Ent. News*, **71**, 156–61.

Crabill, R. E. (1960*c*). On the true nature of the Azygethidae (Chilopoda: Geophilomorpha). *Psyche*, **67**, 76–9.

Crabill, R. E. (1962). Plectrotaxy as a systematic criterion in lithobiomorphic centipedes (Chilopoda: Lithobiomorpha). *Proc. US natn. Mus.*, **113**, 399–412.

Crabill, R. E. (1968). A bizarre case of sexual dimorphism in a centipede, with consequent submergence of a genus (Chilopoda: Geophilomorpha: Mecistocephalidae). *Ent. News*, **79**, 286–7.

Crabill, R. E. (1969). Tracheotaxy as a generic criterion in Himantariidae, with proposal of two new bothriogastrine genera (Chilopoda: Geophilomorpha). *Smithson. contrib. to Zoology* no. 12, 1–9.

Crabill, R. E. (1970). A new family of centipedes from Baja California with introductory thoughts on ordinal revision. *Proc. ent. Soc. Wash.*, **72**, 112–18.

Crawford, C. S. & Riddle, W. A. (1974). Cold hardiness in centipedes and scorpions in New Mexico. *Oikos*, **25**, 86–92.

Crawford, C. S., Riddle, W. A. & Pugach, S. (1975). Overwintering physiology of the centipede *Scolopendra polymorpha*. *Physiol. Zoöl.*, **48**, 290–4.

Crossland, C. (1929). Amphibious centipedes. *Nature. Lond.*, **124**, 794.

Cumming, W. D. (1903). The food and poison of a centipede. *J. Bombay nat., Hist. Soc.*, **15**, 364–5.

Curry, A. (1974). The spiracle structure and resistance to desiccation of centipedes. *Symp. zool. Soc. Lond.*, no 32, 365–82.

Dass, C. M. S. & Jangi, B. S. (1978). Ultrastructural organisation of the poison gland of the centipede *Scolopendra morsitans* Linn. *Indian J. exp. Biol.*, **16**, 748–57.

Davy, J. (1848). Miscellaneous observations on the centipede (*Scolopendra morsitans*) and the large land snail of the West Indies (*Helix oblonga*). *Edinb. Phil. J.*, **45**, 383–8.

DeCastro, A. B. (1921). The poison of the Scolopendridae. Being a special reference to the Andaman species. *Indian med. Gaz.*, **56**, 207.

Delany, M. J. (1975). *The Rodents of Uganda.* London: British Museum.

Demange, J. M. (1942). Remarques sur le système trachéen d'*Hydroschendyla submarina* (Grube) et celui des myriapodes géophilomorphes en général. *Bull. Mus. Hist. nat., Paris*, **14**, 422–7.

Demange, J. M. (1943). Sur le développement post-embryonnaire et la chaetotaxie d'*Hydroschendyla submarina* (Grube) (Myriapodes). *Bull. Mus. Hist. nat., Paris*, **15**, 418–23.

Demange, J. M. (1944). Quelques mots sur la mue de *Lithobius forficatus* L. (Myriapodes, Chilopodes). *Bull. Mus. Hist. nat., Paris*, **16**, 235–7.

Demange, J. M. (1945). Le spermatophore de quelques scolopendrides. *Bull. Mus. Hist. nat., Paris*, **17**, 483–7.

Demange, J. M. (1946). Sur la morphologie comparée des testicules de quelques scolopendrides (Myriapodes, Chilopodes). *Bull. Mus. Hist. nat., Paris*, **18**, 59–64.

Demange, J. M. (1948). Note sur la mue, l'autotomie et la régénération chez une Scutigère (Myriapodes, Chilopodes). *Bull. Mus. Hist. nat., Paris*, **20**, 165–8.

Demange, J. M. (1956). Contributions à l'étude de la biologie, en captivité de *Lithobius piceus gracilitarsus* Bröl. (Myriapodes, Chilopodes). *Bull. Mus. Hist. nat., Paris*, **28**, 388–93.

Demange, J. M. (1958). Contribution à la connaissance de la faune cavernicole de l'Espagne (Myriapodes, Chilopodes: Lithobioidea). *Speleon*, **9**, 27–49.

Demange, J. M. (1961). A propos de la description d'une nouvelle espèce de géophilomorphes de maroc, *Orya panousei* nov. sp. *Soc. Sci nat. phys. Maroc*, **41**, 211–27.

Demange, J. M. (1963). La segmentation dorsale des Myriapodes Chilopodes au niveau de la zone des 7e et 8e segments. *C. r. hebd. Séanc. Acad. Sci., Paris*, **257**, 514–17.

Demange, J. M. (1967). Segmentation du tronc des chilopodes et des diplopodes chilognathes. *Mém. Mus. natn. Hist. nat., Paris,* **44**, 1–188.

Demange, J. M. & Richard, J. (1969). Morphologie de l'appareil génital mâle des scolopendromorphes et son importance en systématique (Myriapodes, Chilopodes). *Bull. Mus. Hist. nat., Paris,* **40**, 968–83.

Descamps, M. (1969*a*). Étude cytologique de la spermatogenèse chez *Lithobius forficatus* L. (Myriapode, Chilopode). *Archs. Zool. exp. gén.,* **110**, 349–61.

Descamps, M. (1969*b*). Étude cytochimique de la spermatogenèse chez *Lithobius forficatus* L. *Histochimie,* **20**, 46–57.

Descamps, M. (1971*a*). Le cycle spermatogénétique chez *Lithobius forficatus* L. (Myriapode, Chilopode) 1. Évolution et étude quantitative des populations cellulaires du testicule au cours du développment post-embryonnaire. *Archs. Zool. exp. gén.,* **112**, 199–209.

Descamps, M. (1971*b*). Les processus de dégénérescence des cellules sexuelles mâles chez *Lithobius forficatus* L., (Myriapode, Chilopode). *Archs. Zool. exp. gén.,* **112**, 439–455.

Descamps, M. (1971*c*). Etude ultrastructurale des spermatogonies de la croissance spermatocytaire chez *Lithobius forficatus* L. *Z. Zellforsch. mikrosk. Anat.,* **121**, 14–26.

Descamps, M. (1971*d*). Le cycle spermatogénétique chez *Lithobius forficatus* L. (Myriapode, Chilopode) II. Influence de facteurs externes sur l'évolution du testicule et des vésicules séminales. *Archs Zool. exp. gén.,* **122**, 731–46.

Descamps, M. (1972*a*). Rôle de la pars intercerebralis dans la régulation du cycle spermatogénétique chez *Lithobius forficatus* L. *Gen. comp. Endocr.,* **18**, 3.

Descamps, M. (1972*b*). Etude ultrastructurale du spermatozoide de *Lithobius forficatus* L. (Myriapode, Chilopode). *Z. Zellforsch.,* **126**, 193–205.

Descamps, M. (1974). Etude du contrôle endocrinien du cycle spermatogénétique chez *Lithobius forficatus* L. (Myriapode, Chilopode). Rôle de la pars intercerebralis. *Gen. comp. Endocr.,* **24**, 191–202.

Descamps, M. (1975). Etude de contrôle endocrinien du cycle spermatogénétique chez *Lithobius forficatus* L. (Myriapode, Chilopode). Rôle du complexe 'cellules neurosécrétrices des lobes frontaux du protcérébron-glands cérébrales'. *Gen. comp. Endocr.,* **25**, 346–57.

Descamps, M. (1977*a*). Recherches expérimentales sur la régulation du cycle spermatogénétique au cours du développement post-embryonnaire chez *Lithobius forficatus* L. (Myriapode, Chilopode). *Archs Biol., Paris,* **88**, 203–15.

Descamps, M. (1977*b*). Influence de la croissance somatique sur le cycle spermatogénétique du *Lithobius forficatus* L. (Myriapode, Chilopode). *Gen. comp. Endocr.,* **33**, 412–22.

Descamps, M. & Herbaut, C. (1971). Sur un cas d'intersexualité chez *Lithobius forficatus* L. (Myriapode, Chilopode). *C. r. hebd. Séanc. Acad. Sci., Paris,* **272**(D), 1648–51.

Descamps, M. & Joly, R. (1971). Rôle de la pars intercerebralis dans le déroulement du cycle spermatogénétique du *Lithobius forficatus* L. *C. r. hebd. Seanc. Acad. Sci., Paris,* **273**(D), 768–70.

Dobroruka, L. J. (1961). *Die Hundertfüssler.* Neue Brehm Bücherlei no. 285. Wittenberg: A. Ziemsem Verlag.

Dohle, W. (1970). Uber Eiablage und Entwicklung von *Scutigera coleoptrata* (Chilopoda). *Bull. Mus. Hist. nat., Paris,* **41**, suppl. no. 2, 1969 (1970), 53–7.

Dorier, A. (1925). Sur la faculté d'enkystement dans l'eau de la larve du *Gordius aquaticus* L. *C. r. hebd. Séanc. Acad. Sci., Paris*, **181**, 1098–9.

Dorier, A. (1929). Sur les gordiaces des myriapodes. *C. r. hebd. Séanc. Acad. Sci., Paris*, **188**, 743–5.

Dowdy, W. W. (1944). A community study of a disturbed deciduous forest area near Cleveland, Ohio, with special reference to invertebrates. *Ecol. Monogr.*, **14**, 193–222.

Drift, J. van der (1951). Analysis of the animal community in a beech forest floor. *Meded. Inst. Toegep. biol. Onderz. Nat.*, **9**, 1–168.

Dubois, M. R. (1886). De la fonction photogénique chez les myriapodes. *C. r. Séanc. Soc. Biol.*, **3**, 518–22, 523–6.

Duboscq, O. (1898). Recherches sur les chilopodes. *Arch. Zool. exp. gén.*, Ser. 3, **6**, 481–650.

Dubuisson, M. (1928). La ventilation trachéene chez les chilopodes et sur la circulation sanguine chez les scutigères. *Archs Zool. exp. gén.*, **67**, 49–63.

Dugès, A. (1838). *Traité de physiologie comparée de l'homme et des animaux*, vols. 1–3. Montpellier.

Dugès, A. (1887). Sur les moeurs d'une grand espèce de Scolopendre mexicaine. *C. r. ent. belg.*, no. 9, 101–4.

Eason, E. H. (1964). *Centipedes of the British Isles*. London: Warne.

Eason, E. H. (1970*a*). On certain characters used to separate species and subspecies of Lithobiidae. *Bull. Mus. Hist. nat., Paris*, **41**, suppl. no. 2, 1969 (1970), 58–60.

Eason, E. H. (1970*b*). A redescription of the species of *Eupolybothrus* s. str. preserved in the British Museum (Natural History) and the Hope Department of Zoology, Oxford (Chilopoda, Lithobiomorpha). *Bull. Br. Mus. nat. Hist. (Zool.)*, **19**(9), 287–310.

Eason, E. H. (1973). The type specimens and identity of the species described in the genus *Lithobius* by R. I. Pocock from 1890 to 1901 (Chilopoda, Lithobiomorpha). *Bull. Br. Mus. nat. Hist. (Zool.)*, **252**, 39–83.

Eason, E. H. (1974). On certain aspects of the generic classification of the Lithobiidae, with special reference to geographical distribution. *Symp. Zool. Soc., Lond.*, no. 32, 65–72.

Easterla, D. A. (1975). Giant desert centipede preys upon snake. *S. West Nat.*, **20**, 411.

Edwards, E. E. (1929). A survey of the insect and other invertebrate fauna of permanent pasture and arable land of certain soil types at Aberystwyth. *Ann. Appl. Biol.*, **16**, 299–323.

Enghoff, H. (1975). Notes on *Lamyctes coeculus* (Brölemann), a cosmopolitic, parthenogenetic centipede (Chilopoda: Henicopidae). *Ent. Scand.*, **6**, 45–6.

Enghoff, H. (1976). The sex ratio of the littoral centipede *Strigamia maritima* (Leach). *Ent. Meddr.*, **44**, 121–2.

Ernst, A. (1971). Licht und elektron mikroskopische Untersuchungen zur Neurosekretion bei *Geophilus longicornis* Leach unter besonderer Berücksichtigung der Neurohämalorgane. *Z. wiss. Zool.*, **182**, 62–130.

Ernst, A. (1976). Ultrastruktur der Sinneshaare auf den Antennen von *Geophilus longicornis* Leach (Myriapoda, Chilopoda). I. Die Sensilla trichoidea. *Zool. Jb. (Anat.)*, **96**, 586–604.

Ewing, H. E. (1928). Observations on the habits and injury caused by the bites or stings of some common North American Arthropods. *Amer. J. Trop. Med.*, **8**, 39–62.

Fabre, J. (1855). Recherches sur l'anatomie des organes reproducteurs et sur le développment des Myriapodes. *Ann. Sci. nat.* (*Zool.*), Ser. 4, **3**, 257–316.

Fachbach, G., Kolossa, J. & Ortner, A. (1975). The nutritional biology of *Salamandra s. salamandra* and *Salamandra atra* (Caudata, Salamandridae). *Salamandra*, **11**, 136–44.

Fahlander, K. (1938). Anatomie und systematische Einteilung der Chilopoden. *Zool. Bidr. Upps.*, **17**, 1–148.

Fairhurst, C. P., Barber, A. D. & Armitage, M. L. (1978). The British myriapod survey – April 1975. *Anh. Verh. naturwiss. Verh. Hamburg* (NF), **21/22**, 129–34.

Faust, E. S. (1928). Vergiftungen durch tierische Gifte. In *Lexicon der Toxicologie* (ed. F. Flury & G. Zanger). Berlin: Springer.

FitzSimmons, V. F. M. (1962). *Snakes of Southern Africa.* London: Macdonald.

Fonesca, F. da (1949). *Animais Peçonhoutas.* Sao Paulo.

Forbes, S. A. (1890). Note on the feeding habits of *Cermatia forceps. Am. Nat.*, **24**, 81–2.

Fuhrmann, H. (1922). Beiträge zur Kenntnis der Hautsinnesorgane der Tracheaten. 1. Die antennalen Sinnesorgane der Myriapoden. *Z. wiss. Zool.*, **119**, 1–52.

Füller, H. (1960). Untersuchungen über den Bau der Stigmen bei Chilopoden. *Zool. Jb.* (*Anat.*), **78**, 129–44.

Füller, H. (1963*a*). Vergleichende Untersuchungen über das Skelettmuskelsystem der Chilopoden. *Abh. dt. Akad. Wiss. Berl. Kl. f. Chemie Geol. u. Biol.* Jahrg., **1962**(3), 1–98.

Füller, H. (1963*b*). Histologische, polarisationsoptische und histochemische Untersuchungen über das bindegewebige Innenskelett der Chilopoden. *Z. wiss. Zool.*, **168**, 184–207.

Füller, H. 1965. Untersuchungen über die Chitintextur des Integuments der Chilopoden. *Zool. Anz.* **175**, 173–181.

Füller, H. (1966). Elektronenmikroskopische Untersuchungen der Malpighischen Gefässe von *Lithobius forficatus* (L.) *Z. wiss. Zool.* **173**, 191–217.

Gabbut, P. D. (1959). The bionomics of the wood cricket *Nemobius sylvestris* (Orthoptera: Gryllidae). *J. Anim. Ecol.*, **28**, 15–42.

Gabe, M. (1952). Sur l'emplacement et les connexions des cellules neuro-sécrétrices dans les ganglions cérébroïdes de quelques Chilopodes. *C. r. hebd. Séanc. Acad. Sci., Paris*, **235**, 1430–2.

Gabe, M. (1956). Contribution à l'histologie de la neuro-sécrétion chez les Chilopodes. In *Bertil Hanström, Zoological papers in honour of his sixty-fifth birthday*, November 20th, 1956 (ed. K. G. Wingstrand). Lund Zoological Institute, pp 163–83.

Gabe, M. (1967). Caractères cytologiques et histochimiques du rein maxillaire des Chilopodes. *C. r. hebd. Séanc. Acad. Sci., Paris* **246**(**D**), 726–9.

Gazagnaire, J. (1888). La phosphorescence chez les myriapodes. *Bull. Soc. zool. Fr.*, **13**, 182–6.

Gazagnaire, J. (1890). La phosphorescence chez les myriapodes de la famille des Geophilidae. *Mém. Soc. zool. Fr.*, **3**, 136–46.

Giard, A. (1893). Sur un diptère parasite des myriapodes du genre *Lithobius. Ann. Soc. ent. Fr.*, **61**, (Bull.), ccxiii–xv.

Goldberg, S. R. (1975). Giant desert centipede preys upon snakes. *S. West Nat.*, **20**, 411–13.

Görner, P. (1959). Optische Orientierungsreaktionen bei Chilopoden. *Z. vergl. Physiol.*, **42**, 1–5.

Grainger, J. P. & Fairley, J. S. (1978). Studies on the biology of the pygmy shrew *Sorex minutus* in the west of Ireland. *J. Zool., Lond.*, **186**, 109–41.

Grassé, P. P. (1953). *Traité de Zoologie* vol. 1, part 2 Protozoaires. Rhizopodes, Actinopodes, Sporozoaires, Cnidosporides. Paris: Masson & Cie.

Grenacher, H. (1880). Über die Augen einiger Myriapoden. *Arch. mikrosk. Anat. EntwMech.*, **18**, 415–67.

Guyetant, R. (1967). Etude de l'alimentation de jeunes batraciens anoures durant la saison estivale. *Ann. Sci. Univ. Besancon Zool. Physical Biol. anim.*, **1967** 69–78.

Haase, E. (1880). Schlesiens Chilopoden 1. Chilopoda Anamorpha. Inaugural Dissertation. Thesis, Breslau.

Haase, E. (1884*a*). Das Respirationssystem der Symphylen und Chilopoda. *Zool. Beitr.*, **1**, 65–96.

Haase, E. (1884*b*). Schlundgerüst und Maxillarorgan von *Scutigera. Zool. Beitr.*, **1**, 97–108.

Hall, R. J. (1976). Summer foods of the salamander *Plethodon wehrlei* (Amphibia, Urodela, Plethodontidae). *J. Herpetol.*, **10**, 129–31.

Haltenorth, T. & Diller, H. (1977). *Säugetiere Afrikas und Madagascars.* Munich: BLV Verlagsgesellschaft.

Haneda, Y. (1939). The terrestrial luminescent animals and plants of Palan and Yap Islands. *Kagaku Nanyo*, **2**, 88–93 (in Japanese).

Hanström, B. (1934). Bemerkungen uber das Komplex-Auge der Scutigeren. *Acta Univ. Lund.*, **30**(6), 1–13.

Harry, O. G. (1970). Gregarines: their effect on the growth of the desert locust. *Nature, Lond.*, **225**, 964–6.

Harvey, E. N. (1952). *Bioluminescence.* New York: Academic Press.

Hemenway, J. (1900). The structure of the eye of *Scutigera* (*Cermatia*) *forceps. Biol. Bull. mar. biol. Lab., Woods Hole*, **1**, 205–13.

Hennings, C. (1903). Zur Biologie der Myriapoden I. Marine Myriapoden. *Biol. Zbl.*, **23**, 720–5.

Hennings, C. (1904*a*). Das Tömösvárysche Organ der Myriapoden I. *Z. wiss. Zool.*, **76**, 26–53.

Hennings, (1904*b*). Zur Biologie der Myriapoden. II. Geruch und Geruchsorgane der Myriapoden. *Biol. Zbl.*, **24**, 274–83.

Hennings, C. (1906). Das Tömösvárysche Organ der Myriapoden. II. *Z. wiss. Zool.*, **80**, 576–641.

Henry, L. M. (1948). The nervous system and the segmentation of the head in annulata. V. Onychophora, VI. Chilopoda, VII. Insecta. *Microentomology*, **13**, 27–48.

Herbaut, C. (1972*a*). Etude cytochimique et ultrastructurale de l'ovogènese chez *Lithobius forficatus* L. (Myriapode, Chilopode). Evolution des constituants cellulaires. *Wilhelm Roux Arch. EntwMech. Org.*, **170**, 115–34.

Herbaut, C. (1972*b*). Nature et origine des réserves vitellines dans l'ovocyte de *Lithobius forficatus* L. (Myriapode, Chilopode). *Z. Zellforsch. mikrosk. Anat.*, **130**, 18–27.

Herbaut, C. (1974). Etude cytochimique et origine des enveloppes ovocytaires chez *Lithobius forficatus* (L.) (Myriapode, Chilopode). *Symp. Zool. Soc., Lond.*, no. 32, 237–47.

Herbaut, C. (1975*a*). Etude expérimentale de la régulation endocrinienne de l'ovogenèse chez *Lithobius forficatus* L. (Myriapode, Chilopode). Rôle de la pars intercerebralis. *Gen. comp. Endocr.* **27**, 34–42.

Herbaut, C. (1975*b*). Influence des facteurs externes sur le cycle ovogenetique chez *Lithobius forficatus* (Myriapode, Chilopode). *Archs Zool. exp. gén.*, **116**, 293–302.

Herbaut, C. (1976*a*). Etude expérimentale de la régulation endocrinienne de l'ovogenèse chez *Lithobius forficatus* L. (Myriapode, Chilopode). Rôle du complexe 'cellules neurosécrétrices protocérébrales-glandes cérébrales'. *Gen. comp. Endocr.*, **28**, 264–76.

Herbaut, C. (1976*b*). Les processus de dégénérescence des cellules sexuelles au cours de l'ovogenèse chez *Lithobius forficatus* (L.) (Myriapode, Chilopode). *Arch. anat. Microsc. Morphol. exp.*, **65**, 175–82.

Herbaut, C. (1977*a*). Evolution du cycle ovogenetique au cours du développement postembryonaire chez *Lithobius forficatus* L. (Myriapode, Chilopode). *Archs Biol., Paris*, **88**, 67–77.

Herbaut, C. (1977*b*). Influence de la croissance somatique sur l'ovogenèse chez *Lithobius forficatus* (L.) (Myriapode, Chilopode). *Archs Zool. exp. gén.*, **118**, 63–72.

Herbaut, C. & Joly, R. (1971). Rôle de la pars intercerebralis dans la croissance ovocytaires chez *Lithobius forficatus* L. *C. r. hebd. Séanc. Acad. Sci., Paris*, **273(D)**, 1515–18.

Herbaut, C. & Joly, R. (1972). Activité ovarienne et cycle ovogenetique chez *Lithobius forficatus* L. (Myriapode, Chilopode). *Archs Zool. exp. gén.*, **113**, 215–25.

Herbst, C. (1891). Beiträge zur Kenntnis der Chilopoden. *Bibl. Zool.*, **3**, 1–42.

Herms, W. B. (1915). *Medical & Veterinary Entomology*. New York: Macmillan.

Herms, W. B. (1939). *Medical Entomology*. New York: Macmillan.

Hesse, R. (1901). Untersuchungen über die Organe der Lichtempfindung bei niederen Thieren VII. Von den Arthropoden Augen, 2. Die Augen der Myriopoden. *Z. wiss. Zool.*, **70**, 347–473.

Heymons, R. (1901). Die Entwicklungsgeschichte der Scolopender. *Zoologica, Stuttg.*, **13**, 1–244.

Hilton, W. A. (1930). Nervous system and sense organs XXXV. Chilopoda. *J. Ent. Zool.*, **22**, 105–15.

Holmgren, N. (1916). Zur vergleichenden Anatomy des Gehirns von Polychaeten, Onychophoren, Xiphosuren, Arachniden, Crustaceen, Myriapoden und Insekten. *K. Sv. Vet. Akad. Handl. NF*, **56**, 1–303.

Holmquist, A. M. (1928). Notes on the life history and habits of the mound-building ant *Formica ulkei* Emery. *Ecology*, **9**, 70–87.

Horne, F. R. (1969). Purine excretion in five scorpions, a uropygid and a centipede. *Biol. Bull. mar. biol. Lab., Woods Hole*, **137**, 155–60.

Horstmann, E. (1968). Die Spermatozoen von *Geophilus linearis* Koch (Chilopoda). *Z. Zellforsch. mikrosk. Anat.*, **89**, 410–29.

Houdemer, E. (1926). Note sur un myriapode vésicant du Tonkin, *Otostigmus aculeatus* Haase. *Bull. Mus. Hist. nat., Paris* **1926**, 213–14.

Hubert, M. (1968). Contribution a l'étude de l'exrétion chez les Myriapodes (Progonéates et Opithsogonéates). *Archs Sci. Physiol.*, **22**, 93–109.
Hubert, M. (1977). Contribution à l'étude des organes excréteurs et de l'excretion chez les Diplopodes et les Chilopodes. Thèse doct. Sc. nat. no. 264. University of Rennes.
Hubert, M. & Razet, P. (1965). Sur les principaux éléments du catabolisme azoté chez les Myriapodes. *C. r. hebd. Séanc. Acad. Sci., Paris*, **261**, 797–800.
Huhta, V. & Koskenniemi, A. (1975). Numbers, biomass and community respiration of soil invertebrates in spruce forests at two latitudes in Finland. *Ann. Zool. Fennici*, **12**, 164–82.
Hyman, L. H. (1951). *The Invertebrates*, vol. 3. Acanthocephala, Aschelminthes and Entoprocta. The pseudocoelomate Bilateria. New York: McGraw-Hill.
Jackson, A. R. (1914). A preliminary list of the Myriapoda of the Chester district. *Lancs. Chesh. Nat.*, **6**, 450.
Jamault-Navarro, C. & Joly, R. (1977). Localisation et cytologie des cellules neurosécrétrices protocérébrales chez *Lithobius forficatus* L. (Myriapode, Chilopode). *Gen. comp. Endocr.*, **31**, 106–20.
Jangi, B. S. (1955). Some aspects of the morphology of the centipede *Scolopendra morsitans* Linn. (Scolopendridae). *Ann. Mag. nat. Hist., Ser.* 12, **8**, 597–607.
Jangi, B. S. (1956). The reproductive system in the male of the centipede, *Scolopendra morsitans* Linn. *Proc. zool. Soc. Lond.*, **127**, 145–59.
Jangi, B. S. (1957). The reproductive system in the female of the centipede, *Scolopendra morsitans* Linn. (Scolopendridae). *Ann. Mag. nat. Hist.*, Ser. 12, **10**, 232–40.
Jangi, B. S. (1959). Further notes on the taxonomy of the centipede *Scolopendra morsitans* Linnaeus (Scolopendridae). *Ent. News.*, **70**, 253–7.
Jangi, B. S. (1960). On the antennal musculature of the centipede, *Scolopendra amazonica* Bucherl. *Bull. Zool. Soc., Coll. of Sci., Nagpur*, **3**, 35–42.
Jangi, B. S. (1961). The skeletomuscular mechanism of the so-called anal legs in the centipede *Scolopendra amazonica* (Scolopendridae). *Ann. ent. Soc. Am.*, **54**, 861–9.
Jangi, B. S. (1964). Sensory physiology of the anal legs of centipedes. *Curr. Sci.*, **33**, 237–8.
Jangi, B. S. (1966). Scolopendra (the Indian centipede). *Indian Zoological Memoirs*, no. 9, Calcutta: Zoological Society of India.
Jangi, B. S. & Dass, C. M. S. (1977). Chemoreceptive function of the poison fang in the centipede *Scolopendra morsitans* L. *Ind. J. exp. Biol.*, **15**, 803–4.
Jangi, B. S. & Dass, C. M. S. (1978). A new species of the genus *Paracryptops* from India, with remarks on the generic and specific characters (Chilopoda: Scolopendromorpha: Cryptopidae). *Zool. J. Linn. Soc.*, **64**, 327–30.
Jeekel, C. A. W. (1964). Beitrag zur Kenntnis der Systematic und Ökologie der Hundertfüsser (Chilopoda) Nordwestdeutschlands. *Abh. Verh. naturwiss. Ver. Hamburg* NF, **8**, 111–53.
Johannsen, O. A. & Butt, F. H. (1941). *Embryology of insects and myriapods.* New York: McGraw-Hill.
Johnson, B. M. (1952). The centipedes and millipedes of Michigan. Ph.D., University of Michigan.
Joly, R. (1962). Les glandes cérébrales, organes inhibiteurs de la mue chez les Myriapodes Chilopodes. *C. r. hebd. Séanc. Acad. Sci., Paris*, **254**, 1679–81.

Joly, R. (1964). Action de l'ecdysone sur le cycle de la mue de *Lithobius forficatus* L. (Myriapode, Chilopode). *C. r. Séanc. Soc. Biol.*, **158**, 548–50.

Joly, R. (1966*a*). Contribution a l'étude du cycle de mue et de son déterminisme chez les Myriapodes Chilopodes. Thèses, Université de Lille.

Joly, R. (1966*b*). Sur l'ultrastructure de la glande cerebrale de *Lithobius forficatus* L. (Myriapode, Chilopode). *C. r. hebd. Séanc. Acad. Sci., Paris*, **263**, 374–7.

Joly, R. (1966*c*). Etude expérimentale du cycle du mue et de sa régulation endocrine chez les Myriapodes Chilopodes. *Gen. comp. Endocr.*, **6**, 519–33.

Joly, R. (1969). Sur l'utrastructure de l'oeil de *Lithobius forficatus* L. (Myriapode, Chilopode). *C. r. hebd. Séanc. Acad. Sci., Paris*, **268**, 3180–2.

Joly, R. (1970). Evolution cyclique des glandes cérébrales au cours de l'intermue chez *Lithobius forficatus* (L.) (Myriapode-Chilopode). *Z. Zellforsch. mikrosk. Anat.*, **110**, 85–96.

Joly, R. (1971). Effet de la destruction de la pars intercerebralis sur l'évolution pondérale chez *Lithobius forficatus* L. *C. r. hebd. Séanc. Acad. Sci., Paris*, **273(D)**, 1208–9.

Joly, R. (1972). Effect de l'injection d'extraits de pédoncules oculaires de crabe sur le cycle de mue de *Lithobius forficatus* L. (Myriapode, Chilopode). *Gen. comp. Endocr.*, **18**, 560–4.

Joly, R. (1976). Influence de quelques interventions expérimentales sur l'activité sécrétoire des glandes cérébrales chez *Lithobius forficatus* L. (Myriapode, Chilopode). *Gen. comp. Endocr.*, **30**, 302–12.

Joly, R. & Descamps, M. (1969). Evolution du testicule, des vésicules séminales et cycle spermatogénétique chez *Lithobius forficatus* L. (Myriapode, Chilopode). *Archs Zool. exp. gén.*, **110**, 341–8.

Joly, R. & Descamps, M. (1970). Le complexe endocrine cephalique des Myriapodes Chilopodes. *Bull. Mus. Hist. nat., Paris*, **41**, suppl. no. 2, 1969 (1970), 75–6.

Joly, R., Descamps, M. Herbaut, C. & Jamault-Navarro, C. (1976). Endocrinologie des Myriapodes Chilopodes. *Bull. Soc. zool. Fr.*, **101**, 867–68.

Joly, R. & Devauchelle, G. (1970). Etude cytochimique de la glande cérébrale de *Lithobius forficatus* L. (Myr., Chil.); nature des secretions. *J. Microscopie*, **9**, 631–42.

Joly, R. & Lehouelleur, J. (1972). Effect de la section antennaire sur le déclenchement de la mue chez *Lithobius forficatus* L. (Myriapode, Chilopode). *Gen. comp. Endocr.*, **19**, 320–4.

Jones, T. H., Conner, W. E., Meinwald, J., Eisner, H. E. & Eisner, T. (1976). Benzoyl cyanide and mandelonitrile in the cyanogenetic secretion of a centipede. *J. Chem. Ecol.*, **2**, 421–9.

Kaestner, A. (1967). *Invertebrate Zoology* (translated and adapted from second German edition by H. W. Levi and L. R. Levi). New York: Wiley, Interscience.

Kaufman, Z. S. (1960). Der Bau der Tracheolen bei einigen Chilopoden. *Dokl. Akad. Nauk. SSSR*, **130**, 693–6 (in Russian).

Kaufman, Z. S. (1961). Development and structure of the tracheal system in *Lithobius forficatus* L. *Zool. Zh.*, **40**, 503–11 (in Russian with English summary).

Kaufman, Z. S. (1962). The structure and development of stigmata in *Lithobius forficatus* L. (Chilopoda, Lithobiidae). *Ent. Obozr.*, **41**, 223–5 (in Russian with English summary).

Kaufman, Z. S. (1964). Structure of the tracheal system in *Cryptops* sp.

(Chilopoda, Scolopendromorpha, Crytopidae). *Ent. Obozr.*, **43**, 167–9 (in Russian with English summary).

Keegan, H. L. (1963). Centipedes and millipedes as pests in tropical areas. In *Venomous and poisonous animals and poisonous plants of the Pacific region* (ed. H. L. Keegan and W. V. Macfarlane), pp. 161–3. London: Pergamon.

Keil, T. (1976). Sinnesorgane auf den Antennen von *Lithobius forficatus* L. (Myriapoda, Chilopoda). 1. Die Funktionsmorphologie der 'Sensilla trichodea'. *Zoomorphol.*, **84**, 77–102.

Keil, T. (1977). Die Antennensinnes- und Hautdrüsenorgane von *Lithobius forficatus* L. Inaugural-Dissertation, Freien Universität, Berlin.

Khanna, V. (1977). Observations on the food and feeding habits of *Scolopendra valida* Lucas. *Geobios*, **4**, 51–3.

Klein, K. (1934). Über die Helligkeitsreaktionen einiger Arthropoden. *Z. wiss. Zool.*, **145**, 1–38.

Klingel, H. (1956). Indirekte Spermatophorenübertragung bei Chilopoden (Hundertfüsser) beobachtet bei der 'Spinnenassel' *Scutigera coleoptrata* Latzel. *Naturwissenschaften*, **43**, 311.

Klingel, H. (1957). Indirekte Spermatophorenübertragung beim Scolopender (*Scolopendra cingulata* Latreille; Chilopoda, Hundertfüsser). *Naturwissenschaften*, **44**, 338.

Klingel, H. (1959). Indirekte Spermatophorenübertragung bei Geophiliden (Hundertfüsser, Chilopoda). *Naturwissenschaften*, **46**, 632–3.

Klingel, H. (1960*a*). Vergleichende Verhaltensbiologie der Chilopoden *Scutigera coleoptrata* L. ('Spinnenassel') und *Scolopendra cingulata* Latreille (Scolopender). *Z. Tierpsychol.*, **17**, 11–30.

Klingel, H. (1960*b*). Die paarung des *Lithobius forficatus* L. *Verh. dt. zool. Ges.*, **23**, 326–32.

Klingel, H. (1962). Das Paarungsverhalten des malaischen Höhlentausendfusses *Thereuopoda decipiens cavernicola* Verhoeff (Scutigeromorpha, Chilopoda). *Zool. Anz.*, **169**, 458–60.

Knoll, H. J. (1974). Untersuchungen zur Entwicklungsgeschichte von *Scutigera coleoptrata* L. (Chilopoda). *Zool. Jb.* (*Anat.*), **92**, 47–132.

Koch, A. (1927). Studien an leuchtenden Tieren 1. Das Leuchten der Myriapoden. *Z. Morph. Ökol. Tiere*, **8**, 241–70.

Kowalewsky, A. (1895). Études des glandes lymphatiques de quelques Myriapodes. *Archs Zool. exp. gén.*, Ser. 3, **3**, 591–616.

Kraus, O. (1957). Eine Gen-Beeinflussung in der Natur als Folge der Radioktivität von Uranlagerstätten? *Natur u. Volk*, **87**, 399–401.

Krishnan, G. (1956). The nature and composition of the epicuticle of some arthropods. *Physiol. Zool.*, **29**, 324–37.

Künckel d'Herkulais, J. (1911). Observations sur les moeurs d'un myriapode, la scutigère coléoptrée. Son utilité comme destructrice des mouches; action de son venin; légende de sa présence accidentale dans l'appareil digestif de l'homme. *C. r. hebd. Séanc. Acad. Sci., Paris*, **153**, 399–401.

Labbé, A. (1899). *Das Tierreich* Lief. 5, Sporozoa. Berlin: Walter de Gruyter.

Latreille, P. A. (1817). Les Myriapodes. In, *Le régne animal*, vol. 3 (ed. G. L. C. F. D. Cuvier). Paris: Deterville.

Latzel, R. (1880). *Die Myriopoden der Österreichisch-Ungarischen Monarchie* I. *Die Chilopoden*. Vienna: Alfred Hölder.

Lawrence, R. F. (1947). Some observations on the post-embryonic development of the Natal forest centipede, *Cormocephalus multispinus* (Kraep.). *Ann. Natal. Mus.*, **11**, 139–56.

Lawrence, R. F. (1953). *The biology of the cryptic fauna of forests with special reference to the indigenous forests of South Africa.* Cape Town: Balkena.

Lawrence, R. F. (1955). Chilopoda, In *South Africa Animal Life*, vol. 2, pp. 4–56. Uppsala: Almqvist & Wiksells Boktryckeri AB.

Lawrence, R. F. (1960). Myriapodes Chilopodes. *Faune de Madagascar*, vol. 12, Tananarive: Institue de Recherche Scientifique.

Lawrence, R. F. (1968). Two new centipedes from southern Africa. *Ann. Cape. Prov. Mus. (Nat. Hist.)*, **6**, 77–9.

Lawrence, T. C. (1934). Notes on the feeding habits of *Scolopendra subspinipes* Leach (Myriopoda). *Proc. Hawaii ent. Soc.*, **8**, 497–8.

Le Moli, F. (1970). Experimental study of predation on wild-type, Oregon and white *Drosophila melanogaster* by *Scutigera coleoptrata. Ann. Inst. Museo Zool. Univ. Napoli*, **19**, 1–18.

Le Moli, F. (1972). Predation on *Drosophila melanogaster* by *Scutigera coleoptrata.* I. Genetic origin of a disadvantageous behaviour. *Atti Accad. Naz. Lincei (Rend. Sc.)*, **53**, 178–85.

Le Moli, F. (1975). Predation on *Drosophila melanogaster* by *Scutigera coleoptrata.* II. Experiments on single mutant genes with possible behavioural implications. *Ateneo Parmense, acta nat.*, **12**, 101–9.

Le Moli, F. (1977). Some behavioural aspects in *Scutigera coleoptrata* (L.). *Ateneo Parmense, acta nat.*, **13**, 669–71.

Léger, L. (1898). Essai sur la classification des Coccidies et description de quelques espèces nouvelles ou peu connues. *Bull. Mus. Nat. Hist., Marseilles*, **1**, 71–123.

Léger, L. & Duboscq, O. (1902). Les grégarines et l'épithelium intestinal chez les tracheata. *Archs Parasitol*, **6**, 377–473.

Léger, L. & Duboscq, O. (1909). Sur les *Chytridiopsis* et leur evolution. *Archs Zool. exp. gén.*, Ser. 5, **1**, Notes et Revue ix–xiii.

Lever, R. A. (1939). Irritant exudation from a centipede. *Nature, Lond.*, **143**, 78–9.

Levieux, J. (1972). Comportement d'alimentation et relations entre les individus chez un fourmi primitive *Amblypone pluto* Gotwald et Levieux. *C. r. hebd. Séanc. Acad. Sci., Paris*, **275**(D), 483–5.

Lévy, R. (1923*a*). Sur le mécanisme de l'hemolyse par le venin de Scolopendre. *C. r. hebd. Séanc. Acad. Sci., Paris*, **177**, 1326–8.

Lévy, (1923*b*). Sur les propriétés hémolytiques du venin de certaines Myriapodes Chilopodes. *Bull. Soc. zool. Fr.*, **48**, 294–7.

Lévy, R. (1927*a*). Intoxication de l'ecrevisse par le venins de deux myriapodes chilopodes. *C. r. Séanc. Soc. Biol.*, **96**, 256–57.

Lévy, (1927*b*). Action antitoxique du sang de *Lithobius forficatus* L. vis-à-vis du venin de la même espèce et vis-à-vis du de *Cryptops anomalans* Newp. *C. r. Séanc. Soc. Biol.*, **96**, 258.

Lewis, J. G. E. (1960). The life history and ecology of the littoral centipede *Strigamia maritima* (Leach). Ph.D. Thesis, University of London.

Lewis, J. G. E. (1961). The life history and ecology of the littoral centipede *Strigamia* (=*Scolioplanes*) *maritima* (Leach). *Proc. zool. Soc. Lond.*, **137**, 221–47.

Lewis, J. G. E. (1962). The ecology, distribution and taxonomy of the centipedes found on the shore in the Plymouth area. *J. mar. biol. Ass. UK*, **42**, 655–64.
Lewis, J. G. E. (1963). On the spiracle structure and resistance to desiccation of four species of geophilomorph centipede. *Ent. exp. appl.*, **6**, 89–94.
Lewis, J. G. E. (1965). The food and reproductive cycles of the centipedes *Lithobius variegatus* and *Lithobius forficatus* in a Yorkshire woodland. *Proc. zool. Soc. Lond.*, **144**, 269–83.
Lewis, J. G. E. (1966). The taxonomy and biology of the centipede *Scolopendra amazonica* in the Sudan. *J. Zool. Lond.*, **149**, 188–203.
Lewis, J. G. E. (1967). The scolopendromorph centipedes of the Sudan with remarks on taxonomic characters in the Scolopendridae. *Proc. Linn. Soc. Lond.*, **178**, 185–207.
Lewis, J. G. E. (1968). Individual variation in a population of the centipede *Scolopendra amazonica* from Nigeria and its implications for methods of taxonomic discrimination in the Scolopendridae. *J. Linn. Soc. (Zool.)*, **47**, 315–26.
Lewis, J. G. E. (1969*a*). Scolopendromorph and geophilomorph centipedes from Eritrea. *J. nat. Hist.*, **3**, 461–70.
Lewis, J. G. E. (1969*b*). The variation of the centipede *Scolopendra amazonica* in Africa. *Zool. J. Linn. Soc.*, **48**, 49–57.
Lewis, J. G. E. (1970). The biology of *Scolopendra amazonica* in Nigerian guinea savannah. *Bull. Mus. Hist. nat., Paris*, **41**, suppl. no. 2, 1969 (1970), 85–90.
Lewis, J. G. E. (1972*a*). The life histories and distribution of the centipedes *Rhysida nuda togoensis* and *Ethmostigmus trigonopodus* (Scolopendromorpha: Scolopendridae) in Nigeria. *J. Zool., Lond.*, **167**, 399–414.
Lewis, J. G. E. (1972*b*). The population density and biomass of the centipede *Scolopendra amazonica* (Bücherl) (Scolopendromorpha: Scolopendridae) in sahel savanna in Nigeria. *Ent. mon. Mag.*, **108**, 16–18.
Lewis, J. G. E. (1973). The taxonomy, distribution and ecology of centipedes of the genus *Asanada* (Scolopendromorpha. Scolopendridae) in Nigera. *Zool. J. Linn. Soc.*, **52**, 97–112.
Lewis, J. G. E. (1978). Variation in tropical scolopendrid centipedes: problems for the taxonomist. *Abh. Verh. naturwiss. Ver. Hamburg (NF)*, **21/22**, 43–50.
Lloyd, M. (1963). Numerical observations on movements of animals between beech litter and fallen branches. *J. Anim. Ecol.*, **32**, 157–63.
Luff, M. L. (1974). Adult and larval feeding habits of *Pterostichus madidus* (F.) (Coleoptera: Carabidae). *J. nat. Hist.*, **8**, 403–9.
Macé, M. (1886). Sur la phosphorescence des gèophiles. *C. r. hebd. Séanc. Acad. Sci., Paris*, **103**, 1273–4.
Macé, M. (1887). Les glandes préanales et la phosphorescence des géophiles. *C. r. Séanc. Soc. Biol.*, Ser. 8, **4**, 37–9.
McIver, S. B. (1975). Structure of cuticular mechanoreceptors of arthropods. *Ann. Rev. entom.*, **20**, 381–97.
MacLeod, J. (1878). Recherches sur l'appareil venimeux des myriapodes chilopodes. *Bull. Acad. roy. Belg.*, Ser. 2, **44**, 1–20.
Manton, S. M. (1950). The evolution of arthropod locomotory mechanisms. Part 1. The locomotion of *Peripatus*. *J. Linn. Soc. (Zool.)*, **41**, 529–70.
Manton, S. M. (1952*a*). The evolution of arthropod locomotory mechanisms. Part 2. General introduction to the locomotory mechanisms of the Arthropoda. *J. Linn. Soc. (Zool.)*, **42**, 93–117.

Manton, S. M. (1952*b*). The evolution of arthropod locomotory mechanisms. Part 3. The locomotion of Chilopoda and Pauropoda. *J. Linn. Soc.* (*Zool.*), **42**, 118–66.

Manton, S. M. (1958). Habits of life and evolution of body design in Arthropoda. *J. Linn. Soc.* (*Zool.*), **44**, 58–72.

Manton, S. M. (1964). Mandibular mechanisms and the evolution of arthropods. *Phil. Trans. R. Soc. Ser. B*, **247**, 1–183.

Manton, S. M. (1965). The evolution of arthropod locomotory mechanisms. Part 8. Functional requirements and body design in Chilopoda, together with a comparative account of their skeletomuscular systems and an appendix on the comparison between burrowing forces of annelids and chilopods and its bearing upon the evolution of the arthropodan haemocoel. *J. Linn. Soc.* (*Zool.*), **46**, 251–483.

Manton, S. M. (1970). Arthropods: Introduction. In *Chemical Zoology* (ed. M. Florin and B. T. Scheer) vol. 5, Arthropoda, Part A, pp. 1–34. London: Academic Press.

Manton, S. M. (1973). The evolution of arthropod locomotory mechanisms. Part 11. Habits, morphology and evolution of the Uniramia (Onychophora, Myriapoda, Hexapoda) and comparisons with the Arachnida, together with a functional review of uniramian musculature. *Zool. J. Linn. Soc.*, **53**, 257–375.

Manton, S. M. (1977). *The Arthropoda. Habits, functional morphology and evolution.* Oxford: Clarendon Press.

Manton, S. M. & Heatley, N. (1937). The feeding, digestion, excretion and food storage of *Peripatopsis. Phil. Trans. R. Soc. Ser. B*, **227**, 411–64.

Marlatt, C. L. (1914). The house centipede (*Scutigera forceps* Raf.). *Circ. US Dept. Agric. Ent.*, Ser. 2, no. 48, pp. 1–4.

Mathur, L. P. (1926). Preliminary observations on the general habits and the so-called poison glands of *Scolopendra morsitans* L. *Proc. Indian Sci. Congr.* **13**, 188.

Matic, Z. (1958). Două Lithobiidae noi pentru fauna R. P. R. şi unele observantii interesant la *Lithobius forficatus. Studii Cerc. Biol.*, **9**, 81–9.

Matic, Z. (1960). Beiträge zur Kenntnis der blinden *Lithobius*-Arten (Chilopoda–Myriapoda) Europas. *Zool. Anz.*, **164**, 443–8.

Matic, Z. (1961). Über die Häutung von *Lithobius forficatus* L. (Chilopoda-Lithobiidae). *Věst. čsl. Společ. zool.*, **25**, 131–4.

Matic, Z. (1962). Notă critica asupra genului *Lithobius* Leach, 1814 (Chilopoda-Lithobiidae) din Europa. *Studii Cerc. Biol.*, **13**, 87–102.

Matic, Z. (1964). Description d'un nouveau lithobiide cavernicole de Bulgarie (Chilopoda). *Annls Speleol.*, **19**, 507–10.

Matic, Z. (1966). *Fauna Republicii Socialiste România*, vol. 6, part 1. Clasa Chilopoda, subclasa Anamorpha. Bucharest: Academia Republicii Socialiste România.

Matic, Z. (1967). Description d'une espèce nouvelle cavernicole de *Lithobius* (Lithobiomorpha, Chilopoda) de l'Algérie. *Annls Speleol.*, **22**, 321–4.

Matic, Z. (1973). Pseudolithobiidae Fam. Nov.: Una nuova famiglia dell'ordine dei Lithobiomorpha (Chilopoda, Anamorpha). *Fragm. Entomologica*, **9**, 134–42.

Mayer, H. (1957). Zur Biologie und Ethologie einheimischer Collembolen. *Zool. Jb.* (*Syst.*), **85**, 501–89.

Mead, M. (1968). Etudes des rythmes d'activité locomotrice chez quelques arthropodes terricoles (Chilopodes, Isopodes). *Rev. Comp. animal*, **2**, 82–101.

Mead, M. (1970). Sur l'obtention d'un rythme d'activité de periode sept jours chez *Scolopendra cingulata* (Chilopodes). *Rev. Comp. animal*, **4**, 75–6.

Mead-Briggs, A. R. (1956). The effect of temperature upon the permeability to water of arthropod cuticles. *J. exp. Biol.*, **33**, 737–49.

Meinert, F. (1870). Myriapoda Musaei Hauniensis 1. Geophili. *Naturh. Tidsskr.*, **3**, 241–68.

Meinert, F. (1883). *Caput Scolopendrae*. Copenhagen: Hagerup.

Meissner, K. (1978). Sequenzen und Zeitmuster in der Putz – und Laufaktivität von Larven des Chilopoden *Lithobius forficatus* (L.). *Biol. Zbl.*, **97**, 265–77.

Meske, C. (1960). Schallreaktionen von *Lithobius forficatus*. *Z. vergl. Physiol.*, **43**, 526–30.

Meske, C. (1961). Untersuchungen zur Sinnesphysiologie von Diplopoden und Chilopoden. *Z. vergl. Physiol.*, **45**, 61–77.

Metschnikoff, E. (1875). Embryologische über *Geophilus*. *Z. wiss. Zool.*, **25**, 313–22.

Minelli, A. (1977). Centipedes. In *Arthropod venoms. Handbook of Experimental Pharmacology* (ed. S. Bettini), vol. 48, pp. 73–85. Berlin: Springer-Verlag.

Misioch, M. (1978). Variation of characters in some geophilid centipedes. *Abh. Verh. naturwiss. Ver. Hamburg (NF)*, **21/22**, 55–62.

Misra, P. L. (1942). On the life history of a new gregarine, *Grebneckiella pixellae* sp. nov., from the centipede *Scolopendra morsitans* Linn., with a note on the family Dactylophoridae Lèger, 1892. *Rec. Indian Mus.*, **44**, 323–37.

Monteith, L. G. (1976*a*). Laboratory feeding studies of potential predators of the apple maggot *Rhagoletis pomonella* (Diptera: Tephritidae) in Ontario. *Proc. entom. Soc. Ont.*, 1975(1976), **106**, 28–33.

Monteith, L. G. (1976*b*). Field studies of potential predators of the apple maggot *Rhagoletis pomonella* (Diptera; Tephritidae) in Ontario. *Proc. entom. Soc. Ont.*, **107**, 23–30.

Morris, H. M. (1922). The insect and other invertebrate fauna of arable land at Rothamsted. *Ann. appl. Biol.*, **9**, 282–305.

Morris, H. M. (1927). The insect and other invertebrate fauna of arable land at Rothamsted. *Ann. appl. Biol.*, **14**, 442–64.

Murakami, Y. (1956*a*). The developmental stadia of *Thereuonema hilgendorfi* Verhoeff (Chilopoda, Scutigeridae). *Zool. Mag., Tokyo*, **65**, 37–41 (in Japanese with English summary).

Murakami, Y. (1956*b*). The life history of *Thereuonema hilgendorfi* Verhoeff (Chilopoda, Scutigeridae). *Zool. Mag., Tokyo*, **65**, 42–6 (in Japanese with English summary).

Murakami, Y. (1958*a*). The food habit of *Thereuonema higendorfi* Verhoeff (Chilopoda, Scutigeridae), *Zool. Mag., Tokyo*, **67**, 138–41 (in Japanese with English summary).

Murakami, Y. (1958*b*). The life history of *Bothropolys asperatus* Koch (Chilopoda, Lithobiidae). *Zool. Mag., Tokyo*, **67**, 217–23 (in Japanese with English summary).

Murakami, Y. (1959*a*). Postembryonic development of the common Myriapoda of Japan I. The anamorphic development of the leg-bearing segments of Scutigeridae and a new aspect on the problem of its tergite. *Zool. Mag., Tokyo*, **68**, 193–9 (in Japanese with English summary).

Murakami, Y. (1959*b*). Postembryonic development of the common Myriapoda of Japan II. *Thereuopoda ferox* Verhoeff (Chilopoda, Scutigeridae). *Zool. Mag., Tokyo*, **68**, 324–9 (in Japanese with English summary).

Murakami, Y. (1960*a*). Postembryonic development of the common Myriapoda of Japan III. *Lithobius pachypedatus* Takakuwa 1. Anamorphic stadia. *Zool. Mag., Tokyo*, **69**, 121–24 (in Japanese with English summary).

Murakami, Y, (1960*b*). Postembryonic development of the common Myriapoda of Japan IV. *Lithobius pachypedatus* Takakuwa 2. Epimorphic stadia. *Zool. Mag., Tokyo*, **69**, 163–6 (in Japanese with English summary).

Murakami, Y. (1960*c*). Postembryonic development of the common Myriapoda of Japan V. *Lithobius pachypedatus* Takakuwa 3. Variation in the number of articles of antennae and coxal pores. *Zool. Mag., Tokyo*, **69**, 169–70 (in Japanese with English summary).

Murakami, Y. (1961*a*). Postembryonic development of the common Myriapoda of Japan VII. *Monotarsobius nihamensis* Murakami (Chilopoda; Lithobiidae) 1. Hemianamorphic stadia of the female. *Zool. Mag., Tokyo*, **70**, 125–30 (in Japanese with English summary).

Murakami, Y. (1961*b*). Postembryonic development of the common Myriapoda of Japan IX. Anamorphic stadia of *Esastigmatobius longitarus* Verhoeff (Chilopoda; Henicopidae). *Zool. Mag., Tokyo*, **70**, 430–4 (in Japanese with English summary).

Murakami, Y. (1963). Postembryonic development of the common Myriapoda of Japan XV. A new species of subgenus *Archilithobius* (Chilopoda, Lithobiidae) and its hemianamorphic stadia. *Zool. Mag., Tokyo*, **72**, 199–203 (in Japanese with English summary).

Murakami, Y. & Paik, K. Y. (1968). Results of the speleological survey in South Korea 1966. XI. Cave-dwelling myriapods from the southern part of Korea. *Bull. Nat. Sci. Mus. Tokyo*, **11**, 363–84.

Muruzesan, R. & Azariah, J. (1972). On the association between the common Indian gerbille *Tatera indica* (Rodentia: Muridae) and the house centipede *Scutigera longicornis*. *Zool. Anz.*, **189**, 326–31.

Narasimhamurti, C. C. (1976). Observations on the morphology and life history of a coccidian *Eimeria mecistophori* n.sp. from the gut of a centipede *Mecistocephalus punctifrons* Newp. *Proc. Indian Acad. Sci., Sect. B*, **84**, 141–7.

Needham, A. E. (1945). On relative proportions in serially repeated structures. *Proc. zool. Soc. Lond.*, **115**, 355–70.

Needham, A. E. (1958). Connective tissue pigment of the centipede, *Lithobius forficatus* (L.). *Nature, Lond.*, **181**, 194–5.

Needham, A. E. (1960). Properties of the connective tissue pigment of *Lithobius forficatus* (L.). *Comp. Biochem. Physiol.*, **1**, 72–100.

Needham, A. E. (1974). *The significance of zoochromes.* Berlin: Springer-Verlag.

Negrea, S. (1969). Aperçu sur les chilopodes cavernicoles de Roumanie. *Bull. Mus. Hist. nat., Paris*, **41**, Suppl. no. 2, 1969(1970), 102–7.

Negrea, S. (1977). Considérations écologiques et biogéographiques sur les Chilopodes de Cuba. In *Résultats des expéditions biospeléologiques Cubano-Roumaines à Cuba*, vol. 2, pp. 303–12. Bucharest: Academia Republicii Socialiste România.

Newman, E. (1867). A *Proctotrupes* parasite of a myriapod. *Entomologist*, **3**, 342–4.

Newport, G. (1843). On the structure, relations and development of the nervous and circulatory systems, and on the existence of a complete circulation of the blood in vessels, in Myriapoda and macrourous Arachnida. *Phil. Trans. R. Soc.*, **133**, 243–302.

Newport, G. (1844). Monograph of the class Myriopoda, order Chilopoda; with observations on the general arrangement of the Articulata. *Trans. Linn. Soc. Lond.* (*Zool.*), **19**, 265–302, 349–439.

Nielsen, C. O. (1962). Carbohydrases in soil and litter invertebrates. *Oikos*, **13**, 200–15.

Norman, W. W. (1897). The poison of centipedes (*Scolopendra morsitans*). *Trans. Texas Acad. Sci.*, **1**, 118 19.

Okeden, W. P. (1903). A centipede eating a snake. *J. Bombay nat. Hist. Soc.*, **15**, 1.

Ormières, R. (1966). Grégarines parasites de Myriapodes Chilopodes: Observations sur les genres *Echinomera* Labbé 1899 et *Acutispora* Crawley 1903. *Protistologica*, **2**, 15–21.

Ormières, R. (1967). Grégarines parasites de Myriapodes Chilopodes: *Rhopalonia stella* Léger, 1899 et un *Actinocephalus*. *Bull. Soc. zool. Fr.*, **92**, 537–541.

Ormières, R. & Marquès, A. (1976). Fixation à leurs hôtes de quelques Dactylophoridae, Eugrégarines parasites de Myriapodes, Chilopodes. Etude ultrastructurale. *Protistologica*, **12**, 415–24.

Ormières, R., Marquès, A. & Puissegur, C. (1977). *Trichorhynchus pulcher* Schneider, 1882, eugrégarine parasite de *Scutigera coleoptrata* L. Cycle, ultrastructure, systématique. *Protistologica*, **13**, 407–17.

Palm, N.-B. (1953). The elimination of injected vital dyes from the blood in myriapods. *Ark. Zool.*, Ser. 2, **6**, 219–46.

Palm, N.-B. (1955). Neurosecretory cells and associated structures in *Lithobius forficatus* L. *Ark. Zool.* **9**, 115–129.

Palmen, E. (1948). The Chilopoda of eastern Fennoscandia. *Ann. Soc. Zool. Fenn. Vanamo*, **13**, 1–45.

Palmen, E. & Rantala, M. (1954). On the life-history and ecology of *Pachymerium ferrugineum* (C. L. Koch) (Chilopoda, Geophilidae). *Ann. Soc. Zool. Fenn. Vanamo*, **16**, 1–44.

Passerini, N. (1882). Sull'organo ventrale de *Geophilus gabrielis* Fabr. *Boll. Soc. entomol. Ital.*, **14**, 323–8.

Paulus, H. F. (1979). Eye structure and the monophyly of the Arthropoda. In *Arthropod Phylogeny* (ed. A. P. Gupta), pp. 299–383. New York: van Nostrand Reinhold.

Pawlowsky, E. (1913). Ein Beitrag zur Kenntnis des Baues der Giftdrüsen von *Scolopendra morsitans*. *Zool. Jb.* (*Anat.*), **36**, 91–112.

Pereira, L. A. & Coscaron, S. (1976). Estudos sobre geophilomorphos neotropicales 1. Sobre dos especies nuevas del genero *Pectiniunguis* Bollman (Schendylidae–Chilopoda). *Revta Soc. ent. argent.*, **35**, 59–75.

Pflugfelder, O. (1933). Über den feineren Bau der Schläfenorgane der Myriapoden. *Z. wiss. Zool.*, **143**, 127–55.

Pianka, E. R. (1968). Notes on the biology of *Varanus eremius*. *West Aust. Natur.*, **11**, 39–44.

Plateau, F. (1878). Recherche sur les phénomènes de la digestion et sur la structure de l'appareil digestif chez les Myriapodes de Belgique. *Acad. r. Sci. Lett. Belg.*, **42**, 1–91.

Plateau, F. (1886). Recherches sur la perception de la lumière par les Myriapodes aveugles. *J. Anat. Physiol., Paris*, **22**, 431–57.

Plateau, F. (1890). Les myriopodes marins et la résistance des arthropodes à respiration aérienne a la submersion. *J. Anat. Physiol., Paris*, **26**, 136–269.

Plowman, T. (1896). A luminous centipede. *Nature, Lond.*, **53**, 249.

Pocock, R. I. (1900). Marine centipede in Somerset. *Zoologist*, **4**, 484–5.

Pocock, R. I. (1902). A new and annectant type of chilopod. *Quart. J. micr. Sci.*, **45**, 417–48.

Pollack, W. (1976). Über die Wirkung von exogenem Ecdysteron auf die Mitose-Aktivität in den Laufbeinknopsen des Chilopoden *Lithobius forficatus* (L.). *Wiss. Z. pädag. Hochsch. 'Dr Theodor Neubauer' Erfurt-Mülhausen.*, **13**, 136–8.

Powders, U. N. & Tietven, W. L. (1974). The comparative food habits of sympatric and allopatric salamanders, *Plethodon glutinosus* and *Plethodon jordani* in eastern Tennessee and adjacent areas. *Herpetologia*, **30**, 167–75.

Prunesco, C. (1963). Anatomical observations of male reproductive system order Lithobiomorpha (Chilopoda, Tracheata). *Rev. Roum. Biol.*, **8**, 357–66 (in Russian).

Prunesco, C. (1964). Anatomie microscopique du systéme génital mâle des lithobiidés (Lithobiomorpha, Chilopoda). *Rev. Roum. Biol., -Zool.*, **9**, 101–7.

Prunesco, C. (1965*a*). Contribution à l'étude anatomique et anatomo-microscopique du système génital femelle de l'ordre Lithobiomorpha. *Rev. Roun. Biol.,-Zool.*, **10**, 10–16.

Prunesco, C. (1965*b*). Contribution à l'étude de l'évolution des Chilopodes. *Rev. roum. Biol.,-Zool.*, **10**, 89–102.

Prunesco, C. (1965*c*). Système génital femelle du genre *Cryptops* (Scolopendromorpha, Chilopoda). *Rev. Roum. Biol.,-Zool.*, **10**, 231–5.

Prunesco, C. (1965*d*). Les systèmes génital et trachéal de *Craterostigmus* (Craterostigmomorpha, Chilopoda). *Rev. Roum. Biol.,-Zool.*, **10**, 309–314.

Prunesco, C. (1965*e*). Le système génital femelle d'*Ethmostigmus trigonopodus* (Otostigmini, Chilopoda). *Rev. Roum. Biol.,-Zool.*, **10**, 407–11.

Prunesco, C. (1967*a*). Le système génital femelle de l'ordre Geophilomorpha. *Rev. Roum. Biol.,-Zool.*, **12**, 251–6.

Prunesco, C. (1967*b*). Le système génital femelle de *Scutigera coleoptrata* L. (Scutigeromorpha, Chilopoda). *Rev. Roum. Biol.,-Zool.*, **12**, 315–20.

Prunesco, C. (1968). Le système génital mâle chez quatre espèces de chilopodes de l'ordre des Geophilomorpha. *Rev. Roum. Biol.,-Zool.*, **13**, 57–62.

Prunesco, C. (1970*a*). Considérations sur l'évolution du système génital des Chilopodes. *Bull. Mus. Hist. nat., Paris*, **41**, suppl. no. 2, 1969(1970), 108–111.

Prunesco, C. (1970*b*). Quelle est la place occupée par *Cermatobius, Craterostigmus,* et *Plutonium* dans la phylogénie des Chilopodes? *Bull. Mus. Hist. nat., Paris*, **41**, suppl. no. 2, 1969(1970), 112–15.

Prunescu, C. (1970*c*). Les cellules neurosécrétrices des ganglions ventraux des Chilopodes anamorphes. *Rev. Roum. Biol.,-Zool.*, **15**, 147–51.

Prunesco, C. (1970*d*). Les cellules neurosécrétrices des ganglions ventraux des Chilopodes épimorphes. *Rev. Roum. Biol.,-Zool.*, **15**, 323–7.

Prunescu, C. & Capuşe, I. (1972). Nouvelles données sur le début de developpment postembryonaire chez les Myriapodes de l'ordre Geophilomorpha. *Trav. Mus. Hist. nat. Gr. Antipa*, **11**, 111–19.

Prunescu, C. & Johns, P. M. (1969). An embryonic gonad in adult males of *Anopsobius neozelandicus* Silv. (Chilopoda). *Rev. Roum. Biol.*, **14**, 407–9.
Rajulu, G. S. (1965), Leeches as endoparasites in centipedes. *Curr. Sci.*, **34**, 408–9.
Rajulu, G. S. (1966). Cardiac physiology of a chilopod *Scolopendra morsitans*. *J. Anim. Morph. Physiol.*, **13**, 114–20.
Rajulu, G. S. (1967*a*). The nature of the proteolytic enzyme systems in *Scolopendra heros*, a chilopod, *Zool. Jb.* (*Physiol.*), **73**, 276–80.
Rajulu, G. S. (1967*b*). Antennal pulsatile organs in *Scolopendra morsitans* (Chilopoda: Myriapoda). *Curr. Sci.*, **36**, 242–3.
Rajulu, G. S. (1968*a*). Neuroendocrine regulation of cardiac activity in *Scolopendra morsitans* (Chilopoda: Myriapoda). *Biol. Zbl.*, **87**, 147–52.
Rajulu, G. S. (1968*b*). On the nature of the pigments of a centipede *Scolopendra morsitans*. *Sci. Cult.*, **34**, 297.
Rajulu, G. S. (1969*a*). Presence of haemocyanin in the blood of a centipede *Scutigera longicornis* (Chilopoda: Myriapoda). *Curr. Sci.*, **38**, 168–9.
Rajulu, G. S. (1969*b*). Blood proteins of *Scolopendra morsitans*, a centipede (Chilopoda: Myriapoda). *Curr. Sci.*, **38**, 472–3.
Rajulu, G. S. (1970*a*). A study of the haemocytes of a centipede *Ethmostigmus spinosus* (Chilopoda: Myriapoda). *Curr. Sci.*, **39**, 324–5.
Rajulu, G. S. (1970*b*). Tracheal pulsation in a marine centipede *Mixophilus indicus*. *Curr. Sci.*, **39**, 397–8.
Rajulu, G. S. (1970*c*). Presence of caecal outgrowths in the alimentary canal of a centipede *Ethmostigmus spinosus* Newport (Chilopoda: Myriapoda). *Curr. Sci.*, **39**, 564–5.
Rajulu, G. S. (1970*d*). Studies on the nature of carbohydrases in a centipede *Scolopendra heros*, together with observations on hydrogen ion concentration of the alimentary tract. *J. Anim. Morph. Physiol.*, **17**, 56–64.
Rajulu, G. S. (1970*e*). Study on the chemo- and mechano-receptors in the last pair of legs of a geophilomorph centipede *Himantarium samuelraji* Rajulu (Chilopoda: Myriapoda) *Monitore zool. ital.* (N.S.), **4**, 55–62.
Rajulu, G. S. (1970*f*). A study on the nature and formation of the spermatophore in a centipede *Ethmostigmus spinosus* (Scolopendromorpha: Myriapoda). *Bull. Mus. Hist. nat., Paris*, **41**, suppl. no. 2, 1969 (1970), 116–21.
Rajulu, G. S. (1971*a*). An electronmicroscopical study of the ultra-structure of the peritrophic membrane of a chilopod *Ethmostigmus spinosus* together with observations on its chemical composition. *Curr. Sci.*, **40**, 134–5.
Rajulu, G. S. (1971*b*). Presence of resilin in the cuticle of the centipede *Scolopendra morsitans* L. *Indian J. exp. Biol.*, **9**, 122–3.
Rajulu, G. S. (1971*c*). X-ray diffraction and electronmicroscopic studies on the fine structure of the tracheae of the centipede *Scutigera longicornis*, together with observations on their chemical composition. *Curr. Sci.*, 40, 467–8.
Rajulu, G. S. (1971*d*). A study of haemocytes in a centipede *Scolopendra morsitans* (Chilopoda, Myriapoda). *Cytologia Jap.*, **36**, 515–21.
Rajulu, G. S. (1972). On the mode of respiration of an estuarine centipede *Mixophilus indicus*. *J. Anim. Morph. Physiol.*, **19**, 181–90.
Rajulu, G. S. (1973*a*). Free amino acids in the haemolymph of Myriapoda. *Curr. Sci.*, **42**, 95–6.
Rajulu, G. S. (1973*b*). Moult cycle of a centipede *Ethmostigmus spinosus* (Chilopoda: Myriapoda). *Curr. Sci.*, **42**, 205–6.

Reinhard, H. J. (1935). New genera and species American muscoid flies (Tachinidae: Diptera). *Ann. ent. Soc. Am.*, **28**, 160–173.

Remmington, C. L. (1950). The bite and habits of the giant centipede (*Scolopendra subspinipes*) in the Philippine Islands. *Am. J. trop. Med.*, **30**, 453–5.

Remy, P. A. (1950). On the enemies of myriapods. *Naturalist, Hull.*, 103–8.

Ribaut, H. (1921). L'armement des pattes chez les lithobies. *Bull. Soc. Hist. nat., Toulouse*, **49**, 312–19.

Richards, O. W. & Davies, R. G. (1977). *Imm's general textbook of entomology* (10th ed) vol. 1, Structure, physiology and development, London: Chapman & Hall.

Riddle, W. A. (1976). Respiratory metabolism of the centipede *Nadabius coloradensis* (Cockerell): influence of temperature, season and starvation. *Comp. Biochem. Physiol.*, **55A**, 147–51.

Ridley, N. N. (1936). The luminous secretion of the centipede *Geophilus* as a defence against the attack of beetles. *Proc. R. ent. Soc. Lond. A*, **11**, 48.

Rilling, G. (1960). Zur Anatomie des braunen Steinläufers *Lithobius forficatus* L. (Chilopoda). Skelettmuskelsystem, peripheres Nervensystem and Sinnesorgane des Rumpfes. *Zool. Jb. (Anat.)*, **78**, 39–128.

Rilling, G. (1968). *Lithobius forficatus.* In *Grosses Zoologisches Praktikum*, part 13b. Stuttgart: Fischer.

Roberts, H. (1956). An ecological study of the arthropods of a mixed beech-oak woodland, with particular reference to Lithobiidae. PhD thesis, University of Southampton.

Rosenberg, J. (1973). Eine bisher unbekannte endokrine Drüse im Kopf von *Scutigera coleoptrata* (Chilopoda, Notostigmophora). *Experientia*, **29**, 690–1.

Rosenberg, J. (1974). Topographie und Ultrastruktur der endokrinen Kopfdrüsen (glandulae capitis) von *Scutigera coleoptrata* L. (Chilopoda; Notostigmophora). *Z. Morph. Ökol. Tiere*, **79**, 311–21.

Rosenberg, J. (1976). Die Ultrastruktur der gabeschen Organe ('Cerebral-drüsen') von *Scutigera coleoptrata* L. (Chilopoda, Notostigmophora). *Zool. Beitr.*, **22**, 281–305.

Rosenberg, J. & Seifert, G. (1975). Ist allein die Glandula ecdysialis die Hätungsdrüse von *Lithobius*? *Experientia*, **31**, 1100.

Rosenberg, J. & Seifert, G. (1977). The coxal glands of Geophilomorpha (Chilopoda): Organs of osmoregulation. *Cell Tiss. Res.*, **182**, 247–51.

Rosenberg, J. & Seifert, G. (1978). Die myelinscheide um Zentralnervensystem und periphere Nerven der Geophilomorpha (Chilopoda). *Zoomorphologie*, **89**, 21–31.

Rossi, G. (1902). Sulla organizzazione dei Myriapodi. *Ricerche Labor. Anat. Roma* **9**, 5–88.

Rubin, D. (1969). Food habits of *Plethodon longicrus* Alder & Dennis. *Herpetologia*, **25**, 102–5.

Rudge, M. R. (1968). The food of the common shrew *Sorex araneus* L. (Insectivora: Soricidae) in Britain. *J. Anim. Ecol.*, **37**, 565–81.

Ryan, G. E. & Croft, J. D. (1974). Observations on the food of the fox, *Vulpes vulpes* (L.) in Kinchega National Park, Menindee, NSW. *Aust. Wildl. Res*, **1**, 89–94.

Salt, G., Hollick, F. S. J., Raw, F. & Brian, M. V. (1948). The arthropod population of pasture soil. *J. Anim. Ecol.*, **17**, 139–50.

Samouelle, G. (1819). *The entomologists useful compendium, or an introduction to the knowledge of British Insects.* London.

Schäfer, M. W. (1972). Reverse turning in *Lithobius forficatus* L. *Monit. Zool. ital.*, **6**, 179–94.

Schäfer, M. W. (1976*a*). Thigmotactic behaviour in *Lithobius forficatus* L. (Myriapoda, Chilopoda). *Monit. Zool. ital.*, **10**, 191–204.

Schäfer, M. W. (1976*b*). Aspects of spontaneous orientation and bias in the maze-behaviour of *Lithobius forficatus* L. (Myriapoda, Chilopoda) and the white mouse (*Mus musculus* L.). *Monit. Zool. ital.*, **10**, 205–17.

Scharmer, J. (1935). Die Bedeutung der Rechts-Links-Struktur und die Orientierung bei *Lithobius forficatus. Zool. Jb. (Physiol.)*, **54**, 459–506.

Schaudinn, F. (1900). Untersuchungen über den Generationswechsel bei Coccidien. *Zool. Jb. (Anat.)*, **13**, 197–292.

Schaufler, B. (1889). Beitrage zur Kenntniss der Chilopoden. *Verh. zool.-bot. Ges. Wien*, **39**, 465–78.

Scheffel, H. (1961). Untersuchungen zur Neurosekretion bei *Lithobius forficatus* L. (Chilopoda). *Zool. Jb. (Anat.)*, **79**, 529–56.

Scheffel, H. (1963). Zur Häutungsphysiologie der Chilopoden. *Zool. Jb. (Physiol.)*., **70**, 284–90.

Scheffel, H. (1965*a*). Über die Wirkung implantierter Cerebraldrüsen auf die Larvenhäutungen bei *Lithobius forficatus* L. (Chilopoda). *Zool. Anz.*, **174**, 173–8.

Scheffel, H. (1965*b*). Der Einfluss von Dekapitation und Schürung auf die Häutung und die Anamorphose der Larven von *Lithobius forficatus* L. (Chilopoda). *Zool. Jb. (Physiol.)*, **71**, 359–70.

Scheffel, H. (1965*c*). Elektronenmikroskopische Untersuchungen über den Bau der Cerebraldrüse der Chilopoden. *Zool. Jb. (Physiol.)*, **71**, 624–40.

Scheffel, H. (1969). Untersuchungen über die hormonale Regulation von Häutung und Anaporphose von *Lithobius forficatus* (L.) (Myriapoda, Chilopoda). *Zool. Jb. (Physiol.)*, **74**, 436–505.

Scheffel, H. & Wilke, C. (1974). Fördernder Einfluss von Actinomycin D auf die Häutunsauslösung durch exogenes Ecdysteron bei Chilopoden-Larven. *Zool. Jb. (Physiol.)*, **78**, 33–9.

Schildknecht, H., Maschwitz, U. & Krauss, D. (1968). Blausäure im Wehrsekret des Erdlaufers *Pachymerium ferrugineum. Naturwissenschaften*, **55**, 230.

Schubart, O. (1929). Thalassobionte und thalassophile Myriopoda. *Tierwelt der Nord und Ostsee*, **9**, 1–20.

Schubart, O. (1960). Die Zahl der in 200 Jahren zoologischer Forschung (1758–1957) beschriebenen Myriapoden-Arten. *Zool. Anz.*, **165**, 84–9.

Seifert, G. (1967*a*). Der pharynxapparat von *Scutigera coleoptrata* L. *Z. Morph. Ökol. Tiere*, **58**, 347–54.

Seifert, G. (1967*b*). Das stomogastrische Nervensystem der Chilopoden. *Zool. Jb. (Anat.)*, **84**, 167–90.

Seifert, G. & Rosenberg, J. (1974). Elektronenmikroskopische Untersuchungen der Häutungsdrüsen ('Lymphstränge') von *Lithobius forficatus* L. (Chilopoda). *Z. Morph. Ökol. Tiere*, **78**, 263–79.

Shinohara, K. (1961). Survey of the Chilopod and Diplopoda of Manazuru seashore Kanagawa Prefecture, Japan. *Science Report of the Yokosuka City Museum*, no. 6, 75–82.

Shinohara, K. (1970). On the phylogeny of Chilopoda. *Proc. Japan. Soc. Syst. Zool.*, no. 6, 35–42.
Shipley, A. E. (1914). Pseudoparasitism. *Parasitology*, **6**, 351–2.
Shrivastava, S. C. (1971). Studies on the cuticle of the Indian common chilopod *Scolopendra morsitans* L. *Indian J. Ent.*, **33**, 183–93.
Shugg, H. B. (1961). Predation on the mouse by centipede. *W. Aust. Nat.*, **8**, 52.
Simon, H. R. (1960). Zur Ernährungsbiologie von *Lithobius forficatus* (Myriapoda, Chilopoda). *Zool. Anz.*, **164**, 19–26.
Sinclair, F. G. (1895). Myriapods. In *The Cambridge Natural History* (ed. S. F. Harmer and A. E. Shipley), vol. 5, pp. 27–80. London: Macmillan.
Snodgrass, R. E. (1952). *A textbook of arthropod anatomy*. New York: Comstock.
Sograff, N. (1882). Zur Embryologie der Chilopoden. *Zool. Anz.*, **5**, 582–5.
Sograff, N. (1883). Materialen zur Kenntnis der Embryonalentwicklung von *Geophilus ferrugineus* L. K. und *Geophilus proximus* L.K. *Nachricht. Ges. Freunde Naturkunde, Anthropol. Ethnolog., Moskau*, **43**, 1–77.
Starling, J. H. (1944). Ecological studies of the pauropods of the Duke Forest. *Ecol. Monogr.*, **14**, 292–310.
Sunderland, K. D. (1975). The diet of some predatory arthropods in cereal crops. *J. appl. Ecol.*, **12**, 507–16.
Suomalainen, P. (1939). Zur Verbreitungsökologie von *Pachymerium ferrugineum* C. Koch im finnischen Schärenhof. *Ann. Zool. Soc. Vanamo*, **7**, 10–14.
Sutcliffe, D. W. (1963). The chemical composition of haemolymph in insects and some other arthropods, in relation to their phylogeny. *Comp. Biochem. Physiol.*, **9**, 212–35.
Sutton, S. L. (1970). Predation on woodlice; an investigation using the precipitin test. *Entomologia exp. appl.*, **13**, 279–85.
Templeton, R. (1846). On the habits and bite of Scolopendrae of Ceylon. *Ann. Mag. nat. Hist.*, Ser. 1, **17**, 65, 495.
Thomas, R. H. (1896). A luminous centipede. *Nature, Lond.*, **53**, 131.
Thomas, R. H. (1902). A luminous centipede. *Nature, Lond.*, **65**, 223.
Thompson, M. (1924). The soil population. An investigation of the biology of the soil in certain districts of Aberystwyth. *Ann. appl. Biol.*, **11**, 249–394.
Thompson, W. D'Arcy (1889). A marine millipede. *Nature, Lond.*, **41**, 176–7.
Thompson, W. R. (1915). Sur le cycle évolutif de *Fortisia foeda*, Diptère parasite d'un *Lithobius*. *C. r. Séanc. Soc. Biol.*, **78**, 413–16.
Thompson, W. R. (1939). Biological control and the theories of the interactions of populations. *Parasitology*, **31**, 299–388.
Tichy, H. (1973). Untersuchungen über die Feinstruktur des Tömösváryschen Sinnesorgans von *Lithobius forficatus* L. (Chilopoda) und zur Frage seiner Funktion. *Zool. Jb. (Anat.)*, **91**, 93–139.
Tobias, D. (1974). New criteria for the differentiation of species within the Lithobiidae. *Symp. zool. Soc. Lond.*, no. 32, 75–87.
Tönniges, C. (1902). Beiträge zur Spermatogenese und Oogenese der Myriopoden. *Z. wiss. Zool.*, **71**, 328–58.
Trauberg, O. (1932). Einige für Lettland neue Arten der der Gattungen *Geophilus* und *Clinopodes*, nebst einigen Bermerkungen über die Variabilität von *Lithobius forficatus* Linné (Chilopoden). *Folia Zool. hydrobiol.*, **4**, 6–12.
Turk, F. A. (1951). Myrapodological notes. III. The iatro-zoology, biology and systematics of some tropical 'myriapods'. *Ann. Mag. nat. Hist.*, Ser. 12, **4**, 35–48.

Turk, F. A. & Turk, S. M. (1958). *The foreshore of Cawsand Bay and district its fauna and flora.* Plymouth.

Tuzet, O. & Manier, J. F. (1953). Les spermatozoides de quelques myriapodes chilopodes et leur transformation dans le réceptacle séminal de la femelle. *Annls Sci. nat. (Zool.)*, Ser. 11, **15**, 221–30.

Vaitilingham, S. (1960). The ecology of the centipede of some Hampshire woodlands. M.Sc. thesis, University of Southampton.

Varma, L. (1972). Muscle receptor organs in the centipede *Scolopendra morsitans* L. *Zool. Anz.*, **188**, 400–7.

Varma, L. (1973). Giant fibres in the ventral nervous system of the centipede *Scolopendra morsitans* L. *Zool. Beitr.* **19**, 63–71.

Vaughan, T. A. (1976). Nocturnal behaviour of the African false Vampire Bat (*Cardioderma cor*). *J. Mammal.*, **57**, 227–48.

Venzmer, G. (1932). *Giftige Tiere und tierische Gifte.* Berlin.

Verhoeff, K. W. (1902–25). Chilopoda. In *Kl. Ordn. Tierreichs* (ed. H. G. Bronn), vol. 5, part 2, book 1. Leipzig: Akademische Verlagsgesellschaft.

Verhoeff, K. W. (1905). Über die Entwicklungstufen der Steinlaüfer, Lithobiiden, und Beiträge zur Kenntnis der Chilopoden. *Zool. Jb., suppl.* 8. *Festchr. Möbius.*, 195–298.

Verhoeff, K. W. (1938). Die europaïsche Spinnen-Assel *Scutigera. Natur Volk*, **68**, 442–8.

Verhoeff, K. W. (1940). Chilopoden-Kieferfuss-Regenerate in freier Natur. *Z. Morph. Ökol. Tiere*, **36**, 645–50.

Verhoeff, K. W. (1941). Zur Kenntnis der Chilopodenstigmen. *Z. Morph. Ökol. Tiere*, **38**, 96–117.

Voinstvenskii, M. A., Petrusenko, A. A. & Boyarchuk, V. P. (1977). Trophic relations of the rook (*Corvus frugilegus* L.) in steppe ecosystem: 1 – Nutrition (diet composition). *Vestn. Zool.*, **6**, 9–17 (in Russian).

Wang, T. H. & Wu, H. W. (1948). On the structure of the Malpighian tubes of centipedes and their excretion of uric acid. *Sinesia*, **18**, 1–11.

Waterhouse, J. S. (1969). An evaluation of a new predaceous centipede *Lamyctes* sp., on the garden symphylan *Scutigerella immaculata. Can. Ent.*, **101**, 1081–3.

Weil, E. (1958). Biologie der einheimischen Geophiliden. *Z. angew. Ent.*, **42**, 173–209.

Wells-Cole, H. (1898). A voracious centipede. *J. Bombay nat. Hist. Soc.*, **12**, 214.

Welsh, J. H. & Batty, C. S. (1963). 5-Hydroxytryptamine content of arthropod venoms. *Toxicon*, **1**, 165–74.

Whitaker, J. O. & Rubin, D. C. (1971). Food habits of *Plethodon jordani metcalfi* and *Plethodon jordani shermani* from North Carolina. *Herpetologica*, **21**, 81–6.

Whitaker, J. O. & Russel, E. M. (1972). Food and ectoparasites of Indiana shrews. *J. Mammal.*, **53**, 329–35.

Whittel, H. R. (1883). On the voracity of a species of *Heterostoma. Proc. Linn. Soc. NSW*, **8**, 33–4.

Wigglesworth, V. B. (1972). *The principles of insect physiology.* (7th edn.) London: Chapman & Hall.

Wignarajah, S. (1968). Energy dynamics of centipede populations (Lithobiomorpha – *L. crassipes* and *L. forficatus*) in woodland ecosystems. Ph.D. thesis, University of Durham.

Willem, V. (1889). Note sur l'existence d'un gésier et sur sa structure dans la famille des Scolopendrides. *Bull. Acad. r. Belg.* (3) **18**, 532–547.

Willem, V. (1892*a*). Les ocelles de *Lithobius* et de *Polyxenus*. *Annls Soc. r. malacol. Belg.*, **27**, lxix-lxxi.

Willem, V. (1892*b*). L'organe de Tömösvary de *Lithobius forficatus*. *Annls Soc. r. malacol. Belg.*, **27**, lxxi-lxii.

Willem, V. (1897). Les glandes filières (coxales) des Lithobies. *Annls Soc. ent. Belg.*, **41**, 87–9.

Wood, D. M. & Wheeler, A. G. (1972). First record in North America of the centipede (Myriap. Chil.) parasite *Loewia foeda* (Dipt. Tachinidae). *Can. Ent.*, **104**, 1363–7.

Wood, H. C. (1865). The Myriapoda of North America. *Trans. Am. phil. Soc.*, **13**, 137–248.

Würmli, M. (1974). Systematic criteria in the Scutigeromorpha. *Symp. zool. Soc. Lond.*, no. 32, 89–98.

Würmli, M. (1975). Systematische Kriterien in der Gruppe von *Scolopendra morsitans* Linné 1758 (Chilopoda, Scolopendridae). *Dt. Ent. Z.* (NF), **22**, 201–6.

Würmli, M. (1976). Contributo alla conoscenza del *Plutonium zwierleini* Cavanno 1881 (Chilopoda: Scolopendromorpha: Cryptopidae). *Animalia Catania*, 1975 (1976) 2, 209–13.

Würmli, M. (1977). Proposed use of the plenary powers to give the family name Henicopidae Pocock, 1901, precedence over the family name Cermatobiidae Haase, 1885 (Myriapoda:Chilopoda). *ZN(S)* 2206. *Bull. Zool. Nomencl.*, **34**, 123–5.

Würmli, M. (1978). Biometrical studies on the taxonomy and the post-embryonic development of some species of *Scolopendra* Linnaeus (Chilopoda). *Abh. Verh. naturwiss. Ver. Hamburg* (NF), **21/22**, 51–54.

Zaleskaja, N. T. (1975). Distribution patterns of Lithobiomorpha on the territory of the USSR. *Progress in Soil Zoology. Proceedings of the 5th International Colloquium on Soil Zoology*, Prague 1973, pp. 163–6.

Zerbib, C. (1966). *Étude descriptive et expérimentale de la differentiation de l'appareil génital du Myriapode Chilopode 'Lithobius forficatus'* L. *Bull. Soc. zool. Fr.*, **91**, 203–16.

INDEX

Page numbers of illustrations are in italic